Titanium Dioxide - Advances and Applications

Edited by Hafiz Muhammad Ali

Published in London, United Kingdom

IntechOpen

Supporting open minds since 2005

Titanium Dioxide - Advances and Applications
http://dx.doi.org/10.5772/intechopen.94670
Edited by Hafiz Muhammad Ali

Contributors
Mouad Dahbi, Nabil El Halya, Karim Elouardi, Abdelwahed Chari, Abdeslam El Bouari, Jones Alami, Haruna Adamu, Barkallah Rachida, Rym Taktak, Noamen Guermazi, Jamel Bouaziz, Tarun Parangi, Manish Kumar Mishra, Xiaoping Wu, Kaman Singh, Ashok Kumar, Rayees Ahamad Bhat, Vibhav Ambardekar, Rajib Das, Partha Pratim Bandyopadhyay, Azrina Abd Aziz, Minhaj Uddin Monir, Fatema Khatun, Sim Lan Ching, Leong Kah Hon, Muhammad Saeed, Atta ul Haq, Samreen Gul Khan, Muhammad Ibrahim, Wafa Selmi, Nabil Hosni, Jamila Ben Naceur, Hager Maghraoui-Meherzi, Radhouane Chtourou, Hafiz Muhammad Ali, Chao Zhou, Tayyab Raza Raza Shah

First published in London, United Kingdom, 2022 by IntechOpen
IntechOpen is the global imprint of INTECHOPEN LIMITED, registered in England and Wales, registration number: 11086078, 5 Princes Gate Court, London, SW7 2QJ, United Kingdom
Printed in Croatia

British Library Cataloguing-in-Publication Data
A catalogue record for this book is available from the British Library

Additional hard and PDF copies can be obtained from orders@intechopen.com

Titanium Dioxide - Advances and Applications
Edited by Hafiz Muhammad Ali
p. cm.
Print ISBN 978-1-83969-475-2
Online ISBN 978-1-83969-476-9
eBook (PDF) ISBN 978-1-83969-477-6

For EU product safety concerns:
IN TECH d.o.o., Prolaz Marije Krucifikse Kozulić 3, 51000 Rijeka, Croatia, info@intechopen.com or visit our website at intechopen.com.

Meet the editor

Dr. Hafiz Muhammad Ali received his doctoral degree in Mechanical Engineering from the School of Engineering and Materials Science, Queen Mary University of London, United Kingdom, in 2011. He worked as a postdoc at the Water and Energy Laboratory, University of California, Merced, United States, in 2016. He is currently an associate professor of Mechanical Engineering at King Fahd University of Petroleum and Minerals, Saudi Arabia. His fields of research include thermal sciences and heat transfer with a focus on electronics cooling, condensation, nanofluids, heat transfer devices, and thermal management. Dr. Ali has supervised numerous undergraduate and postgraduate students. He has more than 230 published journal papers to his credit. He has participated at several international and national conferences as an invited speaker and delivered various keynote talks. Dr. Ali was recognized as a Highly Cited Researcher in the Field of Engineering by the Web of Science in 2021. He was also recognized as being in the top 2% of researchers by Stanford University in 2020 and 2021. He has received numerous awards, including the Best Young Research Scholar Award in Engineering by the Higher Education Commission (HEC) of Pakistan in 2017, HEC's Best Research Paper Award in 2013–2014, and the Research Productivity Award from the Pakistan Council of Science and Technology in 2016–2017. In addition to his academic duties, Dr. Ali is an active editorial member of several international journals, including *Heat Transfer Engineering*, *Journal of Thermal Analysis and Calorimetry*, and *International Journal of Thermofluids.*

Contents

Preface

Titanium dioxide plays an important role in many scientific and engineering applications, thus it is important to further explore the fundamentals and mechanisms of this useful compound.

This book focuses on recent advances and applications of titanium dioxide. It examines the compound's applications in environmental remediation, photocatalytic materials, rechargeable lithium-ion batteries, thin films, energy storage, semiconductors, and much more. It is a useful resource for researchers, scientists, engineers, and students. Chapters include various illustrations and graphs to represent the information in a comprehensive manner.

I would like to extend my appreciation and thanks to the contributing authors. I would also like to express my gratitude and appreciation to my beloved wife Aysha Maryam Ali for her companionship, continuous support, and patience throughout the preparation of this book.

Hafiz Muhammad Ali
Mechanical Engineering Department,
King Fahd University of Petroleum and Minerals,
Dhahran, Saudi Arabia

Acknowledgement

Hafiz Muhammad Ali thankfully acknowledges the support of King Fahd University of Petroleum and Minerals, Dhahran, 31261, Saudi Arabia.

Section 1

Thermal Energy Applications of Titanium Dioxide

Chapter 1

Titanium Dioxide: Advancements and Thermal Applications

Tayyab Raza Shah, Chao Zhou and Hafiz Muhammad Ali

Abstract

Distinctive characteristics of titanium dioxide such as high refractive index, overwhelmingly high melting and boiling point, high toughness, and hardness, photocatalytic nature, ability to absorb or reflect UV-rays, DeNox catalyst, nontoxicity, inert behavior, etc., have brought about the massive use of TiO_2 in a variety of conventional as well as advanced engineering applications. Broad commercial utilization of titanium dioxide in products including paints, anti-air pollutants, cosmetics, skincare and sunblock, pharmaceuticals, surface protection, building energy-saving, etc., accounts for its multibillion dollars market worldwide. Titanium dioxide carries unique thermal and optical characteristics and therefore has gained significance as a potential candidate for advanced applications such as clean hydrogen fuel harvesting, photoelectric solar panels, photothermal conversion, treatment of exhaust gases from combustion engines and power plants, thermal energy storage, thermal management of electronic devices and photovoltaics, and nano-thermofluids. This chapter presents a brief insight into some of the noteworthy characteristics and a comprehensive overview of advanced thermal applications of TiO_2.

Keywords: titanium dioxide, thermal management, energy harvesting, energy storage, nanoparticles

1. Introduction

Titanium dioxide (TiO_2)—a ceramic, commonly known as titania—is a naturally occurring oxide of titanium and is among the most widely used metals. Titania exists in three crystallographic forms, i.e., rutile, anatase, and brookite [1]. Titanium dioxide carries engrossing characteristics, needed to have for a material to be used in a broad range of applications. Some of the key properties of titania have been narrated in **Table 1**. Titania has been used primarily as a pigment in paints for the past hundred-odd years; however, it has been under the spotlight for the past couple of decades as it is deemed as the potential substitute for many of the metals owing to its nontoxic and chemically stable nature. A comprehensive overview of vital features and potential applications of titania has been presented in **Table 2**.

Titanium dioxide is employed in bulk form as well as nanoparticle form. Conventionally, bulk titania (of 0.20 μm general size [4]) has been used; however, avant-garde thermal applications of titania make use of nano-sized particles (100 nm) in medicine, nano-phase changing thermal energy storing materials, and nanofluids. Low cost, durability, and ease to handle are the most fascinating aspects of titania. Moreover, further growth in titania utilization in the future has been projected by the researchers.

Property	Detail
Molecular mass	79.86 (g/mol)
Density	3.9–4.2 (g/cm^3)
Refractive index	2.5–2.75
Mohs hardness	5.5–7
Band gap (rutile)	3.0 eV
Band gap (anatase)	3.2 eV
Structural shape	Tetragonal structure (rutile) Tetragonal structure (anatase) Orthorhombic structure (brookite)

Table 1.
General properties of titanium dioxide [2].

Key feature	Applications	Industrial sectors
Mechanical: • Titania is anticorrosive • Titania has high hardness and toughness • Wear resistant	• Used as an additive for strengthening the material and reducing the wear effects.	• Paint industry • Cosmetics industry • Power sector • Pharmaceuticals • Biomedical • Food industry • Plastic industry
Optical: • High refractive index • UV resistant • UV absorbent	• Used as pigment as it acts as whitener and shimmer brightener as it reflects the light and acts opaque. • Used for scattering UV-rays in skincare and sunblock applications.	
Chemical: • Chemically inert • Insoluble in water • Nontoxic • Photocatalytic • Hydro catalytic • DeNox catalyst	• Used in thin or thick protective paints. • Used for hydrogen extraction from water through hydro catalysis. • Used for decomposing the air pollutants through photocatalysis. • Used for exhaust gas purifier	
Electrical: • Semiconductor • Specific variation in electrical resistance at high temperature in the presence of specific gases	• Due to the semiconductor characteristics, it is used for photoelectric energy harvesting as a solar panel material. • Change in electrical resistance in presence of specific gases is used to detect gases.	
Thermal: • High melting point • High boiling point	• Used for thermal energy storage and energy transportation when used as nanofluid, etc.	

Table 2.
Key in-demand features and applications of titania [3].

State-of-the-art thermal applications of titania include hydrogen fuel extraction, photoelectric energy harvesting, photothermal energy harvesting, thermal energy storage and transportation, thermal management of electronics, automotive engines, electric batteries and photovoltaics, temperature control of buildings, etc. This chapter discusses the mechanism of employing titania in the aforementioned thermal applications, and an in-depth discourse on the performance of titania is presented.

2. Thermal utilization of titanium dioxide

Titania has been utilized for various thermal applications such as heat relieving, photoprotection, storing heat, heat transportation, solar thermal energy cultivation, etc. Utilization of titania in the aforementioned thermal applications has been carried out in the form of bulk titania, titania nanoparticles, titania-based nanofluids, and titania nanoparticle enhancement phase change heat storage materials. The following section contains a detailed discourse on the mechanism of titania-based thermal management and the performance of titania-based nanofluids when employed to perform thermal management of various advanced thermal systems.

2.1 Thermal management

Thermal management stands for the process of relieving excessive heat of components such as electronic devices, solar modules, vehicle engines, electrical batteries, etc., since the prolonged heating of these objects leads to performance deterioration, system size maximization, manufacturing surcharge, operational and handling complications, and system failure.

Thermal management can be carried out either by thermofluids-based cooling or by thermally conductive and solar photoprotection metallic coatings.

2.1.1 Thermal management of photovoltaic modules

Photovoltaic (PV) modules are used to harvest solar photoenergy to generate electrical energy through a photoelectric conversion mechanism. Standalone PV modules are made of semiconducting crystalline materials. There are three categories of PV panels, i.e., monocrystalline, polycrystalline, and amorphous. The photoelectric conversion efficiency of solar panels is quite limited due to certain characteristic constraints of panels' material, and their efficiency range is 10–20% [5]. A substantial component of the solar photoenergy is converted into thermal energy, which gives rise to the heating of panels. Temperature elevation decreases the band gap (energy gap); therefore, the efficiency of solar panels tends to drop by 0.5% when the panel temperature increases by 1°C past 25°C [6]. The relationship between the band gap and the temperature of the semiconductor was presented by Varshni [7] (Eq. (1)). Rodriguez [8] appraised the influence of temperature on the efficiency of PV panels, and he observed a drastic decline in efficiency as evident in **Figure 1**.

$$E_g(T) = E_g(0) - \frac{\sigma T^2}{T + \beta} \tag{1}$$

In Eq. (1), E_g is the energy gap, T is the temperature (in Kelvin), and α and β are the constants of semiconductor materials, and the values of these constants for various semiconducting materials have been presented by Varshni [7].

Complications caused by temperature escalation of PV modules are addressed by heat relieving techniques. Both passive and active methods of cooling have been tested to carry out thermal management of solar modules. Heat absorbed from the cooling fluid is used for thermal applications, and this system is named as photovoltaic thermal (PV/T) system [9]. The efficacy of PV/T systems is evaluated in terms of PV or electrical efficiency, thermal efficiency, and overall efficiency [9].

The electrical efficiency of the PV/T system is appraised by Eq. (2).

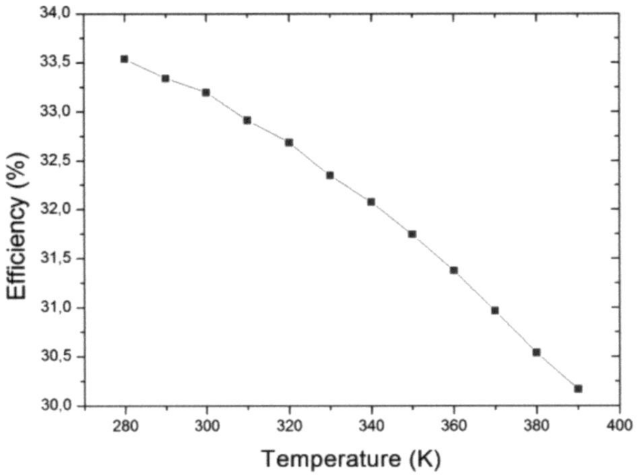

Figure 1.
Influence of temperature variation on the efficiency of PV panels [8].

$$\eta_{el} = \frac{\dot{E}_{el}}{\dot{E}} = \frac{V_{oc} \times I_{sc} \times FF}{G_{eff}} \tag{2}$$

In Eq. (2), $\dot{E}_{el}$ is the output electrical energy, $\dot{E}$ is the incident irradiation, which is converted into electrical energy output, thermal energy output, and energy losses, V_{oc} is the output voltage, I_{sc} is the short-circuit current, G_{eff} represents the absorbed irradiation (effective), and FF stands for fill factor, which is evaluated by Eq. (3), and it depends on the quality of solar cells.

$$FF = f \times \left(\frac{V_{oc}}{T}\right) \tag{3}$$

In Eq. (3), T is the temperature of the solar cell.

The thermal energy output of the PV/T system is appraised by Eq. (4).

$$\dot{E}_{th} = m_f \times C_{pf} \times (T_o - T) \tag{4}$$

In Eq. (4), $\dot{E}_{th}$ is thermal efficiency, m_f is the cooling fluid's mass flow rate, C_{pf} is the specific heat capacity of the cooling fluid, and T_o and T are the fluid outlet and inlet temperature, respectively.

The overall efficiency of the PV/T system is appraised by Eq. (5).

$$\eta_{PV/T} \sim \frac{\dot{E}_{el} + \dot{E}_{th}}{\dot{E}_{th}} \Rightarrow \eta_{PV/T} = \eta_{th} + r \times \eta_{el} \tag{5}$$

In Eq. (5), r is the packing factor, which is calculated by Eq. (6).

$$r = \frac{A_{PV}}{A_c} \tag{6}$$

In Eq. 6, A_{PV} is the photovoltaic area, and A_c is the collector area.

Avant-garde research studies suggest the use of nano-thermofluids for effective thermal management and efficient operation of PV modules. Nanofluids are regarded as the most efficient mean of thermal management that has been widely recommended for temperature regulation of solar panels. Nanofluids are among the most advanced thermofluids, and they carry exceptional thermophysical and optical characteristics. The principal element of nanofluids is the nanoparticles that are uniformly spread in the conventional fluids, e.g., water, ethylene glycol, engine oil, etc. Dispersion of nanoparticles is carried out to tailor the thermophysical and optical characteristics such as thermal conductivity, viscosity, pumping power, and solar light absorption range of conventional thermofluids.

Thermal management of solar modules can be executed by various methods that include rear-end cooling, front-end cooling, double-pass cooling. In rear-end cooling, thermally efficient nanofluid flows beneath the surface of the solar module and takes off the heat of the panel. In front-end cooling, optically efficient nanofluid flows across the frontal surface of the solar panels, and the nanofluid absorbs the UV and IR part of solar radiation that will not be converted into electrical energy and would cause heating of the panel. The front-end method of cooling employs the spectral splitting mechanism. The double-pass method of PV cooling is the combination of rear-end and front-end cooling. The aforementioned PV cooling techniques have been elaborated in **Figure 2**.

Titanium-dioxide-based nanofluids are engineered by dispersing nanoparticles of titania in the base fluid through rigorous nanofluids' preparation techniques. Nanoparticles of titania carry good thermal conductivity and have been reported to have impressive thermohydraulic performance results [10]. The thermal conductivity of titania nanoparticles is 8.4 W/m °C, and they are spherical in general with white color [11].

Rukman et al. [12] appraised the efficiency of a PV/T system using TiO_2/water nanofluid as a coolant. They analyzed the effect of fluid's flow rate (0.012–0.0255 kg/s), irradiation (700 W/m^2 and 900 W/m^2), and nanoparticle concentration (0.5 wt.% and 1.0 wt.%) on the performance of the PV/T system. They reported the overall efficiency of the PV/T system to range from 75–90%, and the electrical efficiency approached 9.9–10.6%. The thermal efficiency of the titania

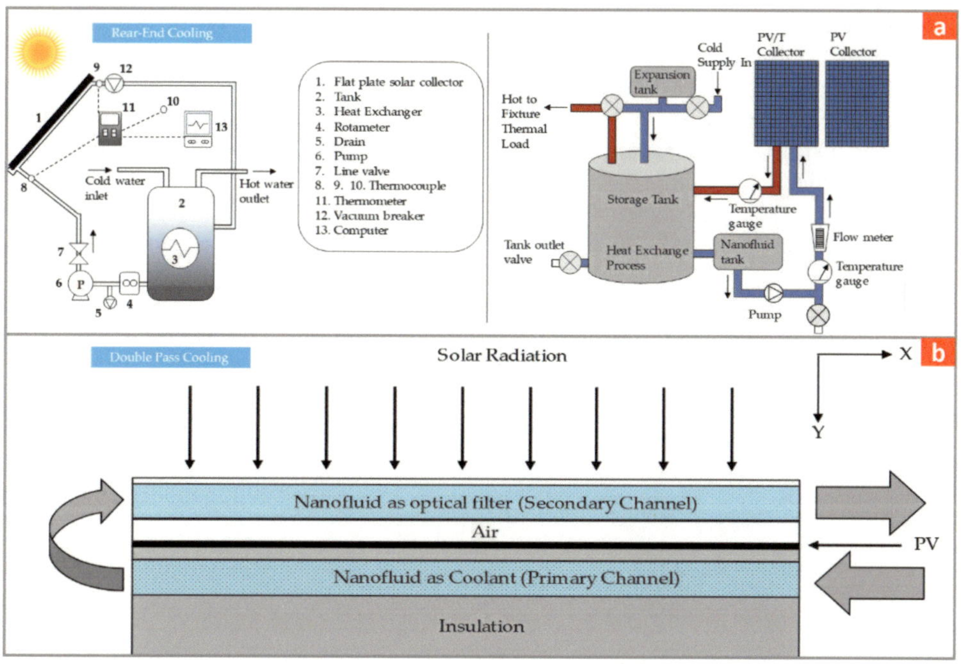

Figure 2.
Methods of PV cooling using nanofluids (a) rear-end cooling, and (b) double-pass cooling [6].

nanofluid-based PV/T system reached 65–80%. At 900 W/m^2 irradiance, electrical efficiency increased by 6.95% and 7.32% at 0.5 wt.% and 1.0 wt.% titania nanoparticle concentration, respectively, as compared with the uncooled PV panel. In another study, Rukman et al. [13] analyzed the performance of the PV/T system using MWCNT and TiO_2_based nanofluids. They concluded that the flow rate increment causes the module surface temperature to decrease, whereas the irradiation increase causes the temperature to increase, and the efficiency gets dropped.

Fadli et al. [14] tested the effect of using TiO_2/water nanofluid on the performance of the PV modules. They observed an increase of 2.6% in PV efficiency and a 10.38 W increase in power output as compared with the reference PV module. At 1100 W/m^2 irradiance, a decline of 11.4°C in PV module temperature was observed. Increased temperature decreased the module efficiency and the use of nanofluid increased the efficiency as compared with the water-cooled module and reference module (**Figure 3**). Maadi et al. [15] analyzed the effect of thermophysical characteristics of nanofluids on the performance of PV/T system using various nanofluids, i.e., 0.2 wt.% Al_2O_3/water, 0.2 wt.% ZnO/water, 0.2 wt.% TiO_2/water, 1.0 wt.% SiO_2/water, and 3.0 wt.% SiO_2/water through comprehensive experimentation and numerical method. They observed that due to greater thermal conductivity, Al_2O_3/water outperformed the rest of the nanofluids in terms of efficiency enhancement of the PV module. Numerical analysis revealed that at 10 wt.% concentration of nanoparticles in the base fluid, the PV efficiency could be improved by 6.23%, 6.02%, 6.88%, and 5.77% by using Al_2O_3/water, ZnO/water, TiO_2/water, and SiO_2/water nanofluids respectively as compared with the water-cooled PV module. Similar analysis was performed by Hasan et al. [16]. They tested SiC/water, TiO_2/water, and SiO_2/water nanofluids and SiC-based nanofluid since superior thermophysical characteristics outperformed TiO_2 and SiO_2-based nanofluids and yielded 12.75% electrical, 85% thermal, and 97.75% combined efficiency. Whereas TiO_2 outperformed SiO_2 as it carries better thermal characteristics such as convective heat transfer coefficient and thermal conductivity. They used the method of jet impingement to achieve a maximum cooling effect (**Figure 4**).

Sardarabadi et al. [17] performed a similar analysis using 0.2 wt.% nanofluids of Al_2O_3/water, TiO_2/water, and ZnO/water. They reported an overall energy efficiency enhancement of 12.34% for water, 15.45% for ZnO/water, 15.93% for TiO_2/water, and 18.27% for Al_2O_3/water. Ebaid et al. [18] and Sardarabadi and Fard [19] performed a similar analysis and reported the same trends as discussed in the previous studies.

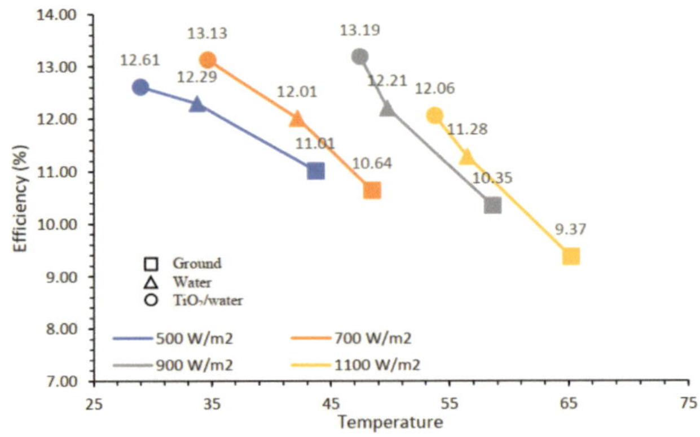

Figure 3.
Effect of temperature, irradiation, and cooling media on the efficiency of PV modules [14].

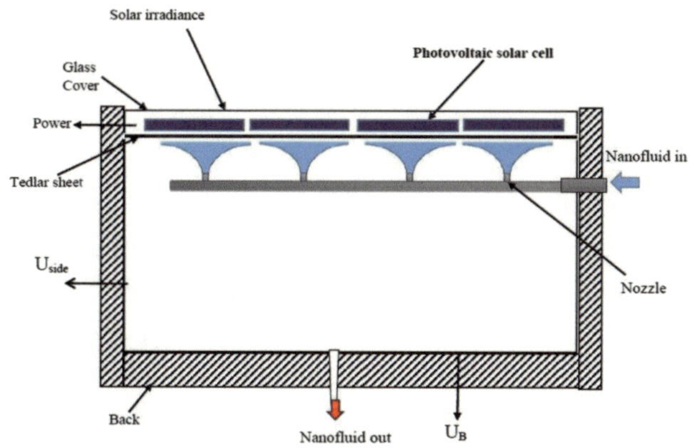

Figure 4.
Nanofluid-assisted PV cooling by jet impingement [16].

Key factors that influence the performance of nanofluids-based PV/T systems are the irradiance, PV cell surface temperature, and flow rate, inlet temperature, and thermophysical characteristics of nanofluids. An increase in irradiation causes higher energy photons to strike the PV surface, which causes the temperature to rise, and therefore, the efficiency tends to drop. The higher flow rate of fluid results in higher convective heat transfer, and therefore, greater cooling of PV panels takes place. Better thermophysical characteristics of nanofluids also intensify the rate of heat transfer and heat storage capacity, which help in achieving higher electrical power output from the PV modules. The bottom surface of solar modules holds considerable significance, and therefore, some studies have also suggested the use of metallic fins to improve temperature regulation [20]. Proper handling of nanofluids is also a crucial aspect of these systems [21].

2.1.2 Thermal management of automotive engines

Automotive engines make use of fuel combustion to generate mechanical energy to power the vehicles. Combustion and the resulting exhaust gases raise the temperature of the combustion chamber up to 2500°C [22]. Extreme temperature can cause the piston to melt, and the molten piston upon solidification causes the welding effect, and it eventually seizes the engine. To prevent failure, automotive engines have a cooling loop in which a coolant is set to flow that takes the heat of the engine and releases the heat in the environment through the radiator heat exchanger. Water, ethylene glycol, and the mixture of water and ethylene glycol have been conventionally utilized as automotive coolants, and the radiator plays a pivotal role in the process of automotive cooling.

State-of-the-art research studies on automotive cooling have extensively tested nanofluids as a potential candidate to replace the conventional automotive coolants to improve the heat transfer performance and system miniaturization [23]. The experimental setup used to simulate the automotive cooling system and test the nanofluids is presented in **Figure 5**.

Titania-based nanofluids have been tested for automotive cooling purposes as well, and quite encouraging results have been reported in recent studies. Chen et al. [25, 26] tested the performance of TiO_2/water nanocoolant for automotive cooling, and they observed 10% enhancement in heat transfer coefficient at 1.0 wt.% concentration of titania nanoparticle. They performed a pump cavitation test and recorded a pump corrosion rating of 10, which means no corrosion/erosion. They

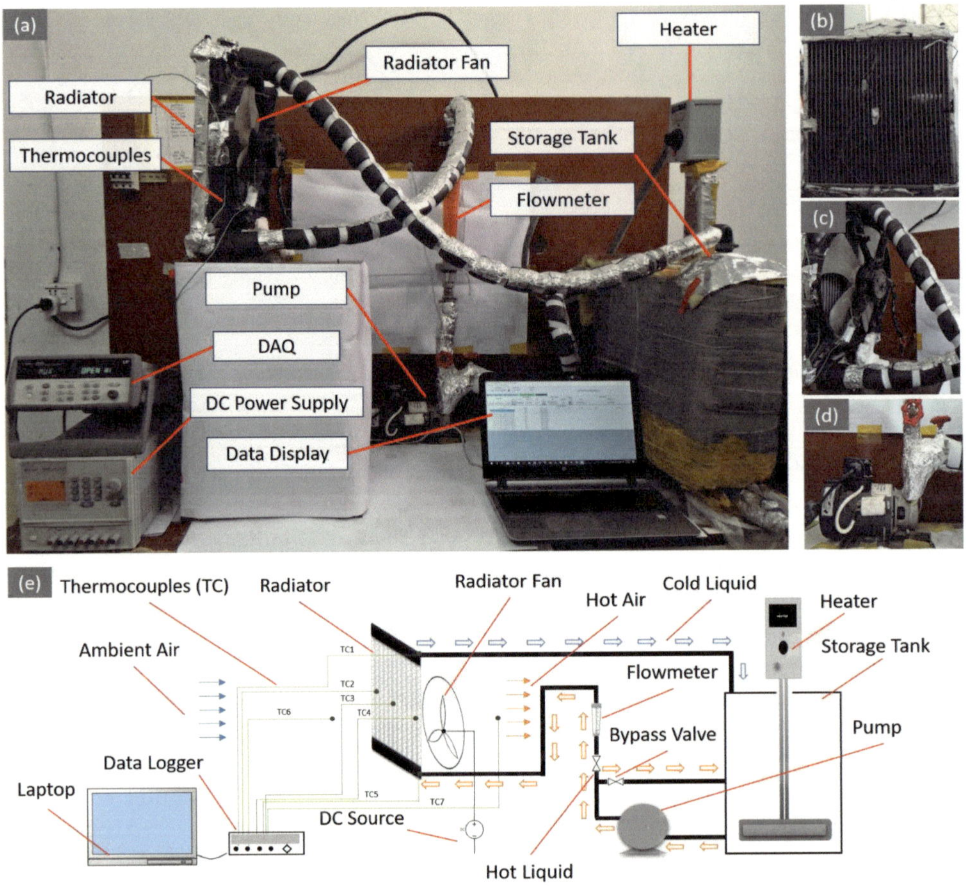

Figure 5.
Automotive cooling experimental setup and components (a–d), and schematic elaboration of the experimental setup (e) [24].

recommended the use of titania in automotive cooling as it has corrosion inhibition characteristics, and therefore, the pumping system and the fluid channels are estimated to operate efficiently and for a longer period. Salamon et al. [27] conducted experiments to analyze the effect of using TiO_2/water-propylene glycol (70:30) nanocoolant of 0.1 vol.% and 0.3 vol.% nanoparticle concentration with flow rate varying from 3 LPM to 6 LPM and inlet temperature ranging from 60°C to 80°C. The use of titania nanofluid resulted in an 8.5% increased heat transfer rate. Moreover, at lower inlet temperature, base fluid outperformed the nanofluids. Usri et al. [28] tested TiO_2/water-ethylene glycol (60:40) nanocoolant and at optimum operating conditions and 1.5 vol.% nanoparticle concentration and 70°C inlet temperature 33.9% increase in heat transfer was reported. An increase in heat transfer rate was recorded as 20.9% at 1.0 vol.% nanoparticle concentration. Titania nanofluids samples were observed to be stable for 2 months after the preparation. Devireddy et al. [29] appraised the performance of TiO_2/water-ethylene glycol (60:40) nanocoolant, and they reported a 37% increase in heat transfer rate at 0.5 vol.% nanoparticle concentration as compared with the base fluid. They observed an increase in Nusselt number (Nu) with an increase in nanoparticle concentration **Figure 6**. Hussein et al. [31] reported a 20% increase in heat transfer efficiency by using TiO_2/water nanocoolant as compared with the water base fluid. Ahmed et al. [32] investigated the automotive cooling potential of TiO_2/water nanocoolant and reported a 47% improvement in radiator effectiveness.

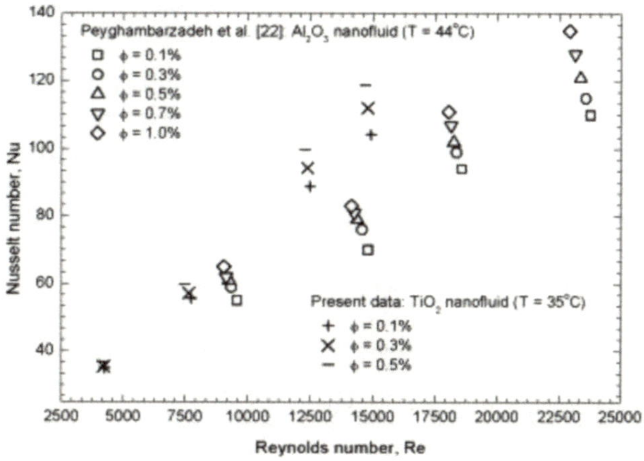

Figure 6.
Effect of titania nanoparticle concentration on Nu, and a comparison of results of Devireddy et al. [29] and Peyghambarzadeh et al. [30].

Titania has also been combined with other nanoparticles to form hybrid nanofluids that carry tunable thermal and optical characteristics. Abbas et al. [33] appraised the performance of avant-garde hybrid nanocoolant having Fe_2O_3-TiO_2 nanoparticles dispersed in water. They reported a 26.7% enhancement in heat transfer rate at a very low nanoparticle concentration of 0.009 vol.% due to the synergistic effect of ferric oxide and titania nanoparticles. The noteworthy finding of their study was that hybrid nanofluids produced quite impressive repeatability tests. They recorded as much as 3% deviation in heat transfer results when the experiments were performed after 12 hours of reference experimentation and 56°C inlet temperature and 15 LPM flow rate (**Figure 7**).

There are some major associated challenges as well when using nanocoolants in automotive cooling. Due to the presence of nanoparticles, there are chances of clogging, and due to high operating temperatures, the nanocoolant can become ineffective as it can cause clustering. High temperature also results in surfactant ineffectiveness. Successive heating and cooling of nanocoolant are also a challenge as it causes the development of thermal stresses.

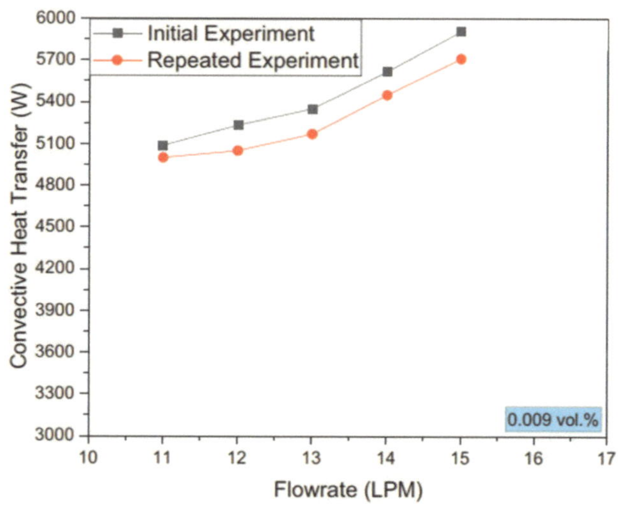

Figure 7.
Heat transfer repeatability test results of 0.009 vol.% Fe_2O_3-TiO_2/water automotive nanocoolant [33].

2.1.3 Thermal management of electronic components

Electronic devices/components produce a considerable extent of heat during their operation. Excessive heat generation causes performance deterioration; therefore, electronic devices/components require temperature regulation for smooth operation. Temperature regulation of these systems is carried out by cooling mechanisms. Conventionally, these components were cooled by air or water; however, modern cooling techniques use metallic heat sinks, nanofluids, heat pipes, and phase change materials [34]. The combined use of metallic heat sinks and nanofluids has been a subject of wide interest by researchers for the past couple of decades. Continuous efforts of researchers have brought massive innovation in heat sink designs. Modern heat sink designs include fins of various shapes and engraved micro/mini channels with different geometries to achieve efficient cooling [35].

Nanofluid and heat-sink-based electronic cooling setup used to evaluate the heat transfer effect has been presented in **Figure 8**. Babar and Ali [36] evaluated the performance of ferric oxide and titania-based nanofluids with pin fin heat sink for the sake of thermal management of electronics. They discussed the performance of nanofluids in terms of Nu and pressure drop/pumping power. They observed Nu improvement of 15.89% for Fe_2O_3/water nanofluid and 14.5% for TiO_2/water nanofluid at nanoparticle concentration of 0.01 vol.% as compared with water. However, the pumping power augmentation for TiO_2/water nanofluid was 30.5% and for Fe_2O_3/water, nanofluid 42.46% enhancement was recorded. Since titania is a corrosion inhibitor and water has no environmental disadvantages and the pumping power enhancement for titania-based nanofluids is also quite low as compared with the ferric oxide, they recommended the use of titania nanofluids for metallic heat sinks. Sajid et al. [37] reported 40.57% augmentation in Nu by using 0.012 vol.% TiO_2/water nanofluid with wavy-mini channel heat sink for electronic cooling. They reported an increase in Nu with increasing Re and nanoparticle loading (**Figure 9**). Arshad and Ali [38] analyzed the performance of TiO_2/water nanofluid in a straight mini channel heat sink. They reported the pressure drop to decrease with an increased heating power of the heating source since the viscosity of the nanofluid tends to decrease with the temperature rise. Qi et al. [39] also reported a drop in surface temperature of the spherically bulging aligned and staggered heat sinks when cooled by a titania-water nanofluid. Similar studies were conducted by Narendran et al. [40], Kumar et al. [41], Tariq et al. [42], and Nitiapiruk et al. [43].

2.1.4 Thermal management of buildings and energy saving

Titania has a high reflection index, and when it is used in paint/coatings for the exterior of buildings, it reflects the solar radiations and prevents the heating of the building. This, in turn, reduces the heating load and saves the cost of air-conditioning. The reflectance of sunlight by the titania nanoparticles has been presented in **Figure 10**. As it is evident in **Figure 10**, UV radiations are reflected by the titania nanoparticles and a minute portion of sunlight is absorbed by the particles, which results in a very minute flow of solar heat inside the buildings. Titania-implanted paints provide thermal resistance, brilliance, protection, and resistance against pollution. Future infrastructure planners are keen to make the modern residential and office buildings energy self-sufficient, which are named as Zero-Energy Buildings or Positive Energy Building with energy surplus [45]. These buildings make use of building integrated photovoltaic panels known as BIPVs [46]. BIPVs have been tested in many countries and have shown positive results. However, the challenge of uneven module temperature could be addressed via titania nanofluids as well [47].

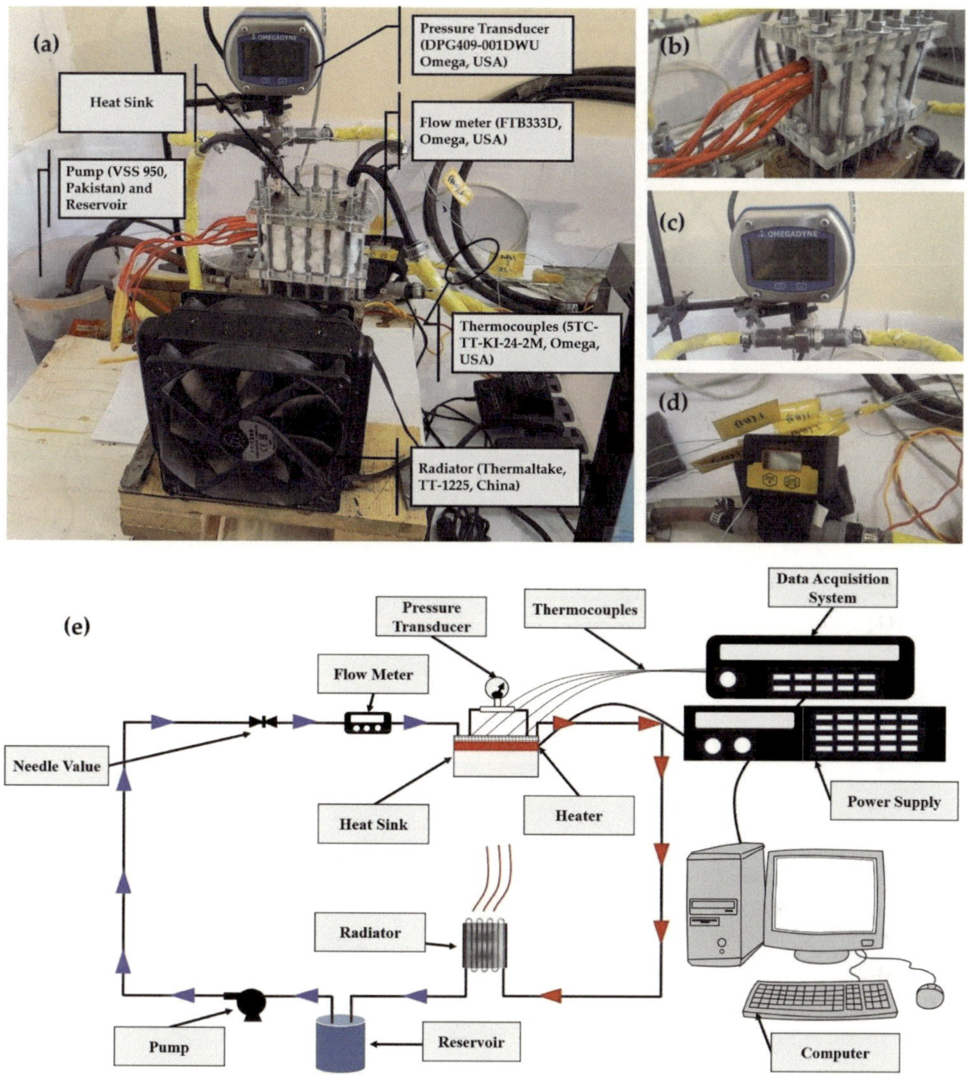

Figure 8.
Heat sink and nanofluid-based electronic cooling experimental setup [36].

2.2 Energy harvesting

Titania is also used for harvesting photoelectric and photothermal energy. The following sections discuss in detail the use of titania in various energy harvesting applications.

2.2.1 Photoelectric energy harvesting

Photoelectric energy is harvested by photovoltaic modules, and titania has also been used to fabricate PV modules for efficient electricity generation from solar energy. Dye-sensitized TiO_2 is used in Grätzel photoelectric solar cells to generate electricity from solar energy. Titania is the essential element of these solar cells. These photovoltaic modules were invented by Grätzel [48] and known as dye-sensitized solar cells (DSSC). The working mechanism of DSSC has been presented in **Figure 11**. These modules are cheaper and produce good efficiency. Initially, the

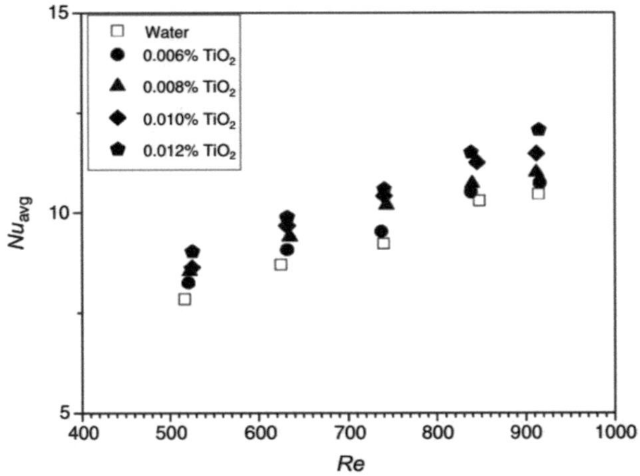

Figure 9.
Effect of Re and nanoparticle concentration on Nu at 35 W source power [37].

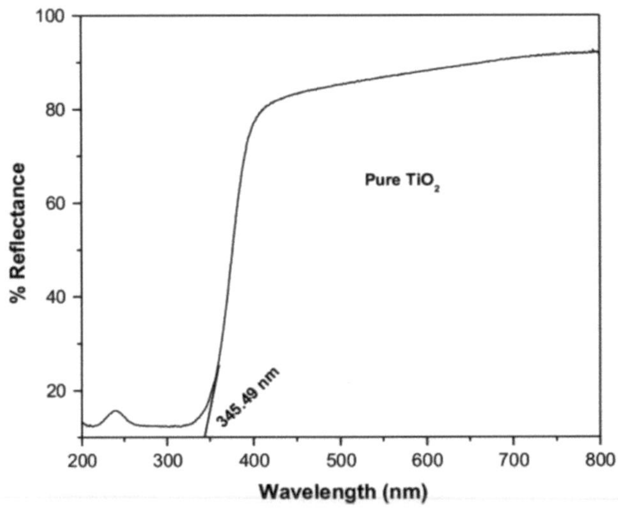

Figure 10.
Reflectance of titania nanoparticles [44].

efficiency of these modules was about 7% as reported by Kay and Grätzel [50]. Later on, the claim of achieving 7%, the photoelectric conversion efficiency was verified by Hagfeldt et al. [51]. The limited efficiency of these modules is attributed to very high resistance and losses due to the cells being in series and resulting in a lower fill factor. Kim et al. [52] conducted a detailed study on power conversion efficiency (PEC) evaluation of these cells and reported 9.4% PEC. Several studies have been conducted on these cells to analyze the performance and environmental and economic aspects [48, 53–56].

2.2.2 Photothermal energy harvesting

Photothermal energy harvesting is carried out by various types of solar collectors (**Figure 12**). The basic element of solar collectors is the receiver tube in which a fluid flows to capture the solar energy by absorbing solar radiations energy and converting it into heat energy through the photothermal conversion process. The

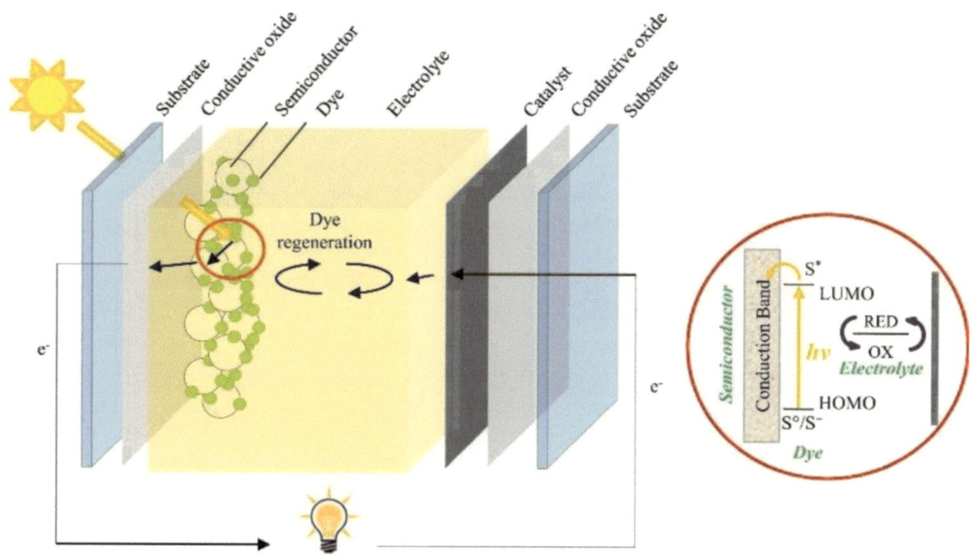

Figure 11.
Schematic illustration of DSSC [49].

photothermal efficiency of these collectors is mainly dependent on the material and geometrical characteristics of the receiver tube and the thermophysical and optical characteristics of the working fluid. Conventionally, water has been utilized as a working fluid in solar collectors; however, water has a very limited range of solar absorbance, which means the rest of the energy is wasted in the form of reflected energy and thermal losses.

In the latest studies, nanofluids have been tested for photothermal conversion since the presence of nanoparticles increases the range of solar absorption, which results in increased thermal energy output and increased efficiency [58]. Moreover, the two types of nanoparticles are also dispersed in the base fluid (called hybrid nanofluids) to tailor the fluid's absorption range through the synergistic effect of different nanoparticle types [59]. Greater thermal conductivity and heat transfer coefficient of nanofluids also intensify the efficiency of solar collectors such as parabolic trough collectors. Since TiO_2 carries good thermal conductivity, titania-based nanofluids have widely been tested for photothermal energy conversion and heat energy transportation in solar collectors [60–62]. Titania depicts higher reflectance and smaller spectral absorbance. However, titania nanoparticles are used to broaden the spectral absorption range of water and other base fluids. Moreover, the combined use of titania nanoparticles with other types of nanoparticles in the base fluid further broadens the solar spectral absorption range and thermal efficiency of solar collectors (**Figure 13**).

Kiliç et al. [63] studied the performance of TiO_2/water nanofluid as a working fluid in a flat plate collector (FPSC). They obtained 48.67% topmost efficiency of the collector when using titania-water nanofluid, and the efficiency of the collector reached 36.20% in the case of water as working fluid. They attributed the efficiency increase to the increased heat transfer surface area and higher heat capacity of the nanofluid. Said et al. [64] also appraised the performance of FPSC using titania-water nanofluid and reported a 76.6% increase in energy efficiency at a 0.5 kg/min flow rate of 0.1 vol.% TiO_2/water nanofluid as compared with water. The topmost energy efficiency was observed to be 16.9%. They reported negligible enhancement in pumping power for nanofluid as compared with the base fluid. Tehrani et al. [65] analyzed the performance of ribbed FPSC using TiO_2/water nanofluid. Ribs

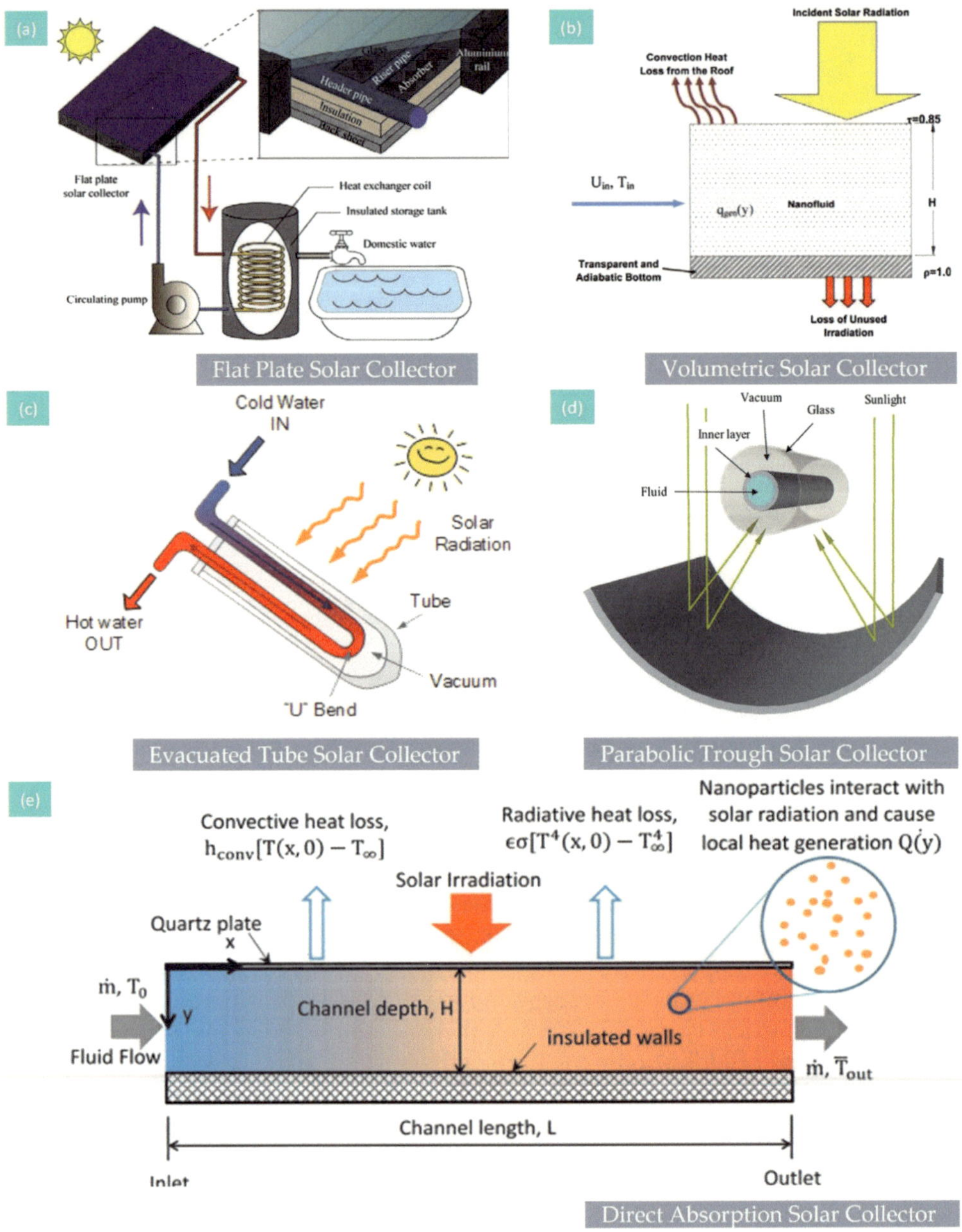

Figure 12.
Photothermal energy-generating solar collector [57].

improved the performance of FPSC by 10%. They reported slight improvement with increasing nanoparticle concentration as well.

Gan et al. [66] appraised the performance of evacuated tube solar collectors ETSC using TiO_2/water nanofluid. They recorded a 16.5% increase in thermal efficiency of the collector by using optimized titania-water nanofluid as compared with the efficiency of the water-based collector. The obtained collector's thermal efficiency was 44.85%. An increase in nanoparticle concentration improved the thermal conductivity of the nanofluid, which resulted in higher thermal efficiency. Hosseini and Dehaj [67] also tested the performance of U-shaped ETSC using titania-water nanofluid. They evaluated the effect of titania particle shape on the collector's efficiency. TiO_2 nanowires (NWs) and nanoparticles (NPs) dispersed in

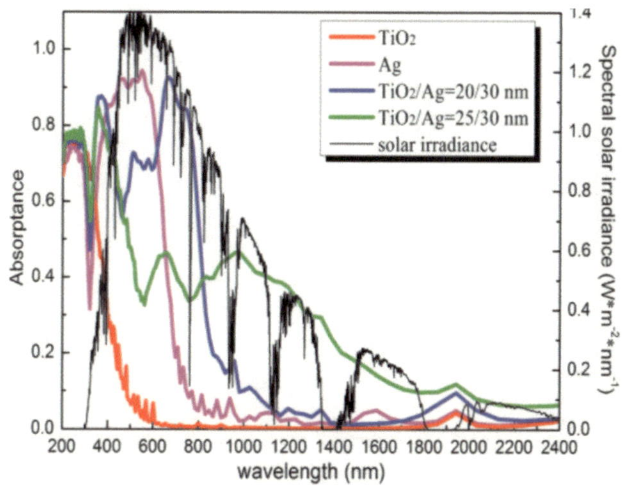

Figure 13.
Absorptance of titania, silver, and combined titania-silver [60].

the base fluid. Efficiency improvement of 21.1% for TiO_2 NWs nanofluid and 12.2% for TiO_2 NPs nanofluid was recorded. However, NWs nanofluid depicted a higher pressure drop.

2.2.3 Thermal energy storage

Titanium holds porous properties and a good ability to stay stable when impregnated in some chemical. Therefore, titania is extensively used in thermal energy storage applications. Thermal energy-storing phase change materials (PCMs) have good thermal storage capacity, but due to their small thermal conductivity, the rate of thermal storage is much slower, which causes them to store a small amount of thermal energy. The addition of titania nanoparticles improves their thermal conductivity, and therefore, the capacity to store thermal energy becomes much higher. Titania also provides chemical stability.

Fikri et al. [68] added titania nano-powder in paraffin wax (PW) PCM to overcome the challenge of low thermal conductivity and small spectral absorption. They also reported the thermal and optical advantages of adding titania in PCMs. Sun et al. [69] reported a 26% elevation in thermal conductivity of polyethylene glycol (PEG) PCM by adding titania nanoparticles.

The challenge associated with nano-enhanced PCMs is the deterioration of thermal conductivity and heat storage capacity due to the thermal cycles. Sami and Etesami [70] reported that the thermal conductivity of nano-titania enhanced PW-PCM dropped below the thermal conductivity of pure PW after several thermal cycles. Deterioration of characteristics takes place due to structural and chemical degradation. Chemically stable nano-PCMs are direly needed for the successful operation of PCMs. Deka et al. [71] impregnated TiO_2 powder in 1-tetradecanoic acid to prepare chemically stable composite-PCM. They observed good chemical stability through rigorous testing techniques. Due to the inclusion of titania, the thermal conductivity of PCM increased by 188% and latent heat storage (LHS) of 97.75 J/g at 52.04°C melting temperature was recorded of composite-PCM. However, after thermal cycling, the LHS declined by only 16.65%.

Another important aspect of PCMs is the time taken for melting and solidification. Common PCMs such as PW take very long to melt and solidify, which reduces the heat storage rate. Studies have suggested the combined use of metallic heat sinks and nanofluids with PCMs to increase the rate of the phase change process. Ding et al. [72] reported a 32.90% reduction in melting time and a 22.57% decrease in solidification time of PW PCM by incorporating a microchannel heat sink with 1.0 wt.% TiO_2/water nanofluid. An increase in nanoparticle concentration increased the rate of melting and solidification due to an increased rate of heat transfer. **Figure 14** represents the effect of varying nanoparticle concentration on Nu and melting and solidification time of PCM.

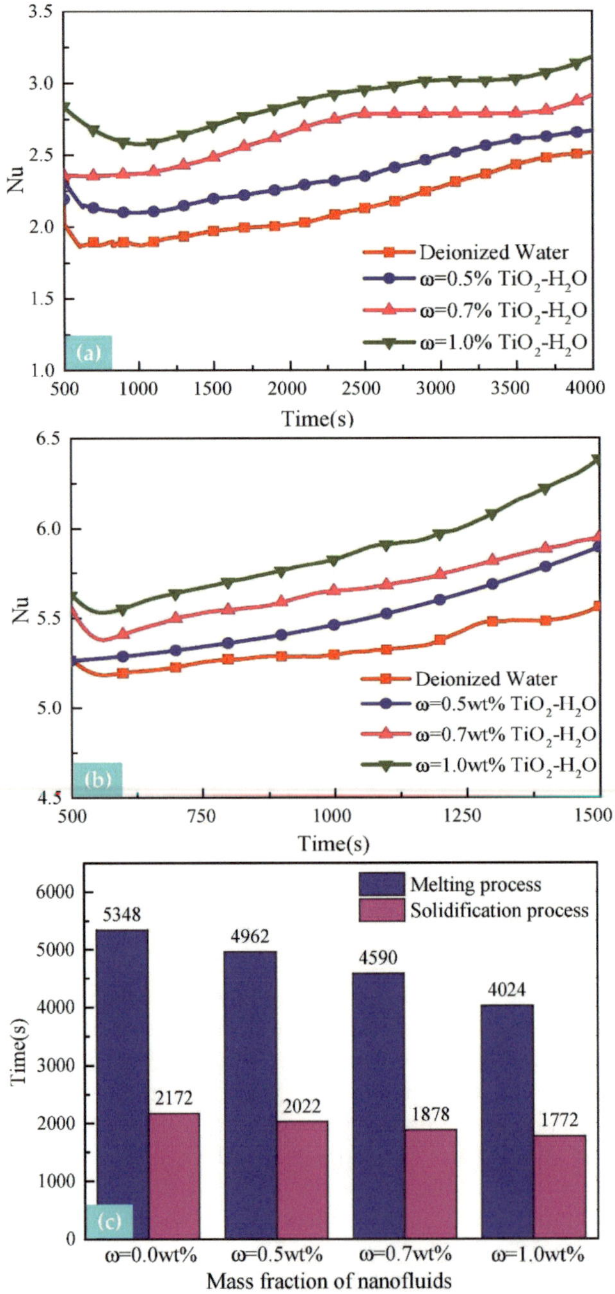

Figure 14.
Nu variation (a) during the melting process, (b) during the solidification process of PCM for different samples of nanofluid, and (c) time of melting and solidification of PCM [72].

3. Conclusions

The chapter presented an insight into the thermal utilization of titanium dioxide. Titania being an abundant metal with unique thermophysical and optical characteristics has been used in many applications for the past hundred years. Titania is expected to attain more advanced applications in the future due to its fascinating environmental and economic aspects. Important findings of the discussed research are as follows.

- Due to corrosion-resistant characteristics, titania is highly recommended for thermal applications such as thermal management systems, thermal energy harvesting systems, and surface coatings.
- High UV-rays reflectance characteristics make the titania best suitable for applications such as sunscreens, cosmetics, textile, and building coating.
- Owing to its semiconducting nature, titania is used in photoelectric solar modules called Grätzel solar panels.
- Titania carries unique optical characteristics; therefore, it is widely used for broadening the spectral absorption range of conventional optical fluids and advanced nanofluids.
- Titania provides chemical stability to the thermal energy storing PCMs and mitigates the effects of thermal cycling.
- Due to excellent thermophysical features, titania is extensively used in thermal transportation, thermal storage rate enhancement, and improving rate melting and solidification of PCMs. The combined use of titania nanofluid and metallic heat sinks is also employed to increase the melting and solidification rate.

Author details

Tayyab Raza Shah[1], Chao Zhou[1] and Hafiz Muhammad Ali[2,3]*

1 College of Engineering, Peking University, Beijing, China

2 Mechanical Engineering Department, King Fahd University of Petroleum and Minerals (KFUPM), Dhahran, Saudi Arabia

3 Interdisciplinary Research Center for Renewable Energy and Power Systems (IRC-REPS), King Fahd University of Petroleum and Minerals, Dhahran, Saudi Arabia

*Address all correspondence to: hafiz.ali@kfupm.edu.sa

References

[1] Ruano P et al. Titanium dioxide and its applications in mechanical, electrical, optical, and biomedical fields. In: Titanium Dioxide. London, UK: IntechOpen; 2021. p. 13

[2] Blanchart P. Extraction, properties and applications of titania. In: Industrial Chemistry of Oxides for Emerging Applications. London, UK: IntechOpen; 2018. pp. 255-309

[3] Haider AJ, Jameel ZN, Al-Hussaini IHM. Review on: Titanium dioxide applications. Energy Procedia. 2019;**157**: 17-29

[4] Song J et al. The effects of particle size distribution on the optical properties of titanium dioxide rutile pigments and their applications in cool non-white coatings. Solar Energy Materials & Solar Cells. 2014;**130**:42-50

[5] Ceylan I, Gürel AE, Demircan H, Aksu B. Cooling of a photovoltaic module with temperature controlled solar collector. Energy and Buildings. 2014;**72**:96-101

[6] Ali HM, Shah TR, Babar H, Khan ZA. Application of Nanofluids for Thermal Management of Photovoltaic Modules: A Review. In: Microfluidics and Nanofluidics. IntechOpen; 2018

[7] Varshni YP. Temperature dependence of the energy gap in semiconductors. Physica. 1967;**34**(1): 149-154

[8] Rodriguez E. Solar cell efficiency vs. module power output: Simulation of a solar cell in a cpv module. In: Solar Cells—Research and Application Perspectives. No. Tourism. IntechOpen; 2013. p. 13

[9] Sardarabadi M, Passandideh-Fard M, Maghrebi MJ, Ghazikhani M. Experimental study of using both ZnO/water nanofluid and phase change material (PCM) in photovoltaic thermal systems. Solar Energy Materials & Solar Cells. 2017;**161**(June 2016):62-69

[10] Ali H, Babar H, Shah T, Sajid M, Qasim M, Javed S. Preparation techniques of TiO_2 nanofluids and challenges: a review. Applied Sciences. 2018;**8**(4):587

[11] Toghraie D, Chaharsoghi VA, Afrand M. Measurement of thermal conductivity of ZnO–TiO_2/EG hybrid nanofluid Effects of temperature and nanoparticles concentration. Journal of Thermal Analysis and Calorimetry. 2016;**125**(1):527-535

[12] Binti Rukman NS, Fudholi A, Mohd Razali NF, Hafidz Ruslan M, Sopian K. Energy and exergy analyses of photovoltaic-thermal (PV/T) system with TiO_2/water nanofluid flow. IOP Conference Series: Earth and Environmental Science. 2019;**268**(1): 012075

[13] Binti Rukman NS, Fudholi A, Mohd Razali NF, Hafidz Ruslan M, Sopian K. Investigation of TiO_2 and MWCNT nanofluids-based photovoltaic-thermal (PV/T) system. IOP Conference Series: Earth and Environmental Science. 2019; **268**(1):012076

[14] Fadli AF, Kristiawan B, Suyitno, Arifin Z. Analysis of TiO_2/water-based photovoltaic thermal (PV/T) collector to improve solar cell performance. IOP Conference Series: Materials Science and Engineering. 2021;**1096**(1):012053

[15] Maadi SR, Kolahan A, Passandideh Fard M, Sardarabadi M. Effects of nanofluids thermo-physical properties on the heat transfer and 1st law of thermodynamic in a serpentine PVT system. In: 17th Conference on Fluid Dynamics. Shahrood, Iran: Shahrood University of Technology; 2017. pp. 27-29

[16] Hasan HA, Sopian K, Jaaz AH, Al-Shamani AN. Experimental investigation of jet array nanofluids impingement in photovoltaic/thermal collector. Solar Energy. 2017;**144**: 321-334

[17] Sardarabadi M, Hosseinzadeh M, Kazemian A, Passandideh-Fard M. Experimental investigation of the effects of using metal-oxides/water nanofluids on a photovoltaic thermal system (PVT) from energy and exergy viewpoints. Energy. 2017;**138**:682-695

[18] Ebaid MSY, Ghrair AM, Al-Busoul M. Experimental investigation of cooling photovoltaic (PV) panels using (TiO_2) nanofluid in water -polyethylene glycol mixture and (Al2O3) nanofluid in water- cetyltrimethylammonium bromide mixture. Energy Conversion and Management. 2018;**155**(November 2017):324-343

[19] Sardarabadi M, Passandideh-Fard M. Experimental and numerical study of metal-oxides/water nanofluids as coolant in photovoltaic thermal systems (PVT). Solar Energy Materials & Solar Cells. 2016;**157**:533-542

[20] Qasim MA, Ali H, Sajid MU, Shah TR. Effect of fins insertion inside phase change material on thermal management of phtovoltaics under climate of Taxila, Pakistan-an experimental study. In: Proceedings Book of 4th International Conference on Advances in Mechanical Engineering (ICAME); Turkey. Istanbul; 2018. pp. 77-84

[21] Ali H, Shah TR, Babar H, Ali AM. Hybrid nanofluids: Techniques and challenges of stability enhancement. In: Proceedings Book of 4th International Conference on Advances in Mechanical Engineering (ICAME); Turkey. Istanbul; 2018. pp. 60-76

[22] Safakish GR. Temperature estimation in the combustion chamber of an internal combustion engine. Advances in Mechanical Engineering. 2012;**2012**:1-6

[23] Abbas F et al. Nanofluid: Potential evaluation in automotive radiator. Journal of Molecular Liquids. 2020;**297**: 112014

[24] Shah TR, Ali HM, Janjua MM. On aqua-based silica (SiO_2–water) nanocoolant: Convective thermal potential and experimental precision evaluation in aluminum tube radiator. Nanomaterials. 2020;**10**(9):1-23

[25] Chen JM, Sun XY, Leng GJ, Feng JH. Performance investigation of TiO_2 nanofluid coolant for automobile cooling applications. Key Engineering Materials. 2015;**645-646**:444-448

[26] Chen J, Jia J. Experimental study of TiO_2 nanofluid coolant for automobile cooling applications. Materials Research Innovations. 2017;**21**(3):177-181

[27] Salamon V, Kumar DS, Thirumalini S. Experimental Investigation of heat transfer characteristics of automobile radiator using TiO_2 -nanofluid coolant. IOP Conference Series: Materials Science and Engineering. 2017;**225**(1): 012101

[28] Usri NA, Azmi WH, Mamat R, Abdul Hamid K. Forced convection heat transfer using water-ethylene glycol (60,40) based nanofluids in automotive cooling system. International Journal of Automotive and Mechanical Engineering. 2015;**11**(June):2747-2755

[29] Devireddy S, Mekala CSR, Veeredhi VR. Improving the cooling performance of automobile radiator with ethylene glycol water based TiO_2 nanofluids. International Communications in Heat and Mass Transfer. 2016;**78**:121-126

[30] Peyghambarzadeh SM, Hashemabadi SH, Hoseini SM, Seifi Jamnani M. Experimental study of heat

transfer enhancement using water/ethylene glycol based nanofluids as a new coolant for car radiators. International Communications in Heat and Mass Transfer. 2011;**38**(9): 1283-1290

[31] Hussein AM, Dawood HK, Bakara RA, Kadirgamaa K. Numerical study on turbulent forced convective heat transfer using nanofluids TiO_2 in an automotive cooling system. Case Studies in Thermal Engineering. 2017;**9** (December 2016):72-78

[32] Ahmed SA, Ozkaymak M, Sözen A, Menlik T, Fahed A. Improving car radiator performance by using TiO_2-water nanofluid. Engineering Science and Technology, an International Journal. 2018;**21**(5):996-1005

[33] Abbas F et al. Towards convective heat transfer optimization in aluminum tube automotive radiators: Potential assessment of novel Fe_2O_3-TiO_2/water hybrid nanofluid. Journal of the Taiwan Institute of Chemical Engineers. 2021; **124**:424-436

[34] Colangelo G, Favale E, Milanese M, de Risi A, Laforgia D. Cooling of electronic devices: Nanofluids contribution. Applied Thermal Engineering. 2017;**127**:421-435

[35] Sajid MU, Ali HM. Recent advances in application of nanofluids in heat transfer devices: A critical review. Renewable and Sustainable Energy Reviews. 2019;**103**(October 2018): 556-592

[36] Babar H, Ali HM. Airfoil shaped pin-fin heat sink: Potential evaluation of ferric oxide and titania nanofluids. Energy Conversion and Management. 2019;**202**(July):112194

[37] Sajid MU, Ali HM, Sufyan A, Rashid D, Zahid SU, Rehman WU. Experimental investigation of TiO_2–water nanofluid flow and heat transfer inside wavy mini-channel heat sinks. Journal of Thermal Analysis and Calorimetry. 2019;**137**(4):1279-1294

[38] Arshad W, Ali HM. Experimental investigation of heat transfer and pressure drop in a straight minichannel heat sink using TiO_2 nanofluid. International Journal of Heat and Mass Transfer. 2017;**110**:248-256

[39] Qi C, Zhao N, Cui X, Chen T, Hu J. Effects of half spherical bulges on heat transfer characteristics of CPU cooled by TiO_2-water nanofluids. International Journal of Heat and Mass Transfer. 2018;**123**:320-330

[40] Narendran G, Bhat MM, Akshay L, Arumuga Perumal D. Experimental analysis on exergy studies of flow through a minichannel using TiO_2/water nanofluids. Thermal Science and Engineering Progress. 2018;**8**(April): 93-104

[41] Kumar A, Narendran G, Perumal DA. Entropy generation study of TiO_2 nanofluid in microchannel heat sink for Electronic cooling application. IOP Conference Series: Materials Science and Engineering. 2018;**376**(1):012013

[42] Tariq HA, Anwar M, Malik A, Ali HM. Hydro-thermal performance of normal-channel facile heat sink using TiO_2-H_2O mixture (rutile–anatase) nanofluids for microprocessor cooling. Journal of Thermal Analysis and Calorimetry. 2021;**145**(5):2487-2502

[43] Nitiapiruk P, Mahian O, Dalkilic AS, Wongwises S. Performance characteristics of a microchannel heat sink using TiO_2/water nanofluid and different thermophysical models. International Communications in Heat and Mass Transfer. 2013;**47**:98-104

[44] Praveen P, Viruthagiri G, Mugundan S, Shanmugam N. Structural, optical and morphological analyses of pristine titanium di-oxide

nanoparticles—Synthesized via sol–gel route. Spectrochimica Acta Part A: Molecular Spectroscopy. 2014;**117** (September):622-629

[45] Kolokotsa D, Rovas D, Kosmatopoulos E, Kalaitzakis K. A roadmap towards intelligent net zero- and positive-energy buildings. Solar Energy. 2011;**85**(12):3067-3084

[46] Norton B et al. Enhancing the performance of building integrated photovoltaics. Solar Energy. 2011;**85**(8): 1629-1664

[47] Hemmat Esfe M, Kamyab MH, Valadkhani M. Application of nanofluids and fluids in photovoltaic thermal system: An updated review. Solar Energy. 2020;**199**:796-818

[48] Grätzel M. Dye-sensitized solar cells. Journal of Photochemistry and Photobiology C: Photochemistry Reviews. 2003;**4**(2):145-153

[49] Parisi ML, Maranghi S, Basosi R. The evolution of the dye sensitized solar cells from Grätzel prototype to up-scaled solar applications: A life cycle assessment approach. Renewable and Sustainable Energy Reviews. 2014;**39**:124-138

[50] Kay A, Grätzel M. Low cost photovoltaic modules based on dye sensitized nanocrystalline titanium dioxide and carbon powder. Solar Energy Materials & Solar Cells. 1996; **44**(1):99-117

[51] Hagfeldt A et al. Verification of high efficiencies for the Grätzel-cell. A 7% efficient solar cell based on dye-sensitized colloidal TiO_2 films. Solar Energy Materials & Solar Cells. 1994; **31**(4):481-488

[52] Kim HS et al. High efficiency solid-state sensitized solar cell-based on submicrometer rutile TiO_2 nanorod and $CH_3NH_3PbI_3$ perovskite sensitizer. Nano Letters. 2013;**13**(6):2412-2417

[53] Munir S, Shah SM, Hussain H, Siddiq M. Adsorption of porphyrin and carminic acid on TiO_2 nanoparticles: A photo-active nano-hybrid material for hybrid bulk heterojunction solar cells. Journal of Photochemistry and Photobiology B: Biology. 2015;**153**:397-404

[54] Huang C, Lv Y, Zhou Q, Kang S, Li X, Mu J. Visible photocatalytic activity and photoelectrochemical behavior of TiO_2 nanoparticles modified with metal porphyrins containing hydroxyl group. Ceramics International. 2014;**40**(5): 7093-7098

[55] Biswas S, Hossain MF, Takahashi T. Fabrication of Grätzel solar cell with TiO_2/CdS bilayered photoelectrode. Thin Solid Films. 2008;**517**(3):1284-1288

[56] Campbell WM, Burrell AK, Officer DL, Jolley KW. Porphyrins as light harvesters in the dye-sensitised TiO_2 solar cell. Coordination Chemistry Reviews. 2004;**248**(13-14):1363-1379

[57] Shah TR, Babar H, Ali HM. Energy harvesting: Role of hybrid nanofluids. In: *Emerging* Nanotechnologies for Renewable Energy. Elsevier; 2021. pp. 173-211

[58] Shah TR, Ali HM. Applications of hybrid nanofluids in solar energy, practical limitations and challenges: A critical review. Solar Energy. 2019;**183** (February):173-203

[59] Shah TR, Koten H, Ali HM. Performance effecting parameters of hybrid nanofluids. In: Hybrid Nanofluids for Convection Heat Transfer. Elsevier; 2020. pp. 179-213

[60] Xuan Y, Duan H, Li Q. Enhancement of solar energy absorption using a plasmonic nanofluid based on TiO_2/Ag composite nanoparticles. RSC Advances. 2014;**4**(31):16206-16213

[61] Said Z, Sajid MH, Saidur R, Mahdiraji GA, Rahim NA. Evaluating

the optical properties of TiO_2 nanofluid for a direct absorption solar collector. Numerical Heat Transfer, Part A: Applications. 2015;**67**(9):1010-1027

[62] Atif M, Farooq WA, Fatehmulla A, Aslam M, Ali SM. Photovoltaic and impedance spectroscopy study of screen-printed TiO_2 based CdS quantum dot sensitized solar cells. Materials (Basel). 2015;**8**(1):355-367

[63] Kiliç F, Menlik T, Sözen A. Effect of titanium dioxide/water nanofluid use on thermal performance of the flat plate solar collector. Solar Energy. 2018;**164** (March):101-108

[64] Said Z et al. Performance enhancement of a flat plate solar collector using titanium dioxide nanofluid and polyethylene glycol dispersant. Journal of Cleaner Production. 2015;**92**:343-353

[65] Bazdidi-Tehrani F, Khabazipur A, Vasefi SI. Flow and heat transfer analysis of TiO_2/water nanofluid in a ribbed flat-plate solar collector. Renewable Energy. 2018;**122**:406-418

[66] Gan YY, Ong HC, Ling TC, Zulkifli NWM, Wang CT, Yang YC. Thermal conductivity optimization and entropy generation analysis of titanium dioxide nanofluid in evacuated tube solar collector. Applied Thermal Engineering. 2018;**145**(April):155-164

[67] Hosseini SMS, Shafiey Dehaj M. Assessment of TiO_2 water-based nanofluids with two distinct morphologies in a U type evacuated tube solar collector. Applied Thermal Engineering. 2021;**182**(September 2020):116086

[68] Fikri MA et al. Thermal stability and light transmission capability of nano TiO_2 enhanced phase change material as thermal energy storage. IOP Conference Series: Materials Science and Engineering. 2021;**1116**(1):012206

[69] Sun X et al. Shape-stabilized composite phase change material PEG@TiO_2 through in situ encapsulation of PEG into 3D nanoporous TiO_2 for thermal energy storage. Renewable Energy. 2021;**170**: 27-37

[70] Sami S, Etesami N. Improving thermal characteristics and stability of phase change material containing TiO_2 nanoparticles after thermal cycles for energy storage. Applied Thermal Engineering. 2017;**124**:346-352

[71] Deka PP, Ansu AK, Sharma RK, Tyagi VV, Sarı A. Development and characterization of form-stable porous TiO_2/tetradecanoic acid based composite PCM with long-term stability as solar thermal energy storage material. International Journal of Energy Research. 2020;**44**(13):10044-10057

[72] Ding M, Liu C, Rao Z. Experimental investigation on heat transfer characteristic of TiO_2-H_2O nanofluid in microchannel for thermal energy storage. Applied Thermal Engineering. 2019;**160**(June):114024

Chapter 2

Titanium Dioxide as Energy Storage Material: A Review on Recent Advancement

Tarun Parangi and Manish Kumar Mishra

Abstract

With the increased attention on sustainable energy, a novel interest has been generated towards construction of energy storage materials and energy conversion devices at minimum environmental impact. Apart from the various potential applications of titanium dioxide (TiO_2), a variety of TiO_2 nanostructure (nanoparticles, nanorods, nanoneedles, nanowires, and nanotubes) are being studied as a promising materials in durable active battery materials. The specific features such as high safety, low cost, thermal and chemical stability, and moderate capacity of TiO_2 nanomaterial made itself as a most interesting candidate for fulfilling the current demand and understanding the related challenges towards the preparation of effective energy storage system. Many more synthetic approaches have been adapted to design different nanostructures for improving the electronic conductivity of TiO_2 by combining with other materials such as carbonaceous materials, conducting polymers, metal oxides etc. The combination can be done through incorporating and doping methods to synthesize TiO_2-based anodic materials having more open channels and active sites for lithium and/or sodium ion transportation. The present chapter contained a broad literature and discussion on the synthetic approaches for TiO_2-based anodic materials for enhancing the lithium ion batteries (LIBs) and sodium ion batteries (SIBs) performance. Based on lithium storage mechanism and role of anodic material, we could conclude on future exploitation development of titania and titania based materials as energy storage materials.

Keywords: Titanium Dioxide, Nanomaterial, Nanostructure, Lithium Ion Battery

1. Introduction

Nowadays, investigation in the field of energy storage and conversion devices with different functionalities is emerging subject for many research investigators [1–3]. Since last two decades, the regular efforts are being implemented towards the development of nanostructured as electrochemical storage materials convenient to access both surface and bulk properties and hence a superior storage and conversion performance. In this view, nanostructured materials are of great interest due to easy availability for modification into degree of crystallinity, phase, particle size, morphology and porosity which are prior characteristics.

As a talk of nanomaterials and nanotechnology, titania (TiO_2) and titania-based materials always been studied first. TiO_2 nanoparticles (NPs) are being widely investigated over the past few decades due to their unique characteristics such as non-toxicity, abundance, thermal and chemical stability, and easy availability. Many more research and progress reports on TiO_2 have shown a great potential in various important applications such as photocatalysis, biomedical, environmental remediation and many more. Beyond these applications, TiO_2 and TiO_2-based nanomaterials also offers novel materials to overcome the energy and environment related challenges. Different TiO_2 nanostructures with high surface area, large pore volumes, tunable pore structures and nano-confined effects have been synthesized and used in various fields with excellent performances [4]. In the past, many comprehensive reviews have been documented on synthesis, properties and applications of TiO_2 and TiO_2-based nanomaterials.

In view of energy storage technologies, recently, lithium-ion batteries (LIBs) are found to be emerging technologies for imperative electric grid applications such as mobile electronics, electric vehicles and renewable energy systems operating on alternating energy sources like wind, tidal, solar and other clean energy sources [5, 6]. The performance of these technologies in terms of capacity, recyclability and rate capability are much more dependent on the characteristics of the active anode and cathode materials. The performance can be improved through fundamental modification with particular strategy with such factors like the power capacity, long term durability and most prior its cost. In this view of direction, finding of energy storage materials with high efficiency and low cost is thrust challenge for the materials scientists. As we talked about various important characteristics of TiO_2, it could be suitable candidate due to its versatile functionalities. The present chapter covering literature on the recent progress of applications of TiO_2 and TiO_2 based materials as energy storage technologies and discussion on the efforts that have been made so far.

Theoretically, the anode part has a crucial role in LIBs and thus, the direction towards development of anode materials is one of the most essential factors which could define the performance of the device [7]. As an ideal anode material, it should possess high specific surface area allowing better insertion for mobile ions (lithium ions for LIBs); large pore size, low volume change and low internal resistance for speedy charging and discharging; low intercalation potential for mobile ions; and operating at moderate condition with economical and environmental benefits.

Among available various suitable anode materials, transition metal oxides in which TiO_2 is following the characteristics of an ideal anode material that makes TiO_2 itself as most promising anode material for LIBs. Apart from these benefits and utilities of TiO_2, some drawbacks still exist like low capacity and poor rate capability [4, 7]. Thus, TiO_2 suffering from poor ionic/electronic conductivity that limits the lithium storage rate. However, the transport of electrons and Li^+ ions can be promoted by engineering of their physicochemical and morphological characteristics as presented in **Figure 1**. In this view, many more researches and efforts have been made to overcome the said disadvantages by designing and adopting different synthetic strategies to obtain various forms of TiO_2 such as zero-dimensional (0D) nanospheres, one dimensional (1D) nanostructures, two-dimensional (2D) nanoarchitectures and three-dimensional (3D) hierarchical nanostructures with different electronic structures. These nanostructures are showing the advantages of providing high contact surface area with the electrolyte as well as short diffusion pathways for electrons and mobile ions such as Li^+ and Na^+. In addition, the adoption like doping of different heteroatoms into TiO_2 lattice which could alter the chemical and physical surface of TiO_2 would open more channels and active sites for transportation of mobile ions due to which electrical conductivity can be increased [7].

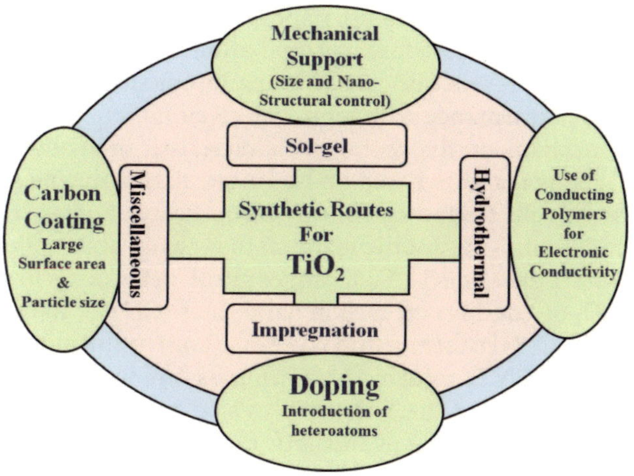

Figure 1.
Synthetic approaches for TiO_2 based anode materials.

2. Synthesis approaches and nanostructural forms

Generally, synthetic strategy is crucial tool to design the structural characteristics due to which materials functionality can be optimized for better performance. Interestingly, TiO_2 can be synthesized in such manner to have different structural forms and hence, versatile functionalities. To summarize recent structural development of titania, this chapter started with a brief introduction on the synthesis and structural characteristics of TiO_2 which has great impact on the performances.

In search of synthesis routes for fabrication of TiO_2-based materials, sol–gel, hydrothermal/solvothermal and impregnation methods of preparation are well recognized. Sol–gel synthesis is mode of preparation offering optimization of various experimental parameters at moderate conditions and allows designing material structure with customized properties. Generally, sol–gel route is widely used to synthesize hybrid and composite types materials by using aqueous and non-aqueous mode in suitable solvent media. In this finding, performance based composite-TiO_2 materials to be used as energy storage materials have been synthesized by sol–gel route. Hydrothermal process consist of treatment of bulk TiO_2 with alkaline solution at high temperature and pressure using Teflon reactor or autoclave instrument. The reaction conditions of pressure and temperature yields selective product which decide the morphology and functionalities of the materials. By use of these useful synthetic techniques, various kind of TiO_2-based materials can be prepared as anode materials in LIBs or/and SIBs applications.

2.1 Carbon coating approach

This approach has received much attention providing superior electrical conductivity, large surface-to-volume ratio, and excellent mechanical and chemical stabilities. In this view many more research works have been done successfully. Kim *et al.* [8] have synthesized anatase TiO_2 nanorods by a hydrothermal method followed by carbon coating in order to improve the electric conductivity through carbonization of pitch used as a carbon source at 700°C for 2 h in Ar flow. The obtained nanorods have been tested for their electrochemical activity in a sodium cell. As results, this anatase TiO_2 nanorod material demonstrates an acceptable cycling performance and a rate capability compared to 1D anatase nanowire TiO_2 and nanowire TiO_2 bulk. In Na cell tests, carbon coated anatase nanorod attain a capacity of ~193 mA·h·g^{-1} on the first charge. The presence of carbon on nanorods surface helps to improve the electrode stable performance with high-rate capability up to the current of 100 C-rate. From the mechanism, the chemical characteristic of anatase nanorod TiO_2 supports a $Ti^{4+}/^{3+}$ redox reaction based on Na^+ intercalation during electrochemical reaction in the Na cell. This study shows carbon-coated anatase nanorod TiO_2 material would be a promising anodic material in rechargeable sodium batteries [8]. Zou *et al.* [9] have developed a facile aqueous sol–gel process to synthesize a porous anatase TiO_2 by coatings on carbon nanotubes (CNTs). In this controlled hydrolysis process, TiO_2 NPs form a thin layer of 20 nm uniformly and continuously covering the surface of CNTs to form an amorphous film without loss of TiO_2 NPs. The capacitive performance of the TiO_2/CNT hybrid electrodes has been evaluated by cyclic voltammetry (CV) and galvanostatic charge–discharge technique using Land battery testing system in the cutoff voltage window of 1–3 V at a scan rate of 50 mV·S^{-1} in 0.5 M H_2SO_4 electrolyte. From this measurement, an ideal rectangular shape was observed in the CV curve of the TiO_2/CNT composite indicating the ideal electrochemical double-layer capacitive behavior of the electrode. The observed good performance and cyclic durability of TiO_2/CNT composite is attributed to the nanometer sized particles facilitate faster charge–discharge rates. In addition, the supported CNTs also has been served as an electron transferring medium and its uniformly coating provides an excellent connection between them to facilitates fast electron exchange, and hence the enhanced electronic conductivity. The calculated average energy density and power density of the TiO_2/CNT composite are found to be ~10.6 W·h·kg^{-1} and 0.954 kW·kg^{-1} at 5 mV·s^{-1} respectively. The porosity of the composite is beneficial for improving contact with electrolyte and reducing the Li^+ ions during application as anode for LIBs. The obtained finding of high specific capacitance and high cycle durability values for the TiO_2/CNT composite makes itself as a promising candidate for supercapacitor application [9].

Zhang *et al.* [10] have adopted a combined sol–gel and hydrothermal routes for the preparation of multi-walled carbon nanotubes (MWCNT) core and mesoporous TiO_2 sheath by using hexadecylamine as a structure directing agent. The prepared CNT@mesoporous TiO_2 hybrid nanocables exhibited with well-crystallized quality, porous feature and large surface area, favoring its electrochemical performance. The introduced CNTs enhance the electronic conductivity. Lithium ion (Li^+) insertion/extraction behavior has been studied to identify the capacity contribution from CNTs in the voltage range of 1–3 V. A stable specific capacity of the CNT@ mesoporous TiO_2 hybrid nanocables could be recovered its initial value after the 30th cycle tested under different current densities even up to current of 10C indicating the high reversibility of lithium ion insertion/extraction in this electrode. The conducted study shows a stable cycling performance with reversible capacities of ~183 mA·h·g^{-1} at current rate of 1C and ~ 112 mA·h·g^{-1} at current rate of 10C retained after 70 cycles. These results suggesting the uses the CNT@mesoporous TiO_2 hybrid nanocables in various fields like photocatalysis and dye-sensitized solar cells. Trang *et al.* [11] have synthesized a stable hybrid mesoporous material through comprised of mesoporous TiO_2 spheres and MWCNTs via a combined sol–gel and solvothermal method. The obtained TiO_2/MWCNT composite has been used as an anode material with a high specific capacity and superior rate capability. In this concept, the MWCNTs served as conducting channels connecting each of the mesoporous TiO_2 spheres in order to form an interconnected network to transport electrons efficiently and therefore the conductivity and stability of the composite can enhance the storage capacity of the lithium ion devices. The composite exhibited with very good LIBs performance with a discharge capacity of ~316 mA·h·g^{-1}

with good reversibility and stability after 100 cycles. The mesoporous nature of the composite offered more open channels for Li^+ ion extraction. The thin walls of the mesoporous TiO_2 reduced the Li ion diffusion length in the solid phase while the pores allowed Li^+ ions to flow smoothly. The mesoporous features allow to TiO_2/MWCNT composite to be used as anode materials significantly for LIBs [11]. Wang *et al.* [12] have implemented a green one-pot hydrothermal process to synthesize hybrid mesoporous CNT@TiO_2-C nanocable using anatase TiO_2, CNT and glucose (as carbon source) as for structure directing agent. The prepared hybrid CNT@TiO_2-C resulted with one-dimensional nanostructure of the CNT with the diameters ranging from 20 to 30 nm. The use of CNT provides high electronic conductivity, high mechanical strength and large-scale availability. The CNT@ TiO_2-C nanocable has been used as anode material for LIBs. The CV analysis curve shows an ideal rectangular shape in the range of 1.0–1.7 V, that are characteristic of charging/discharging of supercapacitance contributed from the anatase TiO_2 and carbon. Two peaks at approximately 1.73 V (cathodic) and 1.98 V (anodic) observed are associated with Li^+ insertion/extraction into/from the anatase TiO_2. These peaks show almost no change in amplitude and voltage positions during the subsequent several cycles, indicating good stability and high reversibility of the electrochemical reaction. The charge–discharge profiles of the nanocable shows a sloped charge–discharge voltage profiles at different rates which is attributed to the decreased crystallite size of TiO_2 and the porous nanostructure of the hybrid material. The observed performance shows an excellent high-rate long-term cycling stability with a high charge capacity of ~187 $mA{\cdot}h{\cdot}g^{-1}$ after 2000 cycles at current of 5 C and ~ 122 $mA{\cdot}h{\cdot}g^{-1}$ after 2000 cycles at current of 50 C. The study reveals about high specific capacity, superior rate capability, and excellent long term cycling stability CNT@TiO_2-C nanocable as anode material for LIBs in half cell [12].

MWCNT array has been grown on a SiO_2 precursor via combustion chemical vapor deposition (CCVD) method and then prepared TiO_2 precursor have been deposited onto MWCNT/SiO_2. The characterization results for the obtained material confirmed positive preliminary data about the capacitance and energy density. The analyzed data proved to be potential as an improvement over comparable electrochemical capacitance device. The present idea of hybridization can be used in current electrochemical theory to create an energy storage device that is both efficient and inexpensive [13]. He *et al.* [14] have synthesized hierarchical rod-in-tube structured TiO_2 with a uniform conductive carbon layer using solvothermal method. With help of modification routes via solvent selection, the morphologies of the prepared TiO_2/C composite can be modified to design various forms like nanoparticles, microrods and microtubes. These modified composites have been tested for their specific capacities for sodium storage capability by using the Na half-cell with a counter electrode consisting of a sodium metal disk, an electrolyte of $NaClO_4$ dissolved in a 1: 1:1 volumetric mixture of dimethyl carbonate, ethylene carbonate (EC), and ethyl methyl carbonate with 5% fluoroethylene carbonate additive and a working electrode of prepared rod-in-tube TiO_2/C composite. As a result, a rod-in-tube TiO_2/C found to be highest electronic conductivity, specific surface area and more porosity compared to other composite materials. These characteristics features with stable crystal texture of a rod-in-tube TiO_2/C composite shows high discharge capacity values of ~277 $mA{\cdot}h^{-1}{\cdot}g^{-1}$ at 50 $mA{\cdot}g^{-1}$ and ~ 153 $mA{\cdot}h^{-1}{\cdot}g^{-1}$ at 5000 $mA{\cdot}g^{-1}$ which are retained over 14000 cycles. The improved electronic conductivity was resulting from the homogeneous carbon layers derived from the carbonization of absorbed solvent which also affects to the electron transfer process, but also suppress the clustering and volume change of TiO_2 nanocrystals during the cycling process. The present synthetic modification routes can lead to a design more advanced materials with better storage performance for SIBs [14].

Liu *et al.* [15] have reported scalable synthesis strategy to fabricate a nanocomposite of anatase TiO_2 NPs anchored on CNTs using ammonia water assisted hydrolysis and *in situ* crystal transformation under calcination in Ar at 500°C. In the fabricated nanostructured TiO_2/CNT, interweaved CNTs has been served as a highway for electrons to facilitate the charge transfer process. The nano-sized TiO_2 found to be useful to improve its contact area with electrolyte and reduced the Li^+ ions diffusion path due to which Li^+ transportation could be accelerated. As results of CV analysis and galvanostatic discharge/charge measurement, the prepared composite exhibited with superior rate capability and cycling stability values of ~92 $mA{\cdot}h{\cdot}g^{-1}$ and retained at a current density of 10 $A{\cdot}g^{-1}$ at current of 60C. The observed results for this hybrid TiO_2/CNT are comparable higher than that of commercial TiO_2 with 80% anatase and 20% rutile phases. The adopted synthetic strategy is eco-friendly, economical and highly scalable for the preparation of anatase-TiO_2/CNTs nanocomposite that holds potential as a superior anode material for high-power LIBs [15]. Cho *et al.* [16] have synthesized phase-pure anatase TiO_2 nanofibers with fiber-in-tube and filled structures by electro-spinning process using tetra-*n*-butyl titanate-polyvinylpyrrolidone (TBT-PVP) composite nanofibers under oxygen and air atmospheres respectively. An intermediate product of carbon free TiO_2 nanofibers with a fiber-in-tube with a core-shell nanostructure was obtained through burning of TiO_2/C composite under oxygen atmosphere during a short time. The repeated combustion and contraction of the TiO_2/C composite core part by burning of the carbon layer lead to formation of the nanofiber with a yolk-shell structure of TiO_2/C@void@TiO_2. These prepared materials have been used as anode materials for LIBs properties using 2032-type coin cells in which electrodes have been prepared using a slurry consisting of 70 wt% active anode material, 20 wt% carbon black as a conductive material, and 10 wt% binder composed of sodium carboxymethyl cellulose (CMC) on a copper foil. As results of electrochemical impedance spectroscopy (EIS) measurements, anatase TiO_2 nanofibers with a fiber-in tube structure exhibited with a superior Li-ion storage capacity (~231 $mA{\cdot}h{\cdot}g^{-1}$) compared to TiO_2 nanofibers with a filled structure (~134 $mA{\cdot}h{\cdot}g^{-1}$) and commercial TiO_2 nanopowder (~223 $mA{\cdot}h{\cdot}g^{-1}$). The TiO_2 nanofibers with the fiber-in-tube structure shows a low charge transfer resistance and structural stability during cycling and rate performances compared to other materials studied in this study [16].

Liao *et al.* [17] have synthesized self-supported, three-dimensional (3D) single-crystalline nanowire array electrodes by using simple hydrothermal process. In this study TiO_2, TiO_2-C and TiO_2-C/SnO_2 have been produced on flexible Ti foil and used as anode materials for electrochemical performance in LIBs using 2032-type coin cells with two electrodes assembled in an Ar filled dry glove box with the prepared materials as working electrodes and Li metal as the counter electrode. The results showing that TiO_2-Sn/C core-shell nanowires exhibit an enhanced electrochemical cycling and rate capability compared to the pure TiO_2 and TiO_2-SnO composite nanowires. A single-crystalline TiO_2-Sn/C core shell nanowire electrodes exhibits a good cycling performance with a discharge capacity higher than ~160 $mA{\cdot}h{\cdot}g^{-1}$ and a capacity retention rate of ~84% even after 100 cycles at a current rate of 10 C. These results could be attributed to such factors like high theoretical capacity of Sn, long cycling stability due to TiO_2 nanowire with self-supported array structures, the space between nanowires is large enough to accommodate the volume changes of Sn during charge/discharge, which can also help to maintain the integrity of the anode during cycling, the carbon shell can suppress cracking and improve the conductivity of the electrode and each nanowire anchored directly to the current collector can help to fast charge transport. With similar intention of successful preparation of carbon containing TiO_2 hybrid composite materials, a single crystalline TiO_2, TiO_2-C and TiO_2-C/MnO_2 core-double-shell nanowire arrays on flexible Ti foil through a layer-by-layer deposition technique has been reported [18]. In this synthetic approach, MnO_2 NPs produced on a TiO_2-C nanowire surface by combination of in situ chemical redox reaction between carbon and $KMnO_4$ using a facile soaking process. The electrochemical characteristics of the prepared materials have been investigated for use in LIBs. The study shows that, TiO_2-C/MnO_2 core double-shell nanowires performed well with enhanced electrochemical cycling and rate properties compared to that of the TiO_2 and TiO_2-C nanowires. For TiO_2-C/MnO_2, a high charge/discharge capacity and stable rate performance of ~332 mA·h·g^{-1} at current rate of 2C. The observed improved electrochemical performance is attributed to the stable structure of the TiO_2 nanowire core, high conductivity of the carbon layer, higher active surface area, and high theoretical capacity of MnO_2 NPs. This study reveals about the adopted layer-by-layer synthetic method proven to be useful technique for preparing unique single crystalline TiO_2-C/MnO_2 nanowire arrays those would be promising anode material for advanced carbon black and binder free LIBs applications. The above two studies [17, 18] can open new ideas for the preparation of self-supported TiO_2/carbon composite based core-shell nanostructures for having better electrochemical performance in the field of LIBs applications.

In addition, graphene oxide (GO) can be considered as another conducting carbon source which has a good ability to form a composite with TiO_2 by using various synthetic techniques. Many more significant research works are in progress to find feasibility for GO through combined with TiO_2 in the field of energy storage materials. Farooq *et al.* [19] have reported a scalable process for *in situ* synthesis of TiO_2 NPs on reduced GO nanosheets by a microwave hydrothermal process to prepared homogenously dispersed and high performance based TiO_2-rGO nanocomposite as anode materials for LIBs. For electrochemical measurement, slurry of the prepared materials was formed and coated onto copper foil, and lithium foil and polypropylene (PP) membrane were used as the counter electrode and a separator respectively. As an electrolyte liquid, 1 M lithium hexaflurophosphate ($LiPF_6$) dissolved in ethylene carbonate (EC) and dimethyl carbonate (DMC) (1: 1 in vol %) have been used. The results of electrochemical impedance spectroscopy reveal that high rate capability and cycle stability have been achieved with developed TiO_2-rGO electrodes. The well-integrated rGO nano-filler enhanced the electronic conductivity enabling the increased rate capability of the electrodes. The present study demonstrates a potential use of the prepared TiO_2-rGO nanocomposite as a viable anode material in advanced Li-ion battery applications which require high power. In the similar way of success, Yue *et al.* [20] have used a simple hydrothermal process to synthesize three dimensional (3D) mesoporous TiO_2 nanocubes grown on RGO nanosheets without use of any surfactants and high temperature of calcination. The prepared TiO_2/RGO composite formed with mesoporous structure contained both rutile and anatase crystalline phases. TiO_2/RGO composite has been exhibited with a very good lithium storage performance as anode materials for LIBs with high specific capacity value of ~180 mA·h·g^{-1} at current rate of 1.2C after 300 cycles. The observed performances is attributed to the relatively large surface area and stable mesoporous structure. The performed above two studies [19, 20] reveal the importance of the synthesis strategy about the formation of desired nanostructure with exploring experimental condition that would give a better performance.

Gao *et al.* [21] have reported sol–gel synthesis of anode materials consisting of mixed (DND) and TiO_2 hollow nanospheres for improving the specific capacity of LIBs. The obtained hybrid material (DND/TiO_2) exhibits high lithium adsorption capacity, large specific surface area and chemical inertness probably due to beneficial characteristics of detonation nanodiamond and porous structure of TiO_2. The electrochemical measurements has been performed using two electrodes of CR2025 button cells in which working electrode made of 80% active material (DND/TiO_2), 10% carbon black and 10% polyvinylidene fluoride (PVDF) and 1 M LiPF6 used as electrolyte. From the capacity performance results, this active material found to be a good anodic material for LIBs with reversible capacity value of ~348 $mA{\cdot}h{\cdot}g^{-1}$ at current of 0.5 C after 100 cycles, high rate performance and long-term cyclic stability value of ~246 $mA{\cdot}h{\cdot}g^{-1}$ at current of 5 C after 800 cycles. The observed performance of this material is found to be superior to that of traditional TiO_2 anode. This study can lead to develop and design anodic materials consisting nanodiomond mixed with other suitable material to have improved and stable performance based LIBs in practical application fields [21].

2.2 Doping approach

This has been well established method for the introduction of impurities into the semiconductor crystal to intentionally change its conductivity due to deficiency or excess of electrons [22]. Tanaka *et al.* [23] have synthesized mesoporous spherical anatase TiO_2 by one-pot solvothermal process using carboxylic acids as organic additives in supercritical methanol medium. The obtained TiO_2 nanostructure can be considered to be roundly integrated metal oxides evaluated as anode materials for Li^+-ion and Na^+-ion batteries. The authors also have developed Nb-doping mesoporous TiO_2 with an aim to improve electrochemical performancefor LIBs. The electrochemical measurements of Li-insertion (charge) and extraction (discharge) has been carried out using 2032-type coin half cells that consisted of Li metal sheets and glass fiber as separator. 1 M bis(trifluoromethanesulfonyl)amide (LiTFSA) dissolved in propylene carbonate has been used as electrolyte. Nb-doped spherical mesoporous TiO_2 exhibited with larger surface area than that of commercial TiO_2. As results of Galvanostatic charge–discharge analysis, the excellent electrochemical performance observed for Nb-doped TiO_2 with the discharge (Li-extraction) capacity of ~147 $mA{\cdot}h{\cdot}g^{-1}$ at the 1000th cycle at a high current rate of 10C. The observed performance can be compared to that $Li_4Ti_5O_{12}$ or a Li^+-ion battery negative electrode. The similar performance observed for Nb-doped for Na-batteries with the discharge (Na-extraction) capacity of ~128 $mA{\cdot}h{\cdot}g^{-1}$ upto 1000 cycles. This performance is attributed to Nb doping into TiO_2 providing large reaction area and it also improved electronic conductivity and broadening the ion-diffusion path in TiO_2 [23]. Synthesis optimization, structural analysis, characterization of the dielectric properties and energy storage density of rutile TiO_2 ceramics co-doped with niobium (Nb^{5+}) and erbium (Er^{3+}) have been investigated [24]. The material $(Er_{0.5}Nb_{0.5})_xTi_{1-x}O_2$ obtained with dense microstructure and smaller grain size distribution giving rise to its high permittivity performance in the subsequent dielectric properties. The observed both intrinsic and extrinsic defect states could be expected to be responsible for the observed high-performance colossal permittivity. As results, the co-doping of Er^{3+} and Nb^{5+} have produced enhanced dielectric properties, where Nb^{5+} ions contribute to a large dielectric permittivity and Er^{3+} decreased dielectric loss. The conducted study suggesting that the prepared co-doped TiO_2 ceramics are favorable for solid-state capacitor and high-energy-density storage applications [24].

2.3 By using conducting polymer

This approach also has gained much attention because of high electrical conductivities and good redox properties of such conducting polymeric materials. Polyoxometalates (POMs), a well-known class of transition metal oxide nanoclusters with interesting structures and diverse properties can give a nanocomposite of TiO_2. In this view, Qu *et al.* [25] have reported synthesis of hybrid bifunctional nanocomposite film of TiO_2 nanowires and POMs, using combination of hydrothermal and layer-by-layer self assembly methods. The prepared nanocomposite film has been tested as anode material for electrochemical performance in which three-electrode system used consisting the as-prepared film served as the working electrodes; Pt plate/Pt wire and Ag/AgCl played the roles of counter electrode and reference electrode, and 1 M $LiClO_4$/propylene carbonate used as electrolyte. The originated electrochemical properties are due to the unique 3D structures of TiO_2 nanowires which provide a larger electrolyte contact area and shorter ion diffusion pathway available for the redox reactions. As results of electrochemical performance, this composite film shows a high coloration efficiency (~150 $cm^2{\cdot}C^{-1}$ at 600 nm) and cyclic stability with its volumetric capacitance about ~172 $F{\cdot}cm^{-3}$ with working voltage window of ~1.77 V, and maintained after 1000 cycles at 0.18 $mA{\cdot}cm^{-2}$. The favorable adsorption of ions and the transport of charges with a very low resistance represents to this composite film as a suitable anodic material for electrochromic and energy storage applications [25].

2.4 Combination with other semiconductors

Ngaotrakanwiwat *et al.* [26] have investigated energy storage capacity of nanocrystalline TiO_2 mixed with V_2O_5 with different compositions using sol–gel synthesis process at different calcination temperatures. The prepared TiO_2-V_2O_5 composite film has been used as a working electrode immersed in a three-electrode electrochemical cell with Ag/AgCl electrode and Pt-wire as a reference electrode and counter electrode, respectively. The obtained TiO_2-V_2O_5 composite with Ti:V (TiO_2:V_2O_5) molar ratio of 1:0.11 calcined at 550°C shows higher energy storage capacities compared to other paired samples and pure WO_3. The performance could be due to the availability of high surface area (~1.75 cm^2) and hence the enhanced ion mobility and intercalation rate.

2.5 Impregnation method

Deka *et al.* [27] have developed a form-stable composite of 1-tetradecanoic acid (TDA) and TiO_2 powder using a facile impregnation method. The long-term chemical and thermal reliability have been investigated by applying 1000 melting-freezing cycles to the composite sample under set temperature range between 20°C and 90°C. The conducted study has demonstrated about a good chemical compatibility and crystalline structure which would be useful for thermal management of electronic devices, automotive components, photovoltaic thermal hybrid designs, solar air/water heating systems, etc.

2.6 Miscellaneous

Hierarchical TiO_2 sphere composed of ultrathin nanotubes consist of both anatase and rutile phases have been synthesized from a low-temperature hydrothermal reaction and calcination process [28]. The obtained hierarchical TiO_2 (~78% anatase and ~ 21% rutile phases) used as-prepared electrode mixed with acetylene black and binder of sodium-CMC in a weight ratio of 7:2:1 using distilled water as solvent. The solution was placed on a copper foil as the current collector and the electrode dried in a vacuum oven at 100°C for 24 h. From this electrochemical tests, prepared material showing the superior performance in Li-storage with a specific capacity of ~167 $mA{\cdot}h{\cdot}g^{-1}$ at 1 $A{\cdot}g^{-1}$ current of 6C and maintained ~187 $mA{\cdot}h{\cdot}g^{-1}$

at rate of 10 A·g^{-1} at current rate of 60C. The capacity retention rate was ~97% at 1 A g and ~ 92% at 2 A·g^{-1} after 500 cycles. The present performance is attributed to the presence of both phases of TiO_2 due to which the synergistic effect has been originated between the defective anatase and rutile renders the shared conduction of electrons through the anatase and Li^+ ions via the rutile at high current rates. In addition, pathway for electron and Li^+ ion conduction was found to be shortened. Also, the interfacial boundaries between the two phases can contribute to extra capacitance at high rates by forming a Li^+/e^- double layer. Overall, the synergy between the anatase and rutile phases, and interfacial storage are revealed to be beneficial for the high-kinetic reaction in lithium-ion batteries [28]. Anatase phase mesoporous TiO_2 has been synthesized using the urea assisted hydrothermal process and used for high power performance for LIBs applications [29]. The material obtained as mono-phasic TiO_2 sub-microspheres of uniform particle size with I41/amd space group. From the rate capability behavior, material has a specific charge capacity of ~162 mA·h·g^{-1} at current rate of 0.5 C and slightly reduced to 160, 154 and 147 mA·h·g^{-1} at 1, 5 and 10C-rates, respectively. The obtained good performance due to the large surface area (~116 m^2·g^{-1}) introduced by the highly porous (pore size of ~7 nm) nano-structured building blocks of each anatase TiO_2 sub-microspheres resulting to make favorable and short diffusion pathway for ionic and electronic diffusion. The near zero strain behavior of this anatase phase mesoporous TiO_2 makes it a suitable anode material for high power lithium applications.

Serrapede *et al.* [30] have compared and quantified the charging mechanisms of defective TiO_2 with respect to stoichiometric anatase. The porous TiO_2 samples have been synthesized by simple thermal oxidation of titanium foil in a 111 volume hydrogen peroxide solution (30% (w/w) in H_2O) at low temperature. Two different thermal treatments have been applied to form a defective TiO_2 by annealing the as-grown material in Ar atmosphere at 150°C and a stoichiometric TiO_2 by air annealing at 450°C. These TiO_2-based electrodes have been used for energy-storage applications by using three electrode cell in de-aerated 1 M $LiPF_6$ in EC:DMC with a Li foil and a Li ring as counter and reference electrodes, respectively. As results, the defective TiO_2 exhibited a higher Li molar fraction with respect to the crystalline one, with a common diffusion-controlled charging-discharging mechanism, initiated by Li^+ ions. A rapid extraction, with respect to Li^+ ion-insertion was observed in both samples. With the theoretical capacity of stoichiometric TiO_2, it was found to be stable for 200 cycles while the defective material exhibited with a lower capacity and it maintained stable even after 1550 cycles. Moretti *et al.* [31] have performed electrochemical test for the nanocrystalline anatase TiO_2 electrodes synthesized by using sodium-CMC binder, alone and in combination with styrene butadiene rubber (SBR) via aqueous process. The electrochemical behavior of these electrodes has been compared with those of electrodes manufactured using traditional polyvinylidene difluoride (PVDF) and its copolymer with hexafluoropropylene (PVDF-HFP) in N-methyl-2-pyrrolidone solution. In case of CMC-based electrodes, a spontaneous reaction of CMC towards the electrolyte probably due to the presence of hydroxide groups at the surface leading to the formation of a surface film, whose presence appears to be beneficial with respect to the first cycle behavior. The study demonstrated that TiO_2 electrodes consist of CMC or without CMC/SBR mixture as binder material exhibit an improved electrochemical performance (energy density of ~ >120 mW·h·g^{-1}) compared to those based fluorinated binders. The adopted aqueous process has been facilitated as environmental friendly and cost-effective fluorine free lithium-ion electrode preparation process.

TiO_2 bulk NPs have been synthesized using modified hydrothermal process under aqueous medium [32]. Their behavior as an intercalation host for Li^+ ions has been explored by performing an electrochemical test in which the active material

Sr. No.	TiO_2 based anode materials and method of synthesis	LIBs or SIBs performance	Ref.
A	**Carbon Coating Approach**		
1	Anatase TiO_2 nanorods synthesized by a hydrothermal method followed by carbon coating	Capacity of ~193 $mA{\cdot}h{\cdot}g^{-1}$ on the first charge in a sodium cell	[8]
2	Porous anatase TiO_2 synthesized by aqueous sol–gel process followed by coatings on carbon nanotubes (CNTs)	Reversible capacity as high as ~200 $mA{\cdot}h{\cdot}g^{-1}$ at a current density of 0.1 $A{\cdot}g^{-1}$ for LIBs	[9]
3	CNT@mesoporous TiO_2 hybrid nanocables prepared by a combined sol–gel and hydrothermal route by using hexadecylamine as a structure directing agent	Discharge capacity as high as ~183 $mA{\cdot}h{\cdot}g^{-1}$ at current of 1 C for LIBs	[10]
4	TiO_2/MWCNT composite prepared using via a combined sol–gel and solvothermal method	Discharge capacity as high as ~316 $mA{\cdot}h{\cdot}g^{-1}$ with good reversibility and stability after 100 cycles	[11]
5	Hybrid mesoporous CNT@TiO_2-C nanocable using anatase TiO_2, CNT and glucose (as carbon source) as for structure directing agent.	charge capacity of ~187 $mA{\cdot}h{\cdot}g^{-1}$ after 2000 cycles at current of 5 C and ~ 122 $mA{\cdot}h{\cdot}g^{-1}$ after 2000 cycles at current of 50 C for LIBs	[12]
6	Prepared TiO_2 precursor deposited onto MWCNT/SiO_2 using combustion chemical vapor deposition (CCVD) process	Performance that is significantly better and more applicable than current electrochemical capacitors for LIBs	[13]
7	Hierarchical rod-in-tube structure TiO_2 with a uniform conductive carbon layer using solvothermal method	Discharge capacity values of ~277 $mA{\cdot}h^{-1}{\cdot}g^{-1}$ at 50 $mA{\cdot}g^{-1}$ and ~ 153 at 5000 $mA{\cdot}g^{-1}$ and retained over 14000 cycles	[14]
8	Anatase TiO_2 NPs anchored on CNTs using ammonia water assisted hydrolysis and *in situ* crystal transformation under calcination in Ar at 500°C	Cycling stability values of ~92 $mA{\cdot}h{\cdot}g^{-1}$ remained at a current density of 10 $A{\cdot}g^{-1}$ at current of 60 C in LIBs	[15]
9	Phase-pure anatase TiO_2 nanofibers with fiber-in-tube and filled structures by electro-spinning process using tetra-*n*-butyl titanate-polyvinylpyrrolidone	Li-ion storage capacity (~231 $mA{\cdot}h{\cdot}g^{-1}$)	[16]
10	Self-supported, three-dimensional single-crystalline nanowire array electrodes by using simple hydrothermal process: TiO_2, TiO_2-C and TiO_2-C/SnO_2 have been produced on flexible Ti foil	Discharge capacity higher than ~160 $mA{\cdot}h{\cdot}g^{-1}$ and a capacity retention rate of ~84% even after 100 cycles at a current rate of 10 C	[17]
11	A single crystalline TiO_2, TiO_2-C and TiO_2-C/MnO_2 core-double-shell nanowire arrays on flexible Ti foil through a layer-by-layer deposition technique	Performance of ~332 $mA{\cdot}h{\cdot}g^{-1}$ at current of 2 C in LIBs	[18]
12	TiO_2 NPs on reduced GO nanosheets by a microwave hydrothermal process	Discharge capacity of ca. 100 $mA{\cdot}h{\cdot}g^{-1}$ with >99% coulombic efficiency at C-rates of up to 20 C in LIBs	[19]
13	Hydrothermal process to synthesize 3D mesoporous TiO_2 nanocubes grown on RGO nanosheets without use of any surfactants and high temperature of calcination	LIBs with high specific capacity value of ~180 $mA{\cdot}h{\cdot}g^{-1}$ at current of 1.2 C after 300 cycles	[20]

Sr. No.	TiO_2 based anode materials and method of synthesis	LIBs or SIBs performance	Ref.
14	Mixed detonation nanodiamond (DND) and TiO_2 hollow nanospheres synthesized by sol–gel route	LIBs with reversible capacity value of ~348 mA·h·g^{-1} at current of 0.5 C after 100 cycles	[21]
B	**Doping Approach**		
15	Nb-doping mesoporous TiO_2 by one-pot solvothermal process using carboxylic acids as organic additives in supercritical methanol medium.	Li-extraction capacity of ~147 mA·h·g^{-1} at the 1000th cycle at a high current rate of 10 C Na-extraction capacity of ~128 mA·h·g^{-1} upto 1000 cycles	[23]
16	TiO_2 ceramics co-doped with niobium (Nb^{5+}) and erbium (Er^{3+})	Better dielectric properties than the singly-doped ones	[24]
C	**By using conducting polymer**		
17	Hybrid bifunctional nanocomposite film based on TiO_2 nanowires and Polyoxometalates (POMs) using combination of hydrothermal and layer-by-layer self assembly methods	Coloration efficiency (~150 cm^2·C^{-1} at 600 *nm*) and cyclic stability with its volumetric capacitance about ~172 F·cm^{-3} with working voltage window of ~1.77 V, and maintained after 1000 cycles at 0.18 mA·cm^{-2}	[25]
D	**Combination with other semiconductors**		
18	Nanocrystalline TiO_2 mixed with V_2O_5 with different compositions using sol–gel synthesis process at different calcination temperatures	Energy storage ability: Initial charging rate of ~3700 C·mol^{-1}·h^{-1}	[26]
E	**Impregnation Method**		
19	Form-stable composite of 1-tetradecanoic acid (TDA) and TiO_2 powder using a facile impregnation method	Superior thermal stability with latent heat storage capacity of ~97 J·g^{-1}	[27]
20	Hierarchical TiO_2 sphere composed of ultrathin nanotubes having mixed anatase and rutile phases synthesized from a low-temperature hydrothermal reaction and calcination	Specific capacity of ~167 mA·h·g^{-1} at 1 A·g^{-1} current of 6 C and maintained ~187 mA·h·g^{-1} at 10 A·g^{-1} at current of 60 C	[28]
21	Anatase phase mesoporous TiO_2 synthesized using the urea assisted hydrothermal process	Specific charge capacity of ~162 mA·h·g^{-1} at current rate of 0.5 C and slightly reduced to 160, 154 and 147 mA·h·g^{-1} at 1, 5 and 10 C-rates, respectively	[29]
22	Porous TiO_2 samples have been synthesized by simple thermal oxidation of titanium foil	LIBs Stable for 200 cycles	[30]
23	Nanocrystalline anatase TiO_2 electrodes synthesized by using sodium carboxymethyl cellulose (CMC) binder, and in combination with styrene butadiene rubber (SBR) via aqueous process	Energy density of ~ >120 mW·h·g^{-1}	[31]
24	TiO_2 bulk NPs synthesized using modified hydrothermal process under aqueous medium	High capacity than other titanate materials at rated above 1000 mA·g^{-1}.	[32]

Table 1.
Synthesis methods and electrochemical performances for TiO_2-based materials.

(TiO_2) mixed with poly(vinylidene fluoride) and Super P carbon in the weight ratio of 70:20:10, then introduced into an electrochemical cell along with a lithium metal counter/reference electrode and liquid electrolyte. The mixture was cast onto copper foil from acetone using the Doctor-Blade technique. From this test, gravimetric capacity for the TiO_2 bulk NPs at all rates up to 18000 $mA \cdot g^{-1}$ was found to be identical to the 6 nm anatase particles which have a much higher proportion of carbon content. In comparison from nanowires to nanotubes and nanoparticles, the amount of Li and hence charge that can be stored, even at low rates, increased with reduced dimensions. The volumetric capacity of composite electrodes with nanoparticulate TiO_2 bulk was found to be notably high than other titanate materials at rated above 1000 $mA \cdot g^{-1}$.

Table 1 summarizes discussed synthetic modification processes for TiO_2-based materials to be used as anode materials for LIBs or SIBs.

3. Role of anode in LIBs: the aging mechanism

The anodic materials is operating in organic electrolyte such as $LiPF_6$ with co-solvents like ethylene carbonate (EC), dimethyl carbonate (DMC), diethyl carbonate (DEC), methyl ethyl carbonate (EMC) [33, 34]. Generally, the most important part of the LIBs cell is the anode/electrolyte interface because of high reactivity of the organic electrolyte with any type of electrode material and lithium ions for LIBs. Due to the interaction occurring between the composite anode and the electrolyte solution would form organic species on the anode surface that can be recognized as solid electrolyte inter-phase (SEI). The produced organic species can undergo reduction with CO_2 and traces of H_2O in the electrolyte to form lithium carbonate which further react with used organic solvents to form transesterification products. On other hand, anion contaminates, such as F^- from HF and PF_5, readily react with lithium to form insoluble reaction products which are non-uniform, electronically insulating, and unstable on the surface of the anodic material. At higher battery potentials, during the intercalation of lithium ions into the anode lattice structure, within the reactions continue, the consumption of lithium ions with co-solvents can precipitate and grow on the anode surface. This process consequently results in the formation of a protective, ionically conductive but electrically insulating passive layer on the surface of the anode during the first charge cycle, the so-called SEI. This SEI creates resistance to lithium ion flow, which results in a rise in the charge transfer resistance and the impedance of the anode. This increase follows charge rate, cycle number, temperature, and surface area/particle size of anodic material. The stable growth of SEI may leads to the loss of active lithium and further decomposition of the electrolyte. This phenomenon is the main degradation mechanism in fully charged batteries at storage conditions [33, 34].

Furthermore, interactions of the anode with the cathode should be included because of the dissolution of the cathode electrode metal from the lattice into the electrolyte solution may occur when batteries stored at voltages greater than ~3.6 V [34]. This may cause deposition of cation contaminates which are incorporated into the SEI layer. This kind of reactions can damage SEI and a short circuit may be generated, which then can lead to thermal runway and battery failure.

Fracture and decrepitating of the electrodes are critical challenges existing in lithium-ion batteries as a result of lithium diffusion during the charging and discharging operations. A large volume change on the order of a few to several hundred percent can be observed due to lithium ions intercalate and de-intercalate. Thus, diffusion-induced stresses (DISs) generate the nucleation and growth of cracks/damages, leading to mechanical degradation of the active electrode materials. Therefore, for nanoscale electrode structures, surface energies and surface stresses can be predictable to have an important impact on the mechanical properties of the electrode materials.

3.1 Effect of particle size, active surface area and porosity of the anode material

Generally, small particles contain short diffusion paths between the anodic particles, which make easy and fast charge and discharge rate the. Similarly, large surface area of the anodic material are prone to higher internal heat generation and lithium ion are consumed during the exothermic reaction at high temperatures (> 60°C) compared to larger particles size, this leads to an enhancement in the irreversible capacity of the anodic material. There is no a direct connection between the porosity of the anode and the reversible capacity of the anode. At high temperature (> 120°C), heat generation from a denser electrode material produces gaseous species through thermal decomposition of the SEI layer. This reveals about the importance of the thermal stability of an anodic material. Overall, the anode of the lithium ion battery undergoes several degradation mechanisms during aging. Ion (Li^+) plating is one aging mechanism which ends the life of a battery more rapidly due to the formation and growth of lithium dendrites. These kinds of degradation mechanisms rarely affect the crystal structure of the anode electrode.

In comparison with Li-ion insertion mechanism where Li_3OCl forms as an intermediate, the presence of Na_3OCl is not observed as an intermediate product after the first conversion due to it is metastable and its limited lifetime [35].

4. Conclusions

In conclusions, the huge amounts of efforts are in progress towards the synthesis, characterization and application of TiO_2 based anode materials in the fields of LIBs. Development can be seen with an aim to gain the superior electrochemical performance and promote the practical utilities of the synthesized TiO_2-based functionalized materials. The synthesis strategies have played a crucial role to establish a material composition with a desired nanostructure and appropriate morphologies including particle size/shape and surface area by adopting different modification methods as discussed in the chapter. The effective synthesis condition can easily provides material informations with unique characteristics and hence high lithium or sodium storage capacity and cyclic stability which are depend on ion flux at the electrode/electrolyte interface, internal resistance and diffusion path.

The energy storage capacity strongly influenced by materials structure and morphologies, thus various structural forms should be explored to enhance the electrochemical performance of modified TiO_2 materials. The chapter providing a bunch of literature reports on how synthetic process can alter the nanostructure that facilitates the electrochemical performance at minimum cost and good durability. The formation of one dimensional (1D), two dimensional (2D) and three dimensional (3D) hierarchical nanostructures have been found to be much stable materials for electrochemical performance. The doping approaches also can open more windows to modified TiO_2 matrices and therefore different structural forms. The uniformly carbon coating also making a huge impact on formation of TiO_2 based modified materials with good surface area and electronic conductivity favoring good energy storage performances. The use of various carbon sources like CNTs, MWCNT and graphene oxide have formed the homogenously dispersed nanocomposites showing a stable cyclic performance for LIBs.

Overall, progressive research works have been well established for TiO_2 to be used as anode materials in the field of energy storage. Although, still challenges are there to improve the Li ion storage performance like low coulombic efficiency, low volumetric energy density etc. To solve the fundamental issues, more development towards material surface alteration or coating to reduce unwanted side reactions and designing hierarchical structural materials by adopting different experimental conditions.

In addition, on other hand, sodium ion battery (SIBs) getting more attention which has to be used as thrust area of research. The prepared materials should be performed for LIBs and SIBs as well.

Acknowledgements

The authors thank the University Grants Commission (UGC), New Delhi, India, for financial assistance under the Dr. D.S. Kothari Post Doctoral Fellowship (No. F.4.-2/2006/(BSR)/CH/16-17/0049).

Conflict of interest

The authors declare no conflicts of interest.

Acronyms and abbreviations

NPs	Nanoparticles
LIBs	Lithium Ion Batteries
SIBs	Sodium Ion Batteries
0D	Zero-dimensional
1D	One-dimensional
2D	Two-dimensional
3D	Three-dimensional
Ar	Argon
RGO & rGO	Reduced Graphene Oxide
CNTs	Carbon Nanotubes
MWCNT	Multi-walled Carbon Nanotubes
CCVD	Combustion Chemical Vapor Deposition
TBT	Tetra-*n*-butyl titanate
PVP	Polyvinylpyrrolidone
EIS	Electrochemical Impedance Spectroscopy
DND	Detonation Nanodiamond
POMs	Polyoxometalates
PVDF	Polyvinylidene Difluoride
HFP	Hexafluoropropylene
SBR	Styrene Butadiene Rubber
CMC	Carboxymethyl Cellulose
LiTFSA	bis(trifluoromethanesulfonyl)amide.
PP	Polypropylene
EC	Ethylene Carbonate
DMC	Dimethyl Carbonate
DEC	Diethyl carbonate
EMC	Methyl Ethyl Carbonate

SEI	Solid Electrolyte Inter-phase
DIS	Diffusion-Induced Stresses

Appendices and nomenclature for scientific units

nm	Nanometer
cm	centimeter
°C	Degree Celsius
C	Coulomb
h	hour
mA	mili Ampere
g	gram
F	Farad
V	Volt
W	Watt
s	Second

Author details

Tarun Parangi* and Manish Kumar Mishra
Department of Chemistry, Sardar Patel University, Vallabh Vidyanagar,
Gujarat, India

*Address all correspondence to: juvenile34@gmail.com

References

[1] Huang Y, Zhu M, Huang Y, Pei Z, Li H, Wang Z, Xue Q, Zhi C. Multifunctional energy storage and conversion devices. Adv. Mater. 2016;28:8344-8364. DOI: 10.1002/adma.201601928.

[2] Zhou F, Ren Z, Zhao Y, Shen X, Wang A, Li Y, Surya C, Chai Y. Perovskite photovoltachromic supercapacitor with all-transparent electrodes. ACS Nano. 2016; 10:6:5900-5908. DOI: org/10.1021/acsnano.6b01202.

[3] Wang K, Wu H, Meng Y, Zhang Y, Wei Z. Integrated energy storage and electrochromic function in one flexible device: An energy storage smart window. Energy Environ. Sci. 2012;5:8384-8389. DOI: org/10.1039/C2EE21643D.

[4] Zhang W, Tian Y, He H, Xu L, Li W, Zhao D. Recent advances in the synthesis of hierarchically mesoporous TiO_2 materials for energy and environmental applications. Nation. Sci. Rev. 2020;7:1702-1725. DOI: 10.1093/nsr/nwaa021.

[5] Deng D, Kim MG, Lee JY, Cho J. Green energy storage materials: Nanostructured TiO_2 and Sn-based anodes for lithium-ion batteries. Energy Environ. Sci. 2009;2:818-837. DOI: org/10.1039/B823474D.

[6] Bruce PG, Scrosati B, Tarascon J-M. Nanomaterials for rechargeable lithium batteries. Angew. Chem., Int. Ed. 2008;47:2930-2946. DOI: org/10.1002/anie.200702505.

[7] Liu Y, Yang Y. Recent progress of TiO_2-based anodes for Li Ion Batteries. J Nanomater. 2016;8123652. DOI: org/10.1155/2016/8123652.

[8] Kim K-T, Sun Y-K, Ali G, Lu J, Chung KY, Amine K, Yoon CS, Myung S-T, Yashiro H. Anatase titania nanorods as an intercalation anode material for rechargeable sodium Batteries. Nano Lett. 2014;14:416-422. DOI: org/10.1021/nl402747x.

[9] Yan L, Xu Y, Zhou M, Chen G, Deng S, Smirnov S, Luo H, Zou G. Porous TiO_2 conformal coating on carbon nanotubes as energy storage materials. Electrochimica Acta. 2015;169:73-81. DOI: org/10.1016/j.electacta.2015.04.061.

[10] He L, Wang C, Yao X, Ma R, Wang H, Chen P, Zhang, K. Synthesis of carbon nanotube/mesoporous TiO_2 coaxial nanocables with enhanced lithium ion battery performance. Carbon. 2014;75:345-352. DOI: org/10.1016/j.carbon.2014.04.013.

[11] Trang NTH, Ali Z, Kang, DJ. Mesoporous TiO_2 spheres interconnected by multi-walled carbon nanotubes as an anode for high-performance lithium ion batteries. ACS Appl. Mater. Interfaces. 2015;18:7(6): 3676-3683. DOI: 10.1021/am508158v.

[12] Wang B, Xin H, Li X, Cheng J, Yang G, Nie F. Mesoporous CNT@ TiO_2-C nanocable with extremely durable high rate capability for lithium-ion battery anodes. Sci. Rep. 2014;4:3729. DOI: 10.1038/srep03729.

[13] Marler S. Hybrid multi-walled carbon nanotube TiO_2 electrode material for next generation energy storage devices. arXiv:1606.00820.

[14] He H, Gan Q, Wang H, Xu G-L, Zhang X, Huang D, Fu F, Tang Y, Amine K, Shao M. Structure-dependent performance of TiO_2/C as anode material for Na-ion batteries. Nano Energy. 2017;44:217-227. DOI: org/10.1016/j.nanoen.2017.11.077

[15] Liu J, Feng H, Jiang J, Qian D, Li J, Peng S, Liu Y. Anatase-TiO_2/CNTs

nanocomposite as a superior high-rate anode material for lithium-ion batteries. J Alloys Compounds 2014;603:144-148. DOI: org/10.1016/j.jallcom.2014.03.089.

[16] Cho JS, Hong YJ, Kang YC. Electrochemical properties of fiber-in-tube- and filled-structured TiO_2 nanofiber anode materials for lithium-ion batteries. Chem. Eur. J. 2015;21: 11082-11087. DOI: org/10.1002/chem.201500729.

[17] Liao JY, Manthiram A. Mesoporous TiO_2-Sn/C core-shell nanowire arrays as high-performance 3D anodes for Li-ion batteries. Adv. Energy Mater. 2014;1400403-1400410. DOI: 10.1002/aenm.201400403.

[18] Liao JY, Higgins D, Lui G, Chabot V, Xiao X, Chen Z. Multifunctional TiO_2—C/MnO_2 core-double-shell nanowire arrays as high-performance 3D electrodes for lithium ion batteries. Nano Lett. 2013;13:5467—5473. DOI: org/10.1021/nl4030159.

[19] Farooq U, Ahmed F, Pervez SA, Rehman S, Pope MA, Fichtner M, Roberts EPL. A stable TiO_2-graphene nanocomposite anode with high rate capability for lithium-ion batteries†. RSC Adv. 2020;10:29975-29982. DOI: 10.1039/d0ra05300g.

[20] Yue Z, Zhen X, Ai J, Shi Q, Mao T, Zhen M, Wang Z, Peng S. Three-dimensional mesoporous nanocube TiO_2/reduced graphene oxide composites with enhanced lithium storage properties. J. Renew. Sustain. Energy. 2020;12,044101-044106. DOI: org/10.1063/5.0013254.

[21] Gao X, Sun X, Jiang Z, Wang Q, Gao N, Li H, Zhang H, Yu K, Su C. Introducing nanodiamond in TiO_2-based anodes for improving the performance of lithium-ion batteries. New J. Chem., 2019;43:3907-3912. DOI: 10.1039/C8NJ06226A.

[22] Parangi T, Mishra MK. Titania nanoparticles as modified photocatalysts: A review on design and development. Comm. Inorg. Chem. 2019;39:2:90-126. DOI: org/10.1080/02603594.2019.1592751.

[23] Tanaka Y, Usui H, Domi Y, Ohtani M, Kobiro K, Sakaguchi H. Mesoporous spherical aggregates consisted of Nb-doped anatase TiO_2 nanoparticles for Li and Na storage materials. ACS Appl. Energy Mater. 2019;2:1:636-643. DOI:10.1021/acsaem.8b01656.

[24] Tse M-Y, Wei X, Hao J. High-performance colossal permittivity materials of (Nb + Er) co-doped TiO_2 for large capacitors and high-energy-density storage devices. Phys. Chem. Chem. Phys. 2016;18:24270-24277. DOI: 10.1039/c6cp02236g.

[25] Qu X, Fu Y, Ma C, Yang Y, Shi D, Chu D, Yu X. Bifunctional electrochromic-energy storage materials with enhanced performance obtained by hybridizing TiO_2 nanowires with POMs. New J. Chem. 2020;44:15475-15482. DOI: org/10.1039/D0NJ02859B.

[26] Ngaotrakanwiwat P, Meeyoo V. TiO_2-V_2O_5 nanocomposites as alternative energy storage substances for photocatalysts. J Nanosci. Nanotechnol. 2012;12:828-833. DOI: 10.1166/jnn.2012.5381.

[27] Deka PP, Ansu AK, Sharma RK, Tyagi VV, Sari A. Development and characterization of form-stable porous TiO_2/tetradecanoic acid based composite PCM with long-term stability as solar thermal energy storage material. Int. J. Energy Res. 2020;1-14. DOI: 10.1002/er.5615.

[28] Song W, Jiang Q, Xie X, Brookfield A, McInnes EJL, Shearing PL, Brett DJL, Xie F, Riley DJ. Synergistic storage of lithium ions in defective anatase/rutile TiO2 for high-rate batteries. Energy Storage

Mater. 2019;22:441-449. DOI: org/10.1016/j.ensm.2019.07.025.

[29] Jung HG, Oh SW, Ce J, Jayaprakash N, Sun Y-K. Mesoporous TiO_2 nano networks: Anode for high power lithium battery applications. Electrochem. Commun. 2009;11: 756-759. DOI:10.1016/j.elecom.2009. 01.030.

[30] Serrapede M, Savino U, Castellino M, Amici J, Bodoardo S, Tresso E, Chiodoni A. Li^+ Insertion in nanostructured TiO_2 for energy storage. Materials. 2020;13,21. DOI: 10.3390/ ma13010021w.

[31] Moretti A, Kim G-T, Bresser D, Renger K, Paillard E, Winter M, Passerini S, Marassi R. Investigation of different binding agents for nanocrystalline anatase TiO_2 anodes and its application in a novel, green lithium-ion battery. J. Power Sources. 2013;221:419-426. DOI: org/10.1016/j. jpowsour.2012.07.142.

[32] Ren Y, Liu Z, Pourpoint F, Armstrong AR, Grey CP, Bruce PG. Nanoparticulate TiO_2(B): An Anode for lithium-ion batteries**. Angew. Chem. Int. Ed. 2012;51:2164-2167. DOI: 10.1002/anie.201108300.

[33] Lin C, Tang A, Mu H, Wang W, Wang C. Aging mechanisms of electrode materials in lithium-ion batteries for electric vehicles. J Chem. 2015;1-11. DOI: org/10.1155/2015/104673.

[34] Agubra V, Fergus J. Lithium ion battery anode aging mechanisms. Materials. 2013;6:1310-1325. DOI:10.3390/ma6041310.

[35] Dutta PK, Myung Y, Venkiteswaran RK, Mehdi L, Browning N, Banerjee P, Mitra S. Mechanism of Na-Ion Storage in BiOCl anode and the sodium-ion battery formation. J. Phys. Chem. C. 2019;123:11500-11507. DOI: 10.1021/acs. jpcc.9b01807.

Section 2

Semiconductor Applications of Titanium Dioxide

Chapter 3

TiO_2: A Semiconductor Photocatalyst

Azrina Abd Aziz, Fatema Khatun, Minhaj Uddin Monir, Sim Lan Ching and Leong Kah Hon

Abstract

Titanium dioxide (TiO_2) is considered as an inert and safe material and has been used in many applications for decades. TiO_2 have been widely studied, due to its interesting general properties in a wide range of fields including catalysis, antibacterial agents, in civil as nano-paint (self-cleaning) and especially photocatalysis, and that affect the quality of life. Thus, the development of nanotechnologies TiO_2 nanoparticles, with numerous novel and useful properties, are increasingly manufactured and used. TiO_2 doped with noble metal are good candidates in the performance these applications. The fascinating physical and chemical features of TiO_2 depend on the crystal phase, size and shape of particles. For example, varying phases of crystalline TiO_2 have different band gaps that rutile TiO_2 of 3.0 eV and anatase TiO_2 of 3.2 eV, determine the photocatalytic performance of TiO_2. This chapter explains basic information on TiO_2 and theoretical concepts of nanostructure of TiO_2 nanoparticles as a semiconductor photocatalyst.

Keywords: TiO_2, nanoparticles, semiconductor, photocatalyst

1. Introduction

An ideal photocatalyst should posseses following nature: biologically and chemically inert, inexpensive, nontoxic, stable towards photocorrosion and certainly highly active and suitable towards the visible/UV light photoreaction. An additional criterion for such ideal photocatalyst is better redox potential of the $H_2O/{}^{\bullet}OH$ couple (EOP = 2.80 V) [1]. Among the available semiconductors, TiO_2, ZnO, ZnS and WO_3 have gap energies sufficient for catalyzing a wide range of chemical reactions [2].

Binary metal sulphide semiconductors such as CdS, CdSe or PbS are regarded as unstable photocatalyst due to easy photoanodic corrosion property and their toxic nature [2, 3]. Furthermore, the iron oxide semiconductors are not suitable as they readily perform the photocathodic reaction and finally produced corrosive materials [1]. ZnO on the other hand is known to have the similar quality of band gap energy like TiO_2 (3.0 eV) but it is highly unstable towards pH condition, where precipitate of Zn $(OH)_2$ is formed on the particle surface and resulted in photocatalyst deactivation [2].

Of these, titanium dioxide (TiO_2) is the most ideal semiconductor for photocatalytic destruction of organic compounds [2, 4]. It also provides an excellent conciliation between catalytic performance and stabilization in aqueous media [5].

 IntechOpen

2. Titanium dioxide (TiO_2)

The strong oxidation and reduction power of photoexcited TiO_2 was successfully discovered by Honda-Fujishima [6, 7]. Since its innovation, extensive efforts have been made in the development of TiO_2 photocatalyst owing to its potential application in the removal of various types of organics and in both aqueous and air phase [8, 9]. It resulted in flexible applicability either in the form of a suspension, or immobilization [10–13].

Regardless the high superior photocatalytic efficiency of TiO_2, it is popular for its unique wider applications as electroceramic, antifogging agent and as therapy for cancer [14, 15]. Moreover, it also acts as an antibacterial agent due to its strong oxidation activity and superhydrophilicity and was employed for electrolysis of water (H_2O) to produce H_2 and for harvesting solar energy as dye-sensitized solar cells [16]. Additionally, it possesses superior pigmentary property, UV light absorption capability and durability [17]. The high refractive indices of rutile and anatase phase of TiO_2 laid a pathway for its versatility nature.

In general, the crystallinity, impurities, specific surface area and density of the surface hydroxy groups are the physical parameters that have influence over the photocatalytic activity of TiO_2. Apart from that the band gap energy, charge-carrier and recombination as well play a crucial role for photoactivation [18, 19]. Among the available phases the anatase is the most stable form with 8–12 kJ/mol. This describes the quantity of energy transferred during phase transformation. The rest phases such as brookite is very uncommon [14, 20–21].

In the terms of density, rutile is greater than anatase, i.e., 4.26 and 3.90 g/mL respectively. However, in terms of utilization and ability, anatase phase is more efficient due to its open structure [14]. The anatase phase remains as most active photocatalyst with greater stability after repeated catalytic cycles illuminated under a photon energy between 300 nm $< \lambda <$ 390 nm [22]. In addition, it is also chemically stable in aqueous media and in broad range of pH ($0 < pH < 14$) [23]. Hence TiO_2 with its major anatase phase had more applicability as photocatalyst [24]. Furthermore, it also exhibits relatively high reactivity and chemical stability only under UV light ($\lambda <$ 387 nm) [7]. Thus, its limited utilization of the activation energy becomes a major drawback. Therefore, to overcome this limitation of TiO_2, an extensive research needs to be emphasized to develop a titania based photocatalyst that can exhibit high reactivity under visible light spectrum ($\lambda >$ 400 nm) and can persuades the utilization of solar spectrum, even under poor luminance of interior lighting [7].

The reduction in energy gap between the CB and VB further leads to the easy recombination of the formed electrons and holes. This recombination further decreases the interfacial charge-carrier transfer. The size of the TiO_2 further causes difficulties in separating them from the waste's stream. All these issues stress the reseachers to find a suitable composition/dopant for its modification [25].

3. Modification of TiO_2 photocatalyst

The photoactivity of TiO_2 is hindered by the narrow UV wavelength spectrum for photoactivation under ambient conditions. The generation of this UV requires additional power source, which shoots the activation cost of the photocatalyst. Therefore, the utilization of renewable energy source could be a better sustainable choice for the photoactivation [26].

Type	Remark	Reference
Chromium (Cr)	This is a study on the effect of the photoreduction of dinitrogen in a gas–solid regime using Cr as a dopant for phenol degradation. The prepared Cr-TiO_2 showed an enhanced activity and displacement of the Fermi level at the TiO_2 interface. But the presence of Cr as donor ion did not favor the charge separation with respect to pure TiO_2.	[34]
Iron (Fe)	Fe^{3+} ion has a unique half-filled electronic structure which resulted in enhanced activity compared to Fe^{4+} and Fe^{2+}. The stable Fe^{3+} ion detrapped the electron and hole to adsorbed oxygen and surface hydroxyl group, thereby surpressed the electron–hole recombinations.	[35]
	This study is about the effect of Fe doping into TiO_2 for the degradation of chloroform ($CHCl_3$). It observed that Fe provided shallow trapping sites for the charge-carriers and increase the photocatalytic efficiency. It also found that Fe can act as trap for both electron and hole, at high dopant concentration.	[36]
	This study is about the Fe-doped with TiO_2 for the degradation of 1, 2-dichloroethane. The study proved that Fe^{3+} alleviates the surface phenomenon and act as both electron–hole traps.	[37]
	The effect of Fe doping for the inactivation of *E. coli* was studied. The Fe-doped TiO_2 proved the enhancement in the inactivation of *E. coli*. Fe played as a source for the inhibition of the anatase crystal growth.	[38]
Molybdenum (Mo)	The Mo as a cationic dopant was studied in 2009. The dopant increased the surface acidity of TiO_2 and accelerated the interfacial charge transfer process. Thus, prepared doped photocatalyst degraded the cationic dyes, rhodamine B (RhB) and methylene blue (MB) and anionic dyes, methyl orange (MO) and congo red (CR) at both alkaline and acidic pH.	[39]
Zinc (Zn)	Similarly, the same group studied the effect of Zn as a dopant in 2010 and applied the prepared photocatalyst for the degradation of CR. They produced a small crystallite size and stable filled electronic configuration of Zn^{2+}-TiO_2. Moreover, this photocatalyst provided a shallow trap for the charge-carriers contributing to the overall activity.	[40]
Manganese (Mn)	They further extended their study with Mn as a potential cationic dopant in 2011. From the study, it showed that Mn^{2+} intimate contact between the mixed anatase and rutile phase of TiO_2. Moreover, this dopant has synergistic effects in the bicrystalline framework of anatase and rutile. The uniqued half-filled electronic structure of Mn^{2+} served as a shallow trap for the charge-carrier contributed for the appreciable degradation of both Indigo carmine dye and 4-chlorophenol.	[41]
Cobalt (Co)	Amadelli et al., 2008 studied the effect of Co-doping. They prepared the photocatalyst by incipient impregnation method and cobalt acetate as a precursor. Co addition brings about conspicuous changes in the point of zero charge and in surface polarity. They found that Co-doped TiO_2 is more active compared to TiO_2. The best photocatalytic result of the prepared photocatalyst is obtained for heat treatment at 400°C for 30 min.	[42]
Thorium (Th)	Thorium a naturaly occurring radioactive element was studied as a dopant in 2009. It observed that Th can contributed well for the BET surface area of photocatalyst. Moreover, the presence of two absorption edges at 460 and 482 nm in visible region enabled the photocatalyst to utilize more visible light. They reported that the strong adsorption of the pollutant was due to the complex formation between the vacant f orbital of Th^{4+}. Besides, they observed that Th induced the oxygen vacancies which served as shallow traps. But thorium could further contribute for radioactive pollution.	[39]

Table 1.
Literature review of selected metal dopant/hybrid on TiO_2.

Type	Remark	Reference
Nitrogen (N)	In 2001, researchers prepared $TiO_{2-x}N_x$ by sputtering the TiO_2 selectively on N_2 (40%)/Ar gas mixture for degradation of MB and acetaldehyde (CH_3CHO) under UV/visible light.	[43]
	Researchers prepared a narrow band gap N-TiO_2 with oxygen vacancies. Thus, prepared N-TiO_2 failed to oxidize formate and NH_3-OH^+ under visible light illumination. Their preparation method resulted the N as a weak anion donor.	[44]
	This is the study about the preparation of N-TiO_2 by thermal decomposition of the Ti^{4+}-bipyridine complex and had moderately better removal of NO_x under artifical visible light.	[45]
	The researchers prepared $TiO_{2-x}N_x$ by solvothermal route, produced violet color of particles and further calcination between 200 and 800°C produced weak violet, bright yellow, weak yellow and gray; no white color of particles. Their findings proved the thermal stability of Ti-N bonding with $TiO_{2-x}N_x$.	[46]
	N-TiO_2 was prepared using Ti-melamine and Ti-salen complex as precursors and reported a higher photoactivity was obtained for the Ti-melamine by degrading MB degradation. The reasoned for the low photoactivity of the Ti-salen complex was due to the low amount of N doping and smaller Ti-N bonds.	[47]
	N-doped layered titanates ($Cs_{0.68}Ti_{1.83}O_{4-x}N_x$ and $H_{0.68}Ti_{1.83}O_{4-x}N_x$) with leciprocate structure was prepared and achieved better photoactivity by degradating RhB under visible light illumination. The unique layered structure of titanates and homogeneous distribution of N dopant contributed for the mobility of the charge-carrier further contributed for the faster and better photoactivity.	[48]
	Multitype N-TiO_2 containing both subtitutional (N-Ti-O and Ti-N-O) and interstitial NO characteristics was prepared. The subtitutional and interstitial technique resulted in intraband states at 0.14 and 0.73 eV above the VB resulting in faster visible light driven photodegradation of gas phase toluene.	[49]
Carbon (C)	A carbonaceous coke-like species embedded in the TiO_2 matrix was obtained by pyrolysis of alcohol in the sol–gel titania processing and calcined at 250 and 400°C. The resulted C-TiO_2 calcined at 250°C exhibited a maximum photoactivity for degradation of 4-chlorophenol (4-CP) under visible light than that the later. Further calcination to the higher tempearature caused the elimination of carbon from the titania matrix. Thus, showed significance of the calcination temperature.	[50]
	A rutile TiO_2 doped with carbon was prepared by a pyrolysis method using combustion products namely CO_2 and H_2O and used for water splitting reaction. Their doping method resulted in a band gap of 2.32 eV against 3.0 eV and showed the supremacy of the preparation method.	[51]
	TiO_2-G (titania-glycine) was successfully prepared through solution combustion route by glycine (G), hexemethylene (H) and oxalyldihydrazide (O) as fuel. The precursors employed contributed for the carbide ion and thus reduced band gap energy and resulted in a superior photoactivity, crystallinity and surface area. The photoactivity was successfully experimented by degrading aqueous phase MB under solar irradiation.	[52]
	TiO_2-C was prepared through a simple carbon sources, which is tetrabutylammonium hydroxide ($C_{16}H_{37}NO$) and glucose ($C_6H_{12}O_6$) as precursors. The adopted precursors contributed in two visible absorption edges in their band gap namely 2.78 and 1.45 eV, respectively.	[53]
	Park et al., 2009 showed the importance of optimum calcination temperature like Lettmann and coworkers in 2001.	[54]

Type	Remark	Reference
Fluorine (F)	The researchers chose F as an anionic dopant and obtained F-TiO_2 via spray hydrolysis from fluorotitanic acid (H_2TiF_6) precursor. Their work showed that F can excited under visible light but does not contributed for the intrinsic absorption properties rather that the extrinsic properties of TiO_2 absorption.	[55]
Boron (B)	Boron was selected as a potential anion dopant and obtained B-TiO_2 which resulted in wider band gap energy, larger thermodynamic and faster charge-carrier transfer.	[56]
Phosphorus (P)	Phosphorus as dopant was experimented via phosphoric acid (H_3PO_4) and had higher photooxidation of *n*-pentane in air. It was due to the formation of Ti ions in tetrahedral coordination of TiO_2.	[57]

Table 2.
Literature review of selected non-metal dopant/hybrid on TiO_2.

Attempt to shift the optical response of TiO_2 from the UV to the visible spectrum will have a profound positive effect on the photocatalytic efficiency [27]. The visible light reception potential of these photocatalysts is either highly unstable under their illumination (e.g. CdS and CdSe) or have limited photoactivity (e.g. WO_3 and Fe_2O_3) [26]. Therefore, doping/hybridizing TiO_2 with transition metals like Fe, Co, Ni, etc. [28], non-metals like N, O, C, B, etc. [29], noble metals like Ag, Pt, Au or its metal ions incorporation [30], incorporating carbon nanotubes [31] and dyed sensitizers [32] have been proposed to improve its photocatalytic efficiency under solar light.

These modifications either by doping or hybridization are based on the concept of balancing both the half-reaction rates of the photocatalytic reaction by adding electron acceptor, or modifying the photocatalyst structure and composition [26]. The enhancement shown by doped ions is by providing charge trapping sites which leads to the reduction of electron–hole recombination [33]. In addition, the excited electron is expelled by the electron acceptors and hence hinders the recombination of electron–hole pairs [26].

In general, parameters such as types of doping/hybrid, its concentration, synthesizing technique and physico-chemical characteristics of the photocatalyst will have control over the photocatalytic activity. It was observed that the formation of both free $^{\bullet}OH$ and active O_2 species is improved with the increase in charge separation efficiency [26]. The detailed literature reviews of these metals and non-metals dopant/hybrid are tabulated in **Tables 1** and **2** respectively.

4. Modification of TiO_2 with graphene oxide

The incorporation of highly conductive carbon materials can also enhance the electron–hole pair separation of the photocatalyst. Graphene oxide commonly known as GO, is one such high conductive carbon materials that can be employed as a dopant/hybrid. Graphene can be regarded as the origin of all graphitic forms and can be curled, rolled or stacked to shape buckyball fullerenes, carbon nanotubes or graphite [58–60]. With free-standing 2-dimensional (2D) crystal and one-atom thickness properties, it has emerged with wide applications in several fields but employed mostly in nanotechnology for the improvisation of materials chemistry [61]. The unique single atom-thick planar sheet of sp^2 hybridized carbon atoms contribute to efficient storing and shuttling of electrons [61]. Moreover, it attracted the scientific community tremendously because of its distinctive electronic properties, superior chemical stability and soaring specific surface area [61].

The exfoliated graphene sheets employ a theoretical surface area of around 2600 m^2/g, and as a result graphene appears as an attractive high-surface area 2D photocatalyst support [62]. Besides it has potential ideal electron sinks or electron transfer bridges [63]. This was attributed to its exceptional structure that allow ballistic transport, in which electrons can travel without scattering at mobilities exceedingly approximately 15,000 $m^2/V/s$ at room temperature [63]. They are also foremost responsive for chemical doping, adsorbed or bound species and structure distortion [64]. Further incorporation of inorganic materials with modified graphene enormously improves their electronic, electrocatalytic and photocatalytic characteristics [32]. Thus, it proves the potential to enhance the fast electron transfer that highly benefits photocatalysis [32].

Recently modifying TiO_2 surface with carbonaceous materials propounded to induce visible-light responsive property. Few types of carbonaceous materials such as graphitic or coke-like carbon [50, 65], or carbonate structural fragments bonding with titanium were employed for this purpose. Graphene oxide supported TiO_2 is expected to create synergistic effect that enhances the solar photocatalytic activity. The synergistic effect attributes to its unique separation efficiency of electrons and holes between TiO_2 and graphene oxide [32, 66]. The photo-reductions initiated in the transformation of graphene oxide to graphene lays a platform for continuous electron conducting network through cross-surface charge percolation and permitted graphene to act as an efficient exciton sink [66].

Nguyen-Phan et al., [32] adopted a simple one-step colloidal blending method as an environmentally friendly that preserves the TiO_2 properties and combines the advantages of graphene oxide. The prepared composites showed superior adsorptivity and photocatalytic activity under both UV and visible light [32]. This was proved through a photocatalytic degradation study by adopting MB as model pollutant excited under artificial solar energy. The study indicated that graphene oxide acted as an adsorbent, electron acceptor and photosensitizer in the process of accelerating photodecomposition [32].

Recently Hu et al., 2012 also reported that graphene oxide/TiO_2 hybrid (GOT) demonstrated an excellent adsorption and photocatalysis performance under visible radiation by degrading MB dye under solar irradiation. The phenomenon was due to the electron sink in GOT that contributed for the photoactivity [67].

5. Noble metal deposited TiO_2 photocatalyst

Apart from these metals, non-metals and graphene oxide as dopant/hybrid, few reseachers have also utilized the noble metals like platinum (Pt), gold (Au) and silver (Ag) as a potential source of dopant. The expensive Pt deposition on both rutile and anatase TiO_2 was performed by Scalfani et al., 1998, Kim et al., 2002, Sun et al., 2003, Bosc and coworkers, 2006 and Hidalgo and coworkers, 2007. All their study showed that Pt had beneficial effects for the photocatalytic oxidation. The Pt as a dopant/hybrid has increased the electron mobility rate of TiO_2. This mobility has contributed for the electron transfer to the adsorbed oxygen (O^-_{ads}) especially in rutile TiO_2. A negligible such effect was observed in the anatase phase [68–72].

Owing to the versatile potential of the Au, it was also chosen as a dopant by the reseachers [73, 74]. The study by reseachers showed that Au can behaves like a semiconductor rather than metallic. The size of Au plays a major role in the interfacial electron transfer to adsorbed oxygen (O^-_{ads}). The Fermi level of the Au particle will be lowered than that of adsorbed oxygen (O^-_{ads}) in TiO_2 and this depends on the particle size of Au. The Au also contributed for the hindering effect of the surface recombination. Thus, they exhibited an improvised visible light photoactivity.

Finally, the influence of Ag as a dopant/hybrid was studied by Szabo et al., 2003 and Christopher et al., 2010. Their study showed that Ag as a dopant contributed for the uniform morphology of the TiO_2 and increased the probability function of excited oxygen atoms via electron transfer from adsorbed oxygen (O^-_{ads}) produced from O_2^{2-} to the hole. The Ag also further exhibited a unique Surface Plasmon Resonance (SPR) property. Thus, exhibited SPR lowered the electron–hole recombination rate. All these potentially improvised the photoactivity of TiO_2 with a better degradation of the pollutants either under natural or artificial visible light illumination [30, 75].

6. Magnetic particles deposited TiO_2 photocatalyst

Though these noble metals dopant/hybrid had a significant contribution on the photocatalytic mechanism of TiO_2, they are highly expensive and further increasing the cost of the prepared photocatalyst. Most of the modification techniques solve the issues of photocatalytic efficiency however, leaving behind the separation difficulty. Such separation could make them reuse and contribute for economy of the treatment cost. Most commonly these photocatalysts are nanoparticles and requires high speed centrifugation or membrane filtration. However, adopting these techniques further burdens the economy of the treatment. Hence doping/hybridizing the photocatalyst with a ferromagnetic material could be a best alternative.

Hence coating the photocatalyst with magnetic particles emerges to be a promising method to separate the photocatalyst from treated stream [76]. Nanoparticles combine with magnetic core and photoactive shell using magnetic granules and semiconductor photocatalyst is reported to possess magnetic property and separation could be achieved easily by applying simple magnetic field [77]. For these purpose magnetic cores such as magnetite (Fe_3O_4), maghemite (Fe_2O_3), nickel ferrite ($NiFe_2O_4$), etc. were chosen. Though separation of photocatalysts was achieved their photocatalytic activity was found to decrease than that of pure TiO_2 [8, 78, 79].

Beydoun et al. prepared the magnetic photocatalyst by coating TiO_2 particles onto Fe_3O_4 particles. They observed that the magnetic core was easily oxidized and suppressed the photoactivity of the TiO_2 [76, 80, 81]. Alternatively, Chen et al., 2001 prepared the magnetically separable photocatalyst by coating TiO_2 particles onto γ-Fe_2O_3 particles. Their preparation method transformed ferromagnetic γ-Fe_2O_3 to α-Fe_2O_3 paramagnetic material and resulted in poor separation efficiency [79]. Such phase transformation from γ-Fe_2O_3 to α-Fe_2O_3 was triggered by annealing temperature. Therefore, difficulties arise to synthesize TiO_2 coated particles with high photoactivity without loss of magnetic property by an iron oxide phase transition, as well as of high crystallinity without agglomeration, or formation of impurities by solid diffusion [9].

Chung et al., 2004 reported a TiO_2-coated $NiFe_2O_4$ photocatalyst by multi-step ultrasonic spray pyrolysis method. Their complicated synthesize method resulted the photocatalyst in micron size. Owing to the micron size of the photocatalyst the activity and the separation efficiency declined drastically [4]. Similarly, Xu et al., 2007 prepared a magnetically separable nitrogen-doped photocatalyst, $TiO_{2x}N_x$/SiO_2/$NiFe_2O_4$ (TSN) by a simple method, which consists of a $NiFe_2O_4$ as magnetic core, a SiO_2 as magnetic barrier and nitrogen as visible-light-active dopant. Their prepared TSN was found to possess a great photocatalytic activity by removing MO in the presence of artificial UV and visible light illumination [9].

In recent years the M type hexaferrites, $MFe_{12}O_{19}$ (M = Ba, Sr., Pb, etc) gained interest over the spinel ferrite ($NiFe_2O_4$), since the magnetic properties of M type hexaferrites allow them to serve as highly stable permanent magnet. One such M type hexaferrites, strontium ferrite ($SrFe_{12}O_{19}$), is regarded as an excellent magnetic material [82]. There was so far no report discussed about nanoscale hexaferrites as carriers for magnetic photocatalyst before Fu et al., 2006 synthesized TiO_2/ $SrFe_{12}O_{19}$ composite nanoparticles with core-shell structure. Despite the fact that the photocatalytic activity of the composite is slightly lower than that of Degussa P25, the separation of composite particles was well achieved with an external magnetic field, thus proved the separation incapability of commercial photocatalyst Degussa P25 [82].

Researches on protective layer-coated permanent magnets nanoparticles have been studied for both fundamental magnetic investigations and practical engineering applications. In such investigations, coated nanoparticles attracted the attention as the coating hinders the nanocomposites from coarsening and agglomeration. In practical engineering applications, coating works well in magnetic applications as an insulate phase to achieve high electric resistivity and behaves as a binder to ease the consolidation of the nanoparticles [18, 19].

Coating magnetic nanoparticles with silica (SiO_2) is becoming a significant topic in the research of magnetic nanocomposites. The formation of SiO_2 interlayer on the surface of magnetic nanoparticles helps to screen the magnetic dipolar attraction between magnetic nanoparticles. It also protects from leaching of the core magnetic materials during the dispersion in the aqueous phase. Moreover, SiO_2 coating could be easily activated to provide its surface with various functional groups due to the presence of abundant silanol groups in it. Finally, SiO_2 interlayer plays a very significant role in providing a chemically inert surface for magnetic layer. Hence inclusion of a protective SiO_2 coating will suppress the electron–hole recombination rate that occurs in the photocatalyst and benefits both the photo and magnetic activity.

7. Doped/hybrid TiO_2 photocatalytic mechanism

The basic mechanism of TiO_2 photocatalyst is described based on **Figure 1**, which inititated by the absorption of the photon hv_1 with energy equal to the band gap of anatase TiO_2 (3.20 eV). Electron–hole pair is produced on the surface of titania as schematized. As known, the electron from the CB is promoted and produced positive hole in VB. The excited state electrons and holes get trap in the metastable states as well as dissipate energy as heat. Besides, they also can react with the electron donor or acceptor adsorbed on its surface. Therefore, the $^{\bullet}OH$ is produced with high oxidation potential which plays an important role in photocatalysis [1, 17].

The interstitial metal doped TiO_2 on the other hand beneficially produced the new energy level in the band gap of TiO_2 by the dispersion of metal nanoparticles in titania matrix. As shown in **Figure 1**, electron from the CB can be excited from the defect state by photon energy equals to hv_2. The improvement of the electrons trapping to inhibit the electron–hole recombination during its photoactivation as well described the additional advantageous of the metal ion doping. Thus, decreasing in the charge-carrier recombination resulted in enhanced photoactivity of the photocatalyst.

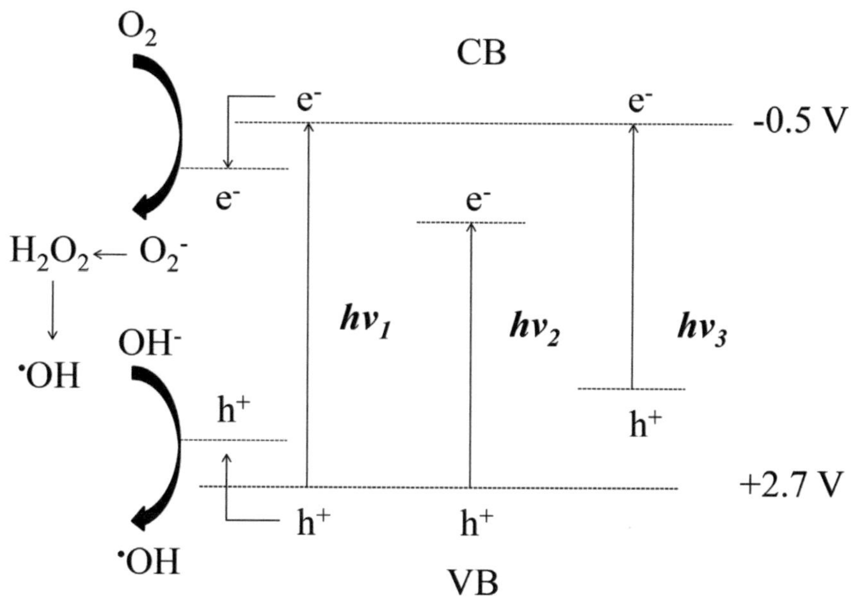

Figure 1.
Mechanism of TiO_2 photocatalysts: hv_1: Pure TiO_2; hv_2: Metal doped TiO_2 and hv_3: Non-metal doped TiO_2 [7].

The modification of TiO_2 with the non-metal ion doping provides four main opinions regarding the changes in the nature mechanism of photocatalyst as described below [7]:

1. Band gap narrowing: Asahi et al., 2001 found that the energies of the CB and VB are very narrow during the absorption of visible light. The N 2p state hybrids with the O 2p state in anatase TiO_2 doped with nitrogen, narrowed the band gap of N-TiO_2 [43].

2. Impurity energy level: Irie et al., 2003 reported that substitutional doping of nitrogen ion into TiO_2 forms an isolated impurity energy level above the VB. Thus, obtained impurity energy level benefits the visible light driven photocatalysis as the visible light only excites electrons in the impurity energy level [83].

3. Oxygen vacancies: Ihara et al., 2003 showed the importance of oxygen deficient sites which act as a blocker for reoxidation. These sites are produced in the grain boundaries in order to emerge visible light activity.

4. Shallowing the acceptor states: Zhao & Liu, 2008 studied the shallowing effect by substitutional doping of nitrogen into TiO_2 [84].

All these obtained modification in the mechanism will contribute for the better and sustainable treatment of the organic pollutants.

8. Conclusions

Though various studies have been carried out to find an ideal semiconductor photocatalyst, TiO_2 however remains as a benchmark and active photocatalyst among them and was proved in both laboratory and pilot studies. Other oxides such as ZrO_2, SnO_2, WO_2 and MoO_3 do not have the similar application prospects as TiO_2 due to the fact that these oxides are much less active, chemically and biologically instable Several commercial TiO_2 photocatalysts are produced worldwide. Among them Degussa P25®, an amorphous TiO_2, emerged as the best photocatalyst due to its better utilization of the UV light. It has a phase ratio of 25:75 (rutile:anatase). It is also considered as standard photocatalyst for environmental applications. The wider band gap, greater recombination of electron–hole pair and low interfacial charger carrier transfer, limit the visible light or sustainable solar energy utilization of TiO_2 photocatalyst. The limitations were successfully achieved by synthesizing a new and modified TiO_2-based composite nanophotocatalysts through a series of simple preparation processes. The nano-size morphology of the composite photocatalysts well created the quantum effect that improved the photocatalytic properties.

Acknowledgements

The authors express gratitude to the Ministry of Higher Education, Malaysia (MOHE) for the Fundamental Research Grant Scheme FRGS (FRGS/1/2017/TK02/UMP/02/20) for the financial support of this book chapter.

Author details

Azrina Abd Aziz[1*], Fatema Khatun[1], Minhaj Uddin Monir[2], Sim Lan Ching[3] and Leong Kah Hon[4]

1 Faculty of Civil Engineering Technology, Universiti Malaysia Pahang, Kuantan, Pahang, Malaysia

2 Faculty of Engineering and Technology, Jashore University of Science and Technology, Department of Petroleum and Mining Engineering, Jashore, Bangladesh

3 Chemical Engineering Department, Lee Kong Chian Faculty of Engineering and Science (LKC FES), UTAR Sungai Long Campus, Selangor, Malaysia

4 Environmental Engineering Department, Faculty of Engineering and Green Technology, UTAR Kampar, Perak, Malaysia

*Address all correspondence to: azrinaaziz@ump.edu.my

References

[1] Hoffmann MR, Martin ST, Choi W, Bahnemann DW. Environmental applications of semiconductor photocatalysis. Chemical Reviews. 1995;95(1): 69-96.

[2] Howe RF, Gratzel M. EPR observation of trapped electrons in colloidal TiO_2. The Journal of Physical Chemistry. 1985; 89(21): 4495-4499.

[3] Fischer CH, Lillie J, Weller H, Katsikas L, Henglein A. Photochemistry of colloidal semiconductors. 29. Fractionation of CdS sols of small particles by exclusion chromatography. Berichte der Bunsengesellschaft für physikalische Chemie. 1989;93(1): 61-64.

[4] Beydoun D, Amal R, Low G, McEvoy S. Role of nanoparticles in photocatalysis. Journal of Nanoparticle Research. 1999;1(4): 439-458.

[5] Aruna ST. Synthesis and properties of nanosized titania. Journal of Materials Synthesis and Processing. 1996; 4(3): 175-179.

[6] Fujishima A, Honda K. Electrochemical photolysis of water at a semiconductor electrode. Nature. 1972;238(5358): 37-38.

[7] Zaleska A. Doped-TiO_2: A review. Recent Patents on Engineering. 2008; 2(3): 157-164.

[8] Chung YS, Park SB, Kang DW. Magnetically separable titania-coated nickel ferrite photocatalyst. Materials Chemistry and Physics. 2004;86(2-3): 375-381.

[9] Xu S, Shangguan W, Yuan J, Chen M, Shi J. Preparations and photocatalytic properties of magnetically separable nitrogen-doped TiO_2 supported on nickel ferrite. Applied Catalysis B: Environmental. 2007; 71(3-4): 177.

[10] Chang H, Wu NM, Zhu FA. Kinetic model for photocatalytic degradation of organic contaminants in a thin-film TiO_2 catalyst. Water Research. 2000; 34(2):407-416.

[11] Huang A, Cao L, Chen J, Spiess FJ, Suib SL, Obee TN, Freihaut JD. Photocatalytic degradation of triethylamine on titanium oxide thin films. Journal of Catalysis. 1999; 188(1): 40-47.

[12] Huan AC, Spiess FJ, Suib S, Obee T, Hay S, J.D. F. Photocatalytic degradation of triethylamine on titanium dioxide thin films. Journal of Catalysis. 1999;188(1): 40-47.

[13] Matthews R. Photooxidation of organic impurities in water using thin films of titanium dioxide. The Journal of Physical Chemistry. 1987; 91(12): 3328-3333.

[14] Al-Rasheed R. Water treatment by heterogeneous photocatalysis an overview. Paper presented at the 4th SWCC Acquired Experience Symposium 2005, Jeddah, Saudi Arabia.

[15] Znad H, Kawase Y. Synthesis and characterization of S-doped Degussa P25 with application in decolorization of Orange II dye as a model substrate. Journal of Molecular Catalysis A: Chemical. 2009; 314(1-2): 55-62.

[16] Hanaor DAH, Sorrell CC. Review of anatase to rutile phase transformation. Journal of Materials Science. 2011; 46(4): 855-874.

[17] Fujishima A, Rao TN, Tryk DA. Titanium dioxide photocatalysis. Journal of Photochemistry and Photobiology C: Photochemistry Reviews. 2000;1(1): 1-21.

[18] Fu W, Yang H, Li M, Chang L, Yu Q, Xu J, Zou G. Preparation and

photocatalytic characteristics of core-shell structure $TiO_2/BaFe_{12}O_{19}$ nanoparticles. Materials Letters. 2006; 60(21-22): 2723-2727.

[19] Fu W, Yang H, Li M, Li M, Yang N, Zou G. Anatase TiO_2 nanolayer coating on cobalt ferrite nanoparticles for magnetic photocatalyst. Materials Letters. 2005; 59(27): 3530-3534.

[20] Bickley R, Gonzalez-Carreno T, Lees J, Palmisano L, Tilley R. A structural investigation of titanium dioxide photocatalysts. Journal of Solid State Chemistry. 1991; 92(1): 178-190.

[21] Cotton F, Wilkinson G, Murillo C, Bochmann M. Advance Inorganic Chemistry. New York, NY: Wiley, 1999.

[22] Malato S, Fernández-Ibáñez P, Maldonado MI, Blanco J, Gernjak W. Decontamination and disinfection of water by solar photocatalysis: Recent overview and trends. Catalyis Today. 2009; 147(1): 1-59.

[23] Herrmann JM. Heterogeneous photocatalysis: Fundamentals and applications to the removal of various types of aqueous pollutants. Catalysis Today. 1999; 53(1):115-129.

[24] Rajeshwar K, Chenthamarakshan CR, Goeringer S, Djukic M. Titania based heterogenenous photocatalysis. Materials, mechanistic issues, and implications for environmental remediation. Pure Applied Chemistry. 2001; 73(12): 1849-1860.

[25] Yu J, Zhang L, Cheng B, Su Y. Hydrothermal preparation and photocatalytic activity of hierarchically sponge-like macro–/mesoporous titania. The Journal of Physical Chemistry C. 2007; 111(28): 10582-10589.

[26] Chong MN, Jin B, Chow CWK, Saint C. Recent developments in photocatalytic water treatment technology: A review. Water Research. 2010; 44(10): 2997-3027.

[27] Liu SX, Chen XY, Chen X, Sun CL. A novel high active TiO_2/AC composite photocatalyst prepared by acid catalyzed hydrothermal method. Chinese Chemical Letters. 2006; 17(4): 529-532.

[28] Litter MI. Heterogeneous photocatalysis transition metal ions in photocatalytic systems. Applied Catalysis B: Environmental. 1999; 23(2-3): 89-114.

[29] Fujishima A, Zhang X, Tryk D. TiO_2 photocatalysis and related surface phenomena. Surface Science Reports. 2008; 63(12): 515-582.

[30] Christopher P, Ingram DB, Linic S. Enhancing photochemical activity of semiconductor nanoparticles with optically active Ag nanostructures: Photochemistry mediated by Ag surface plasmons. The Journal of Physical Chemistry C. 2010; 114(19):9173-9177.

[31] Woan K, Pyrgiotakis G, Sigmund W. Photocatalytic carbon-nanotube-TiO_2 composites. Advanced Materials. 2009; 21(21): 2233-2239.

[32] Nguyen-Phan TD, Pham VH, Shin EW, Pham HD, Kim S, Chung JS, Hur SH. The role of graphene oxide content on the adsorption-enhanced photocatalysis of titanium dioxide/graphene oxide composites. Chemical Engineering Journal. 2011; 170(1): 226-232.

[33] Ahmed S, Rasul MG, Martens WN, Brown R, Hashib MA. Heterogeneous photocatalytic degradation of phenols in wastewater: A review on current status and developments. Desalination. 2010; 261(1-2):3-18.

[34] Palmisano L, Augugliaro V, Sclafani A, Schiavello M. Activity of chromium-ion-doped titania for the

dinitrogen photoreduction to ammonia and for the phenol photodegradation. The Journal of Physical Chemistry. 1988; 92(23): 6710-6713.

[35] Choi W, Termin A, Hoffmann MR. Effects of metal-ion dopants on the photocatalytic reactivity of quantum-sized TiO_2 particles. Angewandte Chemie. 1994; 33(10): 1091-1092.

[36] Zhang Z, Wang CC, Zakaria R, Ying JY. Role of particle size in nanocrystalline TiO_2-based photocatalysts. The Journal of Physical Chemistry B. 1998; 102(52): 10871-10878.

[37] Hung WC, Chen YC, Chu H, Tseng TK. Synthesis and characterization of TiO_2 and Fe/TiO_2 nanoparticles and their performance for photocatalytic degradation of 1, 2-dichloroethane. Applied Surface Science. 2008;255(5): 2205-2213.

[38] Sikong L, Kongreong B, Kantachote D, Sutthisripok W. Photocatalytic activity and antibacterial behavior of Fe^{3+}-doped TiO_2/SnO_2 nanoparticles. Energy Research Journal. 2010; 1(2): 120-125.

[39] Devi LG, Murthy BN, Girish K. Photocatalytic activity of V5+, Mo6+ and Th4+ doped polycrystalline TiO_2 for the degradation of chlorpyrifos under UV/solar light. Journal of Molecular Catalysis A: Chemical. 2009; 308(1-2): 174-181.

[40] Devi LG, Murthy BN, Girish K. Photocatalytic activity of TiO2 doped with Zn2+ and V5+ transition metal ions: Influence of crystallite size and dopant electronic configuration on photocatalytic activity. Materials Science and Engineering: B. 2010; 166(1): 1-6.

[41] Devi LG, Kottam N, Murthy BN, Girish K. Enhanced photocatalytic activity of transition metal ions Mn2+, Ni2+ and Zn2+ doped polycrystalline titania for the degradation of Aniline Blue under UV/solar light. Journal of Molecular Catalysis A: Chemical. 2010; 328(1-2): 44-52.

[42] Amadelli R, Samiolo L, Maldotti A, Molinari A, Valigi M, Gazzoli D. Preparation, characterisation, and photocatalytic behaviour of Co-TiO_2 with visible light response. International Journal of Photoenergy. 2008; 2008:1-9.

[43] Asahi R, Morikawa T, Ohwaki T, Aoki K, Taga Y. Visible-light photocatalysis in nitrogen-doped titanium oxides. Science. 2001;293(5528):269-271.

[44] Mrowetz M, Balcerski W, Colussi AJ, Hoffmann MR. Oxidative power of nitrogen-doped TiO_2 photocatalysts under visible illumination. The Journal of Physical Chemistry B. 2004; 108(45): 17269-17273.

[45] Sano T, Negishi N, Koike K, Takeuchi K, Matsuzawa S. Preparation of a visible light-responsive photocatalyst from a complex of Ti^{4+} with a nitrogen-containing ligand. Journal of Materials Chemistry. 2004; 14(3): 380-384.

[46] Yin S, Aita Y, Komatsu M, Sato, T. Visible-light-induced photocatalytic activity of $TiO_{2-x}N_y$ prepared by solvothermal process in urea–alcohol system. Journal of the European Ceramic Society. 2006; 26(13): 2735-2742.

[47] Sathish M, Viswanathan B, Viswanath RP. Characterization and photocatalytic activity of N-doped TiO_2 prepared by thermal decomposition of Ti–melamine complex. Applied Catalysis B: Environmental. 2007; 74(3-4): 307-312.

[48] Liu G, Wang L, Sun C, Yan X, Wang X, Chen Z, Lu GQ. Band-to-band visible-light photon excitation and

photoactivity induced by homogeneous nitrogen doping in layered titanates. Chemistry of Materials. 2009; 21(7): 1266-1274.

[49] Dong F, Zhao W, Wu Z, Guo S. Band structure and visible light photocatalytic activity of multi-type nitrogen doped TiO_2 nanoparticles prepared by thermal decomposition. Journal of Hazardous Materials. 2009;162(2-3): 763-770.

[50] Lettmann C, Hildenbrand K, Kisch H, Macyk W, Maier W. F.Visible light photodegradation of 4-chlorophenol with a coke-containing titanium dioxide photocatalyst. Applied Catalysis B: Environmental. 2001; 32(4): 215-227.

[51] Khan SUM. Efficient photochemical water splitting by a chemically modified n-TiO_2. Science. 2002; 297(5590): 2243-2245.

[52] Nagaveni K, Hegde MS, Ravishankar N, Subbanna GN, Madras G. Synthesis and structure of nanocrystalline TiO_2 with lower band gap showing high photocatalytic activity. Langmuir. 2004; 20(7): 2900-2907.

[53] Xu C, Killmeyer R, Gray ML, Khan SUM. Photocatalytic effect of carbon-modified n-TiO_2 nanoparticles under visible light illumination. Applied Catalysis B: Environmental. 2006; 64(3-4): 312-317.

[54] Park Y, Kim W, Park H, Tachikawa T, Majima T, Choi W. Carbon-doped TiO_2 photocatalyst synthesized without using an external carbon precursor and the visible light activity. Applied Catalysis B: Environmental. 2009; 91(1-2): 355-361.

[55] Li D, Haneda H, Labhsetwar NK, Hishita S, Ohashi N. Visible-light-driven photocatalysis on fluorine-doped TiO_2 powders by the creation of surface oxygen vacancies. Chemical Physics Letters. 2005; 401(4-6): 579-584

[56] Chen D, Yang D, Wang Q, Jiang Z. Effects of boron doping on photocatalytic activity and microstructure of titanium dioxide nanoparticles. Industrial & Engineering Chemistry Research. 2006;45(12): 4110-4116.

[57] Yu JC., Zhang L, Zheng Z, Zhao J. Synthesis and characterization of phosphated mesoporous titanium dioxide with high photocatalytic activity. Chemistry of Materials. 2003; 15(11): 2280-2286.

[58] Akhavan O. Graphene nanomesh by ZnO nanorod photocatalysts. ACS Nano. 2010; 4(7): 4174-4180.

[59] Akhavan O, Abdolahad M, Esfandiar A, Mohatashamifar M. Photodegradation of graphene oxide sheets by TiO_2 nanoparticles after a photocatalytic reduction. The Journal of Physical Chemistry C. 2010; 114(30):12955-12959.

[60] An X, Yu J. C.Graphene-based photocatalytic composites. RSC Advances. 2011; 1(8):1426-1434.

[61] Dreyer DR, Park S, Bielawski CW, Ruoff RS. The chemistry of graphene oxide. Chemical Society Reviews. 2010;39(1): 228-240.

[62] Lee JS, You KH, Park CB. Highly photoactive, low bandgap TiO_2 nanoparticles wrapped by graphene. Advanced Materials. 2012; 24(8): 1084-1088.

[63] Huang X, Yin Z, Wu S, Qi X, He Q, Zhang Q, Zhang H. Graphene-based materials: Synthesis, characterization, properties, and applications. Small. 2011; 7(14): 1876-1902.

[64] Kim F, Cote LJ, Huang J. Graphene oxide: Surface activity and two-dimensional assembly. Advanced Materials. 2010; 22(17): 1954-1958.

[65] Zhang J, Xiong Z, Zhao XS. Graphene-metal-oxide composites for

the degradation of dyes under visible light irradiation. Journal of Materials Chemistry. 2011; 21(11): 3634-3640.

[66] Manga KK, Wang S, Jaiswal M, Bao Q, Loh KP. High-gain graphene-titanium oxide photoconductor made from inkjet printable ionic solution. Advanced Materials. 2010; 22(46): 5265-5270.

[67] Hu Z, Huang Y, Sun S, Guan W, Yao Y, Tang P, Li C. Visible light driven photodynamic anticancer activity of graphene oxide/TiO2 hybrid. Carbon. 2012; 50(3): 994-1004

[68] Bosc F, Ayral A, Keller N, Keller, V. Room temperature visible light oxidation of CO by high surface area rutile TiO_2-supported metal photocatalyst. Applied Catalysis B: Environmental. 2007;69(3-4): 133-137.

[69] Hidalgo MC, Maicu M, Navío JA, Colón G. Photocatalytic properties of surface modified platinised TiO_2: Effects of particle size and structural composition. Catalysis Today. 2007; 129(1-2): 43-49.

[70] Kim S, Choi W. Dual photocatalytic pathways of trichloroacetate degradation on TiO_2: Effects of nanosized platinum deposits on kinetics and mechanism. The Journal of Physical Chemistry B. 2002; 106(51): 13311-13317.

[71] Sclafani A, Herrmann JM. Influence of metallic silver and of platinum-silver bimetallic deposits on the photocatalytic activity of titania (anatase and rutile) in organic and aqueous media. Journal of Photochemistry and Photobiology A: Chemistry. 1998; 113(2): 181-188.

[72] Sun B, Vorontsov AV, Smirniotis PG. Role of platinum deposited on TiO_2 in phenol photocatalytic oxidation. Langmuir. 2003; 19(8): 3151-3156.

[73] Orlov A, Jefferson DA, Tikhov M, Lambert RM. Enhancement of MTBE photocatalytic degradation by modification of TiO2 with gold nanoparticles. Catalysis Communications. 2007; 8(5): 821-824.

[74] Tian B, Zhang J, Tong T, Chen F. Preparation of Au/TiO_2 catalysts from Au(I)–thiosulfate complex and study of their photocatalytic activity for the degradation of methyl orange. Applied Catalysis B: Environmental. 2008; 79(4): 394-401.

[75] Szabó-Bárdos E, Czili H, Horváth A. Photocatalytic oxidation of oxalic acid enhanced by silver deposition on a TiO_2 surface. Journal of Photochemistry and Photobiology A: Chemistry. 2003; 154(2-3): 195-201.

[76] Beydoun D, Amal R, Low G, McEvoy S. Occurrence and prevention of photodissolution at the phase junction of magnetite and titanium dioxide. Journal of Molecular Catalysis A: Chemical. 2002;180(1-2): 193-200.

[77] Ao Y, Xu J, Zhang S, Fu D. Synthesis of a magnetically separable composite photocatalyst with high photocatalytic activity under sunlight. Journal of Physics and Chemistry of Solids. 2009; 70(6):1042-1047.

[78] Beydoun D, Amal R. Novel photocatalyst: Titania-coated magnetite. Activity and photodissolution. The Journal of Physical Chemistry B. 2000;104(18): 4387-4396.

[79] Chen F, Xie Y, Zhao J, Lu G. Photocatalytic degradation of dyes on a magnetically separated photocatalyst under visible and UV irradiation. Chemosphere. 2001; 44(5):1159-1168.

[80] Beydoun D, Amal R, Low GKC, McEvoy S. Novel photocatalyst: titania-coated magnetite. Activity and photodissolution. The Journal of Physical Chemistry B. 2000;104(18): 4387-4396.

[81] Beydoun D, Amal R, Scott J, Low G, McEvoy S. Studies on the mineralization and separation efficiencies of a magnetic photocatalyst. Chemical Engineering & Technology. 2001;24(7): 745-748.

[82] Fu W, Yang H, Chang L, Hari B, Li M, Zou G. Anatase TiO_2 nanolayer coating on strontium ferrite nanoparticles for magnetic photocatalyst. Colloids and Surfaces A: Physicochemical and Engineering Aspects. 2006; 289(1-3): 47-52.

[83] Iri H., Watanabe Y, Hashimoto K. Nitrogen-concentration dependance on photocatalytic activity of $TiO_{2-x}N_x$ powders. The Journal of Physical Chemistry B. 2003; 107(23): 5483-5486.

[84] Zhao Z, Liu Q. Mechanism of higher photocatalytic activity of anatase TiO_2 doped with nitrogen under visible-light irradiation from density functional theory calculation. Journal of Physics D: Applied Physics. 2008; 41: 1-10.

Chapter 4

Photocatalytic Applications of Titanium Dioxide (TiO_2)

Atta ul Haq, Muhammad Saeed, Samreen Gul Khan and Muhammad Ibrahim

Abstract

Water pollution is one the fundamental problems that have got the serious concerns of the researchers. Water poluution arises due to a number of reasons including domestic, industrial, agricultural, scinec and technology. The textile industry is the main industry that releases the dyes contaminated wastewater to the environment. A varities of protocols have been attempeted for the removal of dyes from aqueous body. Photocatalysis is one of the effective techniques which offer opportunities to overcome the aqueous pollution caused by rapid industrialization and urbanization. The semiconductor metal oxides used as photocatalysts are capable to provide a sustainable and clean ecosystem due to the tunable physiochemical characteristics of semiconductor metal oxides. Titanium dioxide (TiO_2) is one of the metal oxides that can be effectively employed as a photocatalyst in the abatement of aqueous pollution due to organic compounds. The catalytic performance of titanium dioxide depends on several parameters like its crystallinity, surface area, and morphology. Titanium dioxide has shown good performance in the different photocatalytic systems, however, the characteristics like wide band gap and low conductivity limit the photocatalytic performance of titanium dioxide. Various attempts have been made to improve the photocatalytic performance of titanium dioxide. Herein, we summarize the various attempts to improve the photocatalytic performance of titanium dioxide in the abatement of aqueous pollution. The attempts made for the improvement of photocatalytic performance of titanium dioxide include modifications in composition, doping of other metal, and formation of heterojunctions with other metal oxides.

Keywords: Titanium dioxide, Photocatalysis, Photodegradation, Wastewater

1. Introduction

Water pollution is one the fundamental problems that have got the serious concerns of the researchers. Water poluution arises due to a number of reasons including domestic, industrial, agricultural, scinec and technology. The contribution of industrial sector in water pollution is significant. A number of industries like textile industry, leather industry, food industry, pharmaceutical industry, printing industries etc. release their wastewater to the environment. These wastewater of these industries contain a variety of organic and inorganic pollutants. Hence, the wastewater released from these induries cause water pollution. The water pollution due to dyes is very serious issue. The textile industry is the main industry that releases the dyes contaminated wastewater to the environment. Textile industry uses a huge amount of dyes annualy. It is estimated that about 700000 tons of the various dyes are produced annualy around the world. These dyes have been classifies as azo dyes, basic dyes, direct dyes, vat dyes, reactive dyes etc. More than 15% of the globally produced dyes are released to the environment in wastewater from textile and other industries. These dyes released in aqueous system are highly toxic and carcinogeneic in nature. These dyes badly affect the living organisms. Dyes are highly colored substances that impart intense color to the aqueous body when dyes contaminated wastewater is released into it. Due to intense color of aqueous body, the sunlight is inhibited by penetration into the interior of the aqueous body. As a result of non-penetration of sunlight, the aqueous ecosysytem is badly affected by dyes [1–6].

A varities of protocols have been attempeted for the removal of dyes from aqueous body. Techniques like biological degradation technique, membrane filteration technique, adsorption atechniques, sedimentation technique, physical coagulation technique and chemical coagulation technique has been attempted for the elimination of dyes from aqueous medium. However, these techniques are not successful in the removal of dyes. As these techniques do not degrade the dyes molecules, therefore, these techniques fail to remove the dyes from aqueous medium. These techniques only transform the dyes pollutants from one form to another form. Furthermore, these techniques give rise secondary pollution [7–10]. Hence, there is an urgent need to develop effective techniques for the removal of these dyes and other organic pollutants from aqueous medium.

The advanced oxidation processes are considered as effective alternative technique for the removal of dyes from aqueous medium. Advanced oxidation processes are chemical processes which are based on production of hydroxyl radicals. These hydroxyl radicals are highly reactive spectices. These species take part in complete degradation of organic pollutants molecules. The advanced oxidation processes include heterogeneous photocatalysis, electrochemical oxidation and Fenon reactions. **Table 1** shows various advanced oxidation proceses. The advanced oxidation processes have many advantages compared to conventional techniques as follows [11–13].

1. The pollutants molecules are directly converted to simple inorganic molecules like water and carbon dioxide by advanced oxidation processes.

2. The advanced oxidation processes can degradade a wide range of organic pollutant with out any selectivity.

3. There is no formation of hazardous products in advanced oxidation process as the organic pollutants are converted into simple inorganic molecules like water and carbon dioxide.

Due to the above mentioned merits of advanced oxidation processes, the advanced oxidation processes have got considerable interest of the scientists. The photocatalysis using heterogeneous metal oxide semicondutors as photocatalyst in the presence of ultra violet or visible light is one of the techniques of advanced oxidation proesses. The photocatalytic treatment of organic pollutants has got significant attention of the researchers recently. The photocatalytic technique doesnot need any extra ordinary energy requirement. Furthermore, there is no formation of hazardous by-products in photocatalysis. A large number of reports are available of photocatalytic degradation of dyes in aqueous medium. **Figure 1** shows the total number of photocatalytic reports for various dyes reported during 2000–2017.

No	AOP	Oxidatant	Benifits	Drawbacks
1	Fenton's reaction	$^{\cdot}O_2^-$ $^{\cdot}OH$	No formation of bromate. No off-gas treatment. Less energy-intensive than ozone and UV alone.	Maintenance costs. Works in low pH only. Requirement of Iron extraction system.
2	Photocatalyst	UV $^{\cdot}O_2^-$ $^{\cdot}OH$ $^{\cdot}HO_2$	Low energy requirement than other AOPs. No formation of bromate. Use of solar irradiation. No off-gas treatment.	Requirement of pre-treatment. Loss of catalytic activity with time. Sensitivity to pH.
3	H_2O_2/O_3	UV O_3 $^{\cdot}O_2^-$ $^{\cdot}OH$	Higher generation of OH. More effective than O3 or UV alone. Supplementary disinfectant.	Requirement of high cost and energy. Formation of bromate. Inhibition to penetration of UV light due to turbidity. Contamination due to failure of the UV lamp.
4	O_3	O_3 $^{\cdot}OH$	The established technique for the treatment of wastewater. Supplementary disinfectant.	Requirement ozone off-gas treatment.
5	Electron beam	e_{aq} H_2O_2 $\cdot OH$ $\cdot H$	No formation of bromate. Minimal effect of turbidity. No off-gas treatment. Can help in disinfection.	Requirement of skilled professionals. Not scalable for practical application. Energy and cost intensive.
6	Cavitation	$^{\cdot}H$ $^{\cdot}OH$	Less energy consumption. No formation of bromate. Low maintenance cost. No off-gas treatment.	Not scalable for practical application. Efficiency is low.
7	O_3/UV	O_3 UV $^{\cdot}O_2^-$ $\cdot OH$	Higher production of OH than H_2O_2/UV. More effective than UV or O_3 alone.	Inhibition to penetration of UV light due to turbidity. Contamination due to failure of UV lamp. Formation of bromate. Comparatively costly.
8	H_2O_2/UV	H_2O_2 UV $^{\cdot}O_2^-$ $\cdot OH$	No formation of bromate. No off-gas treatment is required. Can oxidize more MTBE compared to H_2O_2 or UV alone.	Inhibition to penetration of UV light due to turbidity. Contamination due to failure of the UV lamp. Absorption of UV by interfering compounds.

Table 1.
Various advanced oxidation processes.

Figure 2 shows the total number, percentage and number of reports under visible and ultra violet radiation for various dyes.

The data given in **Figure 2** shows that researchers have studidied the tiazine dyes mostly. The xanthenes dyes are at second position in the mostly investigated dyes in the subject of photocatalytic degradation.

The applications and fundamentals of photocatalysis developed tremendously during the last four decades. Photocatalysis can be defined as a reaction induced by irradiation of light in the presence of a substance called a catalyst. The photocatalytic reactions are initiated by the absorption of light having energy greater than or equal to the band gap energy of the photocatalyst. The energy difference between the highest filled energy level (valence band) and lowest vacant

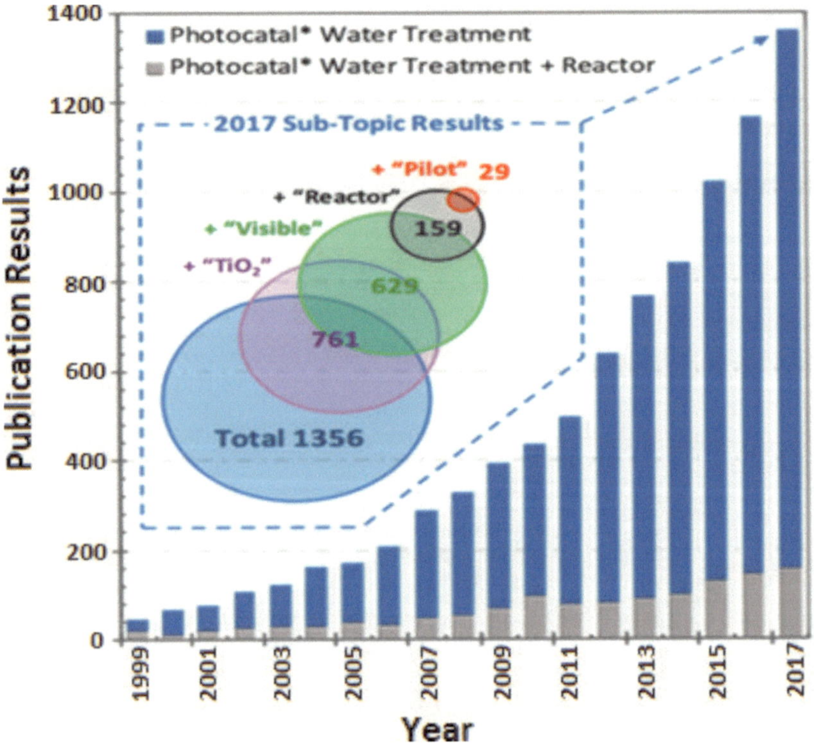

Figure 1.
Number of publications on photocatalytic treatment of wastewater. This figure is reproduced with permission from Environ. Sci. Technol. 2019, 53, 2937–2947 Copyright 2019 American Chemical Society.

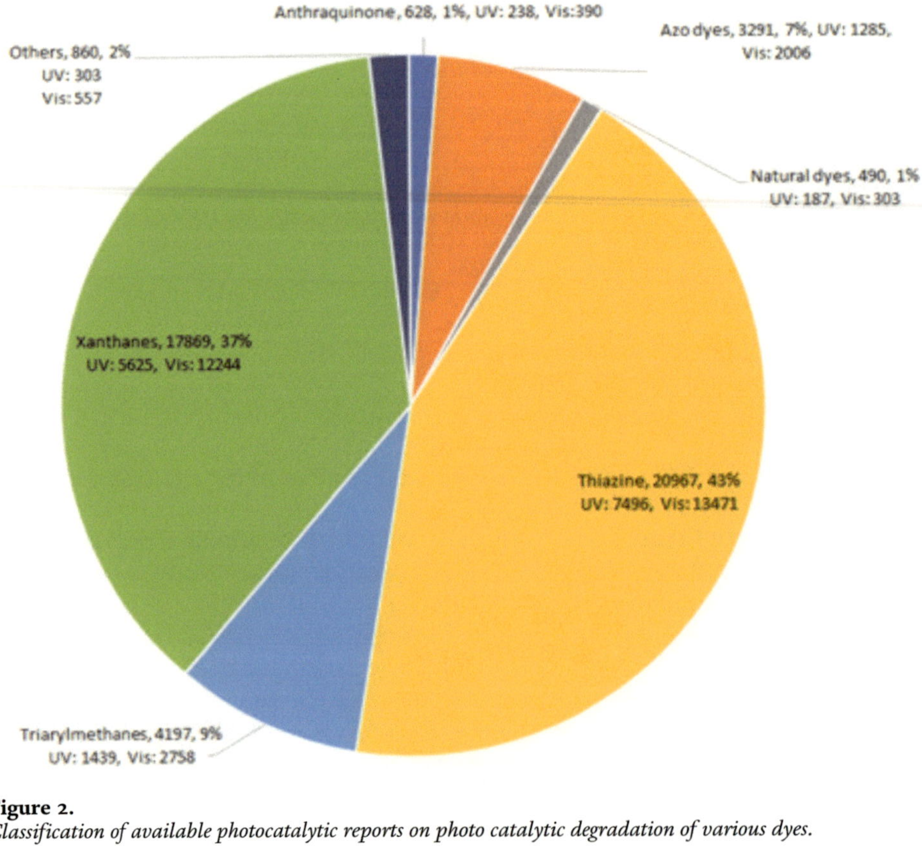

Figure 2.
Classification of available photocatalytic reports on photo catalytic degradation of various dyes.

energy level (conduction band) of photocatalytic substance is called band gap energy of the cited substance [14, 15]. The absorption of light causes the excitation of electrons from the highest filled energy level (valence band) to the lowest vacant energy level (conduction band) of the photocatalyst. This photoinduced excitation creates a positive hole (h^+) in the valence band and electrons (e^-) in conduction band of the substance. After formation of positive holes and electrons, two types of processes may proceed further.

1. The positive holes and electron recombine, and energy is released in the form of heat.

2. The positive holes and electrons take part in reactions and initiates a series of redox reactions called photocatalytic reactions.

Hence, for photocatalytic reactions, process 1 mentioned above (recombination of positive holes and electrons) must be prevented to favor the photocatalytic reactions. The goal of photocatalysis is to initiate reactions of positive holes and electrons with the reductants and oxidants to produce oxidized and reduced products, respectively [16].

Presently, photocatalysis is used in several emerging fields like photodegradation of aqueous organic pollutants, production of hydrogen by water splitting, treatment of gaseous environmental pollutants like NOx, treatment of halogenated hydrocarbon, inactivation of microorganisms, treatment of pesticides and organohalide compounds, oxidation of micropollutants and many more [17–21].

2. Titanium dioxide (TiO_2) as a catalyst

Generally, the semiconductor metal oxides are used as photocatalysts in photocatalysis. Fujishima et al., the pioneer of photocatalysis, employed titanium dioxide (TiO_2) as a catalyst for the production of hydrogen gas by splitting water for the first time [22]. Water can not be decomposed by visible light because it is transparent to it. Water can only be decomposed if it is irradiated with light having a wavelength less than 190 nm. Fujishima and co-workers electrochemically decomposed water using TiO_2 electrode. They reported that water can be electrochemically decomposed if a potential difference of 1.23 V is applied between anode and cathode. The potential difference of 1.23 V is equivalent to the energy of photons having wavelength about 1000 nm. Water can be electrochemically decomposed under any one of the following conditions.

1. The production of oxygen takes place at a potential which is more negative as compared to potential at which production of hydrogen takes place under normal conditions.

2. The production of hydrogen takes place at potential which is more positive as compared to potential at which production of oxygen takes place under normal conditions.

3. The potential at which production of oxygen takes place is made more negative and the potential at which production of hydrogen takes place is made more positive.

Fujishima and co-workers used TiO_2 electrode for electrothermal decomposition of water. They investigated the current–voltage curves under light condition and dark condition. They found that anodic current flowed under the irradiation of light having wavelength less than 415 nm. The energy of these radiation is equivalent to 3.0 eV. This energy is equal to the band gap of TiO_2. On the basis of these observations, it was suggested that irradiation of light produced holes in the valence band of TiO_2. Similarly, the production of oxygen at −0.5 V was also confirmed by various electrochemical measurement. They constructed an electrochemical cell. The TiO_2 was used as electrode which was connected to a Pt electrode. The irradiation of surface of TiO_2 electrode caused a current to flow from Pt electrode to TiO_2 electrode. The flow of current from Pt electrode to TiO_2 electrode suggested that production of oxygen takes place TiO_2 electrode by oxidation reaction and production of hydrogen takes place at Pt electrode by reduction reaction. It was suggested irradiation caused decomposition of water in the absence of any external potential. The decomposition of water took place according to following reactions.

1. Production of hole and electron by excitation of TiO_2

$$TiO_2 \overset{2\,h\vartheta}{\rightarrow} 2h^+ + 2e^- \tag{1}$$

1. Production of oxygen by oxidation reaction at TiO_2

$$H_2O + 2h^+ \rightarrow \frac{1}{2}O_2 + 2H^+ \tag{2}$$

1. Production of hydrogen by reduction at Pt

$$2H^+ + 2e^- \rightarrow H_2 \tag{3}$$

The net decomposition reaction is

$$H_2O + 2h\vartheta \rightarrow H_2 + \frac{1}{2}O_2 \tag{4}$$

Since the work of Fujishima et al., TiO_2 gained the attention of many researchers and presently it is the most used substance in the field of photocatalysis.

Titanium dioxide (TiO_2) exists in three crystalline phases: Brookite, Anatase, and Rutile phase. The Brookite phase of titanium dioxide is unstable and therefore it is not used in photocatalytic applications. Anatase and Rutile phases of titanium dioxide are thermodynamically stable phases. The rutile phase is mostly used in photocatalytic applications due to its easy preparation and higher catalytic performance. Studies have shown that mixed phases of titanium dioxide are used as catalysts for higher photocatalytic performances. It is believed that mixed phases titanium dioxide exhibits higher catalytic performance due to the movement of photoinduced electrons from Rutile to Anatase phase of titanium dioxide. This movement of electrons prevents the recombination of positive holes and electrons and ultimately enhances the photocatalytic performance [23–25]. However, it has also been shown that electrons move from Anatase to Rutile phase in mixed-phase photocatalysts [26]. The band gap energy of the anatase and rutile phase of titanium dioxide is 3.2 eV and 3.0 eV, respectively. Irradiation of titanium dioxide with photons havening energy equal to or greater than 3.2 eV results in the excitation of an electron from the valence band to the conduction band. This excitation results in the formation of an electron–hole pair. These photo-induced charges move to the surface of titanium dioxide and promote a series of redox reactions. The positive holes lead to the formation of vacancies in titanium dioxide as well as excite the reduced species. The photoinduced electrons produce $O_2^{\bullet}$ free radicals. These free radicals are highly reactive and unstable species, so they react further [27, 28]. This whole process can be summarized as follows.

1. Production of positive hole and electron by excitation of TiO_2

$$TiO_2 + Irradiation \rightarrow h^+(TiO_2) + e^-(TiO_2) \tag{5}$$

2. Recombination of positive hole and electron

$$h^+(TiO_2) + e^-(TiO_2) \rightarrow Heat \tag{6}$$

3. Production of OH radical by oxidation of water by reaction with positive hole

$$h^+(TiO_2) + H_2O \rightarrow OH^\bullet \tag{7}$$

4. Reduction of oxygen by reaction with electron

$$e^-(TiO_2) + O_2 \rightarrow O_2^- \tag{8}$$

5. Production of OH radical by reaction of super oxide anion with water

$$O_2^- + H_2O \rightarrow OH^\bullet \tag{9}$$

6. Reaction of OH radicals with reactants

$$Reactants + OH^\bullet \rightarrow Products \tag{10}$$

The mechanism given above has been proposed based on electron spin resonance (ESR) and spin trapping studies. However, Ângelo [29] reported 82%

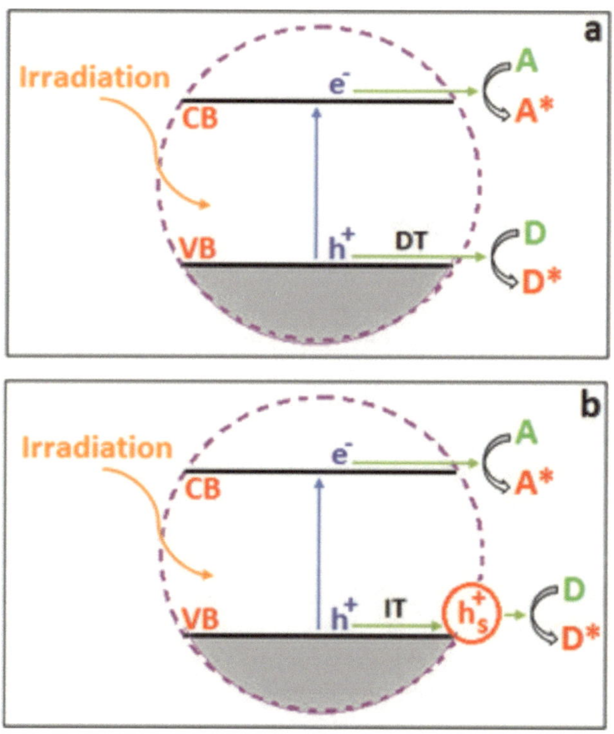

Figure 3.
Graphical representation of direct–indirect mechanism; a) direct transfer of positive holes, b) indirect transfer of positive holes.

conversion of NO for a feed containing 75% NO and 25% RH with −20°C as dew point; the same work indicates that the water-adsorbed monolayer is reached for a relative humidity of 25%. If OH• were the main species in the conversion of NO, then the conversion of NO for dry feed would be much smaller. Hence, this study questions Eq. (3) in the above mechanism or otherwise the role of OH• radicals in photocatalysis. Montoya et al., [30] reported against the formation of OH• radicals by direct reaction of positive holes with water. They proposed a direct–indirect model (D-I) for titanium dioxide catalyzed reactions. The proposed mechanism is given and explained in **Figure 3**. They proposed two mechanisms for the transfer of interfacial charges.

1. In the case of stronger electronic interaction, there is a direct transfer (DT) of the photo-induced positive holes for the reactions with adsorbed species.

2. In the case of weaker interaction, there is an indirect transfer (IT) of the photo-induced positive holes. The transfer of positive holes takes place in two steps. In the first step, the positive holes are trapped by lattice oxygen of TiO_2 and generate lattice oxygen radicals. In the second step, the trapped holes are transferred to the adsorbed reactants through tunneling.

3. How to improve the catalytic performance of titanium dioxide

Although titanium dioxide has been used successfully as a photocatalyst for several reactions. However, its wide band gap (3.2 eV) limits its photocatalytic applications. Due to its wide band gap, titanium dioxide can be used as a photocatalyst under ultraviolet radiation only. The best photocatalyst is one that can be used under visible light as well as ultraviolet light because the sunlight is mainly composed of visible light. Being longer wavelength, the visible light can not excite electrons from the valence band to the conduction band of the titanium dioxide. The photocatalytic performance of titanium dioxide can be improved in two ways.

1. By prevention of recombination of photoinduced positive holes and electrons

2. By narrowing the band gap of the TiO_2

The prevention of recombination of photoinduced positive holes and electrons will permit the occurrence of redox reactions of the positive holes and electrons. These redox reactions will generate hydroxyl radicals and ultimately the photocatalytic performance will be enhanced.

The narrowing of band gap will allow the absorption of visible light by photocatalyst, so the photocatalyst will absorb a wide range of the solar spectrum. The solar spectrum consists of 47% infrared, 46% visible, and 4–5% ultraviolet light. Hence the absorption of visible light will also improve the photocatalytic performance. It has been reported that recombination of the photoinduced positive holes and electrons has a significant contribution to the photocatalytic performance of the catalysts. Most of the researchers work to develop methods and techniques to prevent the recombination of these positive holes and electrons [31–33]. The researchers have proposed that crystal structure significantly affects the photocatalytic performance of the photocatalyst as it plays a significant role in the recombination of positive holes and electrons. Hence, attempts have been made for modifications in the crystal structure to improve the photocatalytic performance of titanium dioxide.

Furthermore, the low photo conversion efficiency of the TiO_2 is also a challenge for the researchers. It has been reported that loss of efficiency is associated with each photo catalytic step. Due to loss of efficiency at each step, the observed photo catalytic efficiency with TiO_2 reported in literature is very low. This low photo catalytic efficiency of TiO_2 is the critical drawback associated with TiO_2 catalyzed treatment. Hence, the photocatalytic efficiency is expected to be low compared to other advanced oxidation processes. However, the TiO_2 has the ability to absorb less energetic photons. This characteristic of TiO_2 enable the researcher to develop TiO_2 based photo catalysts that can absorb light in visible region. The presence of hydrogen peroxide is not needed in visible light photocatalysis. The presenece of hydrogen peroxide is needed in other advanced oxidation processes.

4. Structural modifications of TiO_2

Various structural modifications in the crystal structure of titanium dioxide have been proposed for the improvement of its photocatalytic performance. All these modifications enhance the photocatalytic performance by decreasing the rate of recombination of positive holes and electrons. These modifications allow the absorption of visible light as well. Two types of structural modifications are commonly used for enhancement in the photocatalytic performance of titanium dioxide. These modifications are doping of titanium dioxide with metals or nonmetals and formation of heterojunctions with other semiconductors.

4.1 Doping

Doping is the addition of elements (impurities) into the inner structure of titanium dioxide. The doping of elements causes a red shift in the absorption of light. It also causes the prevention of recombination of photo-induced positive holes and electrons. Hence, doping improves the photocatalytic performance by absorbing of wide range of radiations and separation of charge carriers. Silver (Ag) is one of the elements used for doping of titanium dioxide. The silver (Ag) doping is used for improvement of photocatalytic performance of titanium dioxide towards photo degradation of organic pollutants [34–36]. Saeed et al., [37] have reported the improvement in photocatalytic performance of titanium dioxide for photodegradation of methylene blue and rhodamine B dyes by incorporation of silver (Ag) in the structure of titanium dioxide. They investigated the effect of silver (Ag) on the photocatalytic performance of titanium dioxide using 2, 4, 6, and 8% loading of Ag on titanium dioxide. They reported 65, 84, 97, and 78% photodegradation of methylene blue over 2, 4, 6, and 8% Ag-TiO_2 as a catalyst, respectively. It was found that doping of Ag enhanced the photocatalytic performance of titanium dioxide significantly. **Figure 4** shows the comparison of photocatalytic activity of TiO_2 and 6% Ag-TiO_2 for photodegradation of methylene blue and rhodamine B. The data given in **Figure 4** was obtained by performing degradation experiments with 0.1 g of Ag-TiO_2. A 50 mL solution of methylene blue and/or rhodamine B having concentration of 100 mg/L was used asmodel solution separately for degradation study.

It was found that higher loading of Ag decreased the photocatalytic performance of titanium dioxide. Higher loading of Ag blocks the active sites of the catalyst, therefore the photocatalytic performance decreased [38].

Similarly, Ag-TiO_2 with 1, 3, 5, 7, and 10% Ag have been also reported for photodegradation of methylene blue and methyl orange dyes [39]. In this study, the Ag-TiO_2 loaded with 5% Ag showed the highest photocatalytic performance for

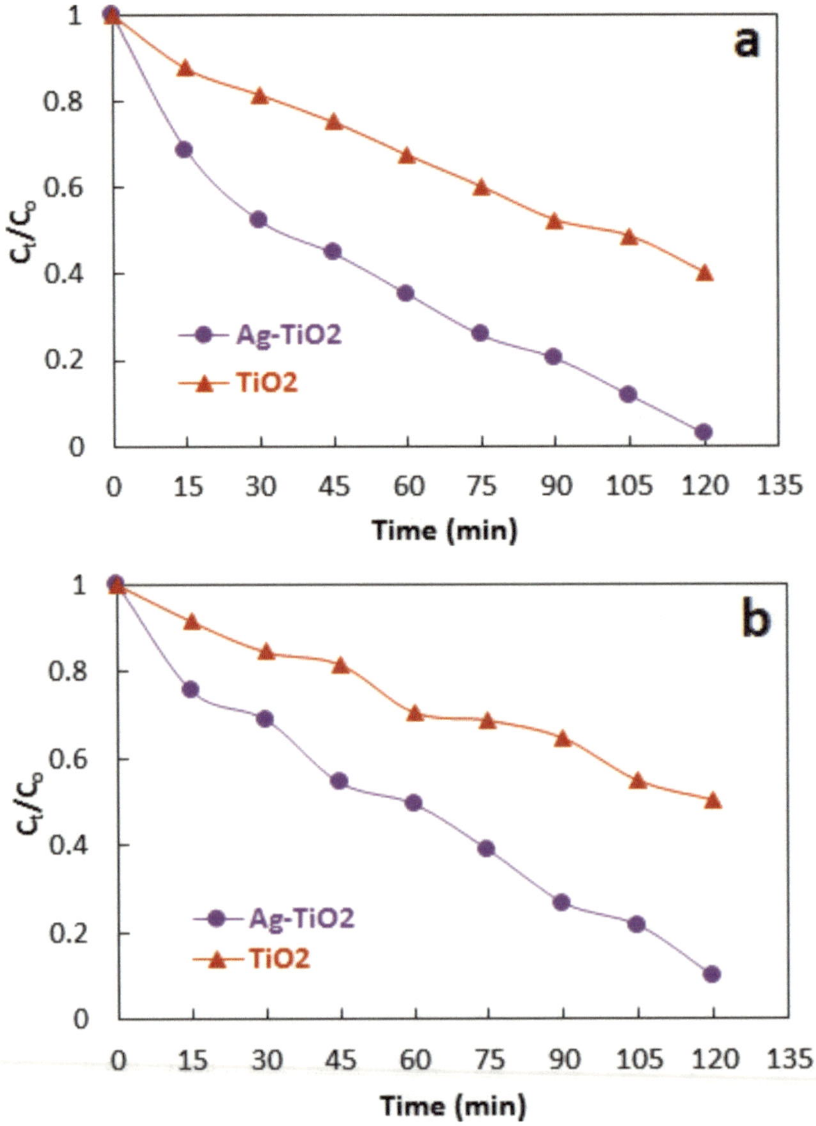

Figure 4.
Comparison of photocatalytic performance of TiO_2 *and Ag-*TiO_2 *towards photodegradation of methylene blue dye (a) and rhodamine B dye (b) [24].*

photodegradation of dyes. it was reported that doping of silver generates the surface defects by the creation of oxygen vacancies and Ti (III) sites in the structure of TiO_2. The formation of Ti (III) has been proposed by the flow of electrons from Ag to Ti (IV). The generation of oxygen vacancies leads to the formation of defect energy levels below the conduction band of TiO_2. Also, the incorporation of Ag narrows the band gap of titanium dioxide due to the formation of Ag 4 d states. These modifications favor the excitation of electrons from the valence band to the conduction band under visible light irradiations. The photogenerated positive holes and electrons flow to the surface of titanium dioxide. The positive holes initiate oxidation by reaction with H_2O or OH ions and generate reactive hydroxyl radicals ($OH^•$). Similarly, the photogenerated electrons initiate reduction by reaction with adsorbed oxygen and give rise to super oxide anion radicals. The photo-generated positive holes are trapped by Ag as well and produce Ag (II). The Ag (II) also initiate oxidation reactions by reaction with H_2O or OH ions and ultimately produce hydroxyl radicals ($OH^•$). Similarly, the Ag (I) traps the photo-generated electrons and produces Ag (0) and then these trapped electrons are transferred to oxygen or Ti (IV). Hence, the recombination of positive holes and electrons is decreased and ultimately the photocatalytic performance is increased [40–44]. Chemical reactions 11 to 18 explain the whole mechanism.

$$Ag - TiO_2 + Irradiation \rightarrow Ag - TiO_2(h^+) + Ag - TiO_2\ (e^-) \quad (11)$$

$$H_2O\ or\ OH^- + Ag - TiO_2(h^+) \rightarrow OH^\bullet \quad (12)$$

$$O_2(ads) + Ag - TiO_2(e^-) \rightarrow O_2^- \quad (13)$$

$$Ag\ (I) + Ag - TiO_2(h^+) \rightarrow Ag(II) \quad (14)$$

$$H_2O\ or\ OH^- + Ag\ (II) \rightarrow OH^\bullet + Ag\ (I) \quad (15)$$

$$Ag\ (I) + Ag - TiO_2(e^-) \rightarrow Ag\ (0) \quad (16)$$

$$Ag\ (0) + O_2(ads) \rightarrow O_2^- + Ag\ (I) \quad (17)$$

$$Ag\ (0) + Ti\ (IV) \rightarrow Ag\ (I) + Ti\ (III) \quad (18)$$

$$Ti\ (III) + O_2(ads) \rightarrow Ti\ (IV) + O_2^- \quad (19)$$

$$Ag - Ti_2(h^+) + O_2^- \rightarrow O^- \quad (20)$$

$$O^- + H_2O \rightarrow OH^- + OH^\bullet \quad (21)$$

$$Pollutants + OH^\bullet \rightarrow Degradation\ products \quad (22)$$

Similarly, other elements can also be used for doping TiO_2. Doping of nitrogen is also one of the most studied approaches for the enhancement of photocatalytic performance of titanium dioxide. The doping of nitrogen has been used to extend the absorption of light towards the visible wavelength side. Some of the others have reported that doping of nitrogen results in narrowing of the band gap of titanium dioxide. Some researchers have argued the interaction between valance band, conduction band, and energy states of doping element cause the narrowing of band gap [13, 45, 46]. Di Valentin and co-workers [47] have reported the density functional theory (DFT) study for the evaluation of the photocatalytic performance of N-doped-TiO_2. They predicted that nitrogen atoms occupy either interstitial or substitutional sites in the lattice of titanium dioxide. As a result, localized energy states are generated. In the case of the interstitial position of nitrogen, discrete energy states are formed above the valence band. In the case of the substitutional position of nitrogen atoms, energy levels are formed in extension to the valence band. In the same way, the doping of other elements like carbon can also improve the catalytic performance of titanium dioxide by narrowing its band gap [48]. It has also been reported that modifications in the (101) plane of titanium dioxide take place with doping of elements. The modifications resulted from the doping of elements enhance the movement of photo-generated electrons to other places within the structure. This flow of electrons to other places increases the lifetime of photo-induced charge carriers and ultimately causes an improvement in photocatalytic performance [49].

Although the doping of elements in the lattice of titanium dioxide has been used for enhancement in the photocatalytic performance of titanium dioxide, however, these dopants may also decrease the photocatalytic performance as these dopants can promote the recombination of photogenerated positive holes and electron. Therefore, the doping of elements in high concentrations must be avoided [50].

The photo catalytic performance of TiO_2 can be enhanced by doping of iron as well. The doping ot TiO_2 with iron produces mixed oxides as well as mixture composed of mixed oxides and simple oxides. The Fe (III) and Ti (IV) have almost similar radii, therefore, Fe(III) occupies the substitutional positions. The presence of Fe (III) decrease the rate of recombination of positive holes and electrons by separating them and hence ultimately increases the catalytic performance. The Fe (III) traps the positive hole and produces Fe(IV). Then, the Fe (IV) reacts with hydroxyl ions and produce the hydroxyl radicals and O_2^- [51, 52]. The doping of TiO_2 with Fe shifts the light absorption ability towards visible light region. Under irradiation by visible light, excitation takes place (Fe(III)/Fe(IV) to conduction band of TiO_2. By irradiation, the Fe(III) changes to Fe(IV) by absorption of visible radiation because the t2g level of d orbital of Fe(III) is above the valence band of TiO_2. The electron released from Fe(III) is shifted to conduction band of TiO_2. The shifted electron produces hydroxyl radicals by further reactions.

4.2 Heterojunction

A heterojunction is an interface that occurs between two layers or regions of dissimilar crystalline semiconductors. As stated in an earlier section that titanium dioxide is very important in photocatalysis. However, two factors limit the photocatalytic activity of titanium dioxide. These factors include the wide band gap and fast recombination of positive holes and electrons. The formation of heterojunction of titanium dioxide with other semiconductor metals oxide is also an attempt to improve the photocatalytic performance by separation of positive holes and electrons and narrowing the band gap. The synthesis of heterojunction or composite of titanium dioxide with other semiconductors has gained much attention [53]. The formation of heterojunction shifts the absorption capacity of titanium dioxide towards the visible wavelength side and thus improves the catalytic performance. Different semiconductor metal oxides can be used for the formation of heterojunction. Zinc oxide (ZnO) is one of the semiconductors that can be used for the formation of heterojunction of titanium dioxide. The zinc oxide has band gap similar to titanium dioxide and it possesses good catalytic activity. Therefore, the TiO_2-ZnO heterojunction is expected to show good photocatalytic performance under visible light irradiation [54–56]. Saeed and his coworkers [57] have reported the synthesis of ZnO-TiO_2 heterojunction as an efficient photocatalyst for the photodegradation of methyl orange. They found that photodegradation of methyl orange was 98% with ZnO-TiO_2 catalyst. The photocatalytic performance was much higher than the photocatalytic performance of ZnO and TiO_2 alone having 75 and 60% activity, respectively. **Figure 5** shows the comparison of photocatalytic performance of ZnO-TiO_2 heterojunction with pure semiconductor ZnO and TiO_2. The data given in **Figure 5** was obtained by performing degradation experiments with 50 mg of ZnO or TiO_2 or ZnO-TiO_2. A 50 mL solution of methyl orange having concentration of 100 mg/L was used asmodel solution for degradation study.

The ZnO-TiO_2 exhibited higher photocatalytic performance due to the synergistic effect between zinc oxide and titanium dioxide. This synergetic effect arises due to the formation of heterojunction. When ZnO-TiO_2 heterojunction is irradiated with light, positive holes and electrons are formed in valence band and conduction band respectively. The positive hole flows from the titanium dioxide valence band to the zinc oxide valence band. At the same time, the electrons flow from the zinc oxide conduction band to the titanium dioxide conduction band. This flow of positive holes and electrons has been explained in **Figure 6**. The flow of positive holes and electrons separates the positive holes and electrons from one another. As a result, the recombination of positive holes and electrons is suppressed. Therefore,

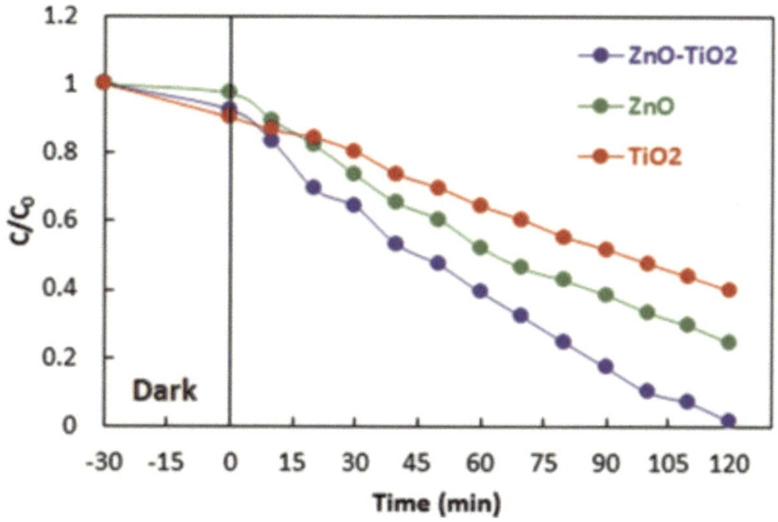

Figure 5.
Comparison of photo catalytic performance of ZnO-TiO_2, ZnO and TiO_2 towards photodegradation of methyl orange [43] (this figure has reproduced with the permission of DeGruyter).

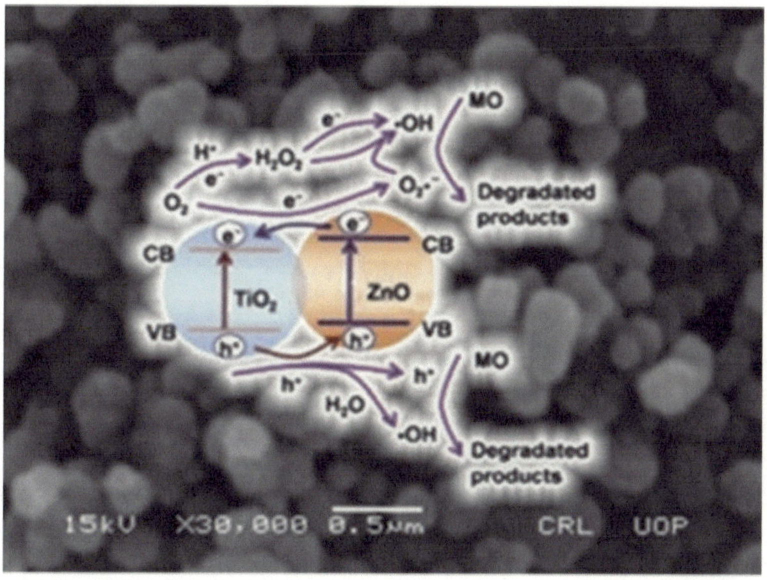

Figure 6.
Process of separation of positive holes and electrons for the improvement of photocatalytic performance towards photodegradation of methyl orange [43] (this figure has reproduced with the permission of DeGruyter).

these positive holes and electrons take proceed the redox reactions. Hence, the photocatalytic performance is increased.

Similarly, heterojunctions of titanium dioxide with other semiconductors have also been reported. For example, Abd-Rabboh and his co-workers [53] have reported the synthesis of $BiVO_4$-TiO_2 heterojunction as an effective photocatalyst for photodegradation of rhodamine B dye. They reported the heterojunction between $BiVO_4$ and TiO_2 for the production of hydrogen gas and photo degradation of rhodamine B dye. it was found that formation of heterojunction shifted that absorption of radiation by TiO_2 towards visible light region. The prepared heterojunction was tested as catalysts for degradation of rhodamine B dye. It was

No	Catalyst	Substrate	Comments	Reference
1	TiO_2-SiO_2	Acid Orange 7 (AO7)	TiO_2-SiO_2 was 12.3 and 2.3 times efficient than TiO_2 and P-25	[61]
2	TiO_2-zeolite	C.I. Basic Violet 10	Photodegradation followed 1st order kinetics	[62]
3	TiO_2-zeolite	Methyl Orange (MO)	TiO_2-zeolite exhibited higher performance irrespective of concentration of MO	[63]
4	TiO_2 Degussa P-25	Malachite Green (MG)	99.9% degradation achieved with 0.5 g/L catalyst, Catalytic performance was higher at pH higher than ZPC pH	[64]
5	TiO_2-graphitic carbon (TiO_2-GC)	Rhodamine B	TiO_2-GC-950 showed rate of 0.012 per min compared to rate of 0.006 per min with TiO_2	[65]
6	TiO_2-graphitic carbon (TiO_2-GC)	Phenol	TiO_2-GC-950 showed rate of 0.012 per min compared to rate of 0.008 per min with TiO_2	[65]
7	TiO_2 Degussa P-25	Triphenylmethane dye	Degussa P-25 was much active than than other TiO_2	[66]
8	TiO_2 Degussa	C.I. Reactive Red	Formation of electrons and holes by irradiation were confirmed by persulfate ions and ethaol	[67]
9	TiO_2 nanoparticles	Reactive Red	Optimum pH 3, increase in dye concentration decreased color removal	[68]
10	TiO_2-glass photoreactor	Methyl Red	TiO_2-glass photoreactor showed lower catalytic performance than TiO_2, because immobilization reduced active surface area	[69]
11	TiO_2-Carbon	Phenol	Both TiO_2-Carbon catalysts with pellet and powder carbon showed good performance	[70]
12	TiO_2 Degussa	Phenol	TiO_2 combined with hydrogen peroxide and ultra violet light showed good catalytic performance	[71]
13	TiO_2 prepared by sol–gel method	Phenol	The operating conditions significantly affected the catalytic activity	[72]
14	Commercial TiO_2	Bisphenol A (BPA)	Complete degradation achieved after 20 h under UV irradiation	[73]
15	TiO_2-MWCNTs	2,4-dinitrophenol	0.05% MWCNTs:TiO_2 was best combination	[74]
16	TiO_2-Activated carbon	2,4-dichlorophenol (DCP)	Improved catalytic performance was due to synergistic effect	[75]
17	Cerium-doped TiO_2	Phenol	The optimum doping was found as 0.4 wt %	[76]
18	Iron-doped TiO_2	Methylene blue	The optimum doping was found as 0.1 mole %	[77]
19	Vanadium-doped TiO_2	2,4-dichlorophenol and Methylene Blue	The optimum doping was found as 1 mole %	[78]
20	Bismuth-doped TiO2	Methylene blue	The optimum doping was found as 0.05 mole %	[79]

Table 2.
A summary of photocatalytic degradation of pollutant in the presence of TiO_2 based photocatalysts.

found that $BiVO_4$-TiO_2 heterojunction showed a photo catalytic performance of ten times greater than bare TiO_2. The rate constant for photodegradation of rhodamine B were 0.021 and 0.0023 per minute with $BiVO_4$-TiO2 and TiO_2 as catalyst respectively. Mousavi and Ghasemi [58] have reported TiO_2-$CoTiO_3$ heterojunction as photocatalyst for photodegradation of different dyes. They reported 99% photodegradation of methyl orange, methylene blue, and rhodamine b dye over TiO_2-$CoTiO_3$ heterojunction as photocatalyst under visible light irradiation. Another research group [59] has used TiO_2-Ti_3C_2 heterojunction as a catalyst for photodegradation of methyl orange with 99% performance under sunlight irradiation. CuO-TiO_2 heterojunction has also been reported for photodegradation of phenol with excellent photocatalytic performance [60]. Hence, it is concluded that the formation of heterojunction for titanium dioxide with other semiconductor metal oxide enhanced the photocatalytic performance of titanium dioxide.

Although a lot of literature is available on photocatalytic degradation of dyes and other organic pollutants in the presence of TiO_2 based photocatalysts, a summary is given in **Table 2**.

5. Conclusions

Water pollution is one the fundamental problems that have got the serious concerns of the researchers. Water poluution arises due to a number of reasons including domestic, industrial, agricultural, scinec and technology. The textile industry is the main industry that releases the dyes contaminated wastewater to the environment. A varities of protocols have been attempeted for the removal of dyes from aqueous body. Photocatalysis plays a significant role in various applications. Semiconductor metal oxides are used as photocatalysts in photocatalysis. Titanium dioxide is one of the metal oxides that has been used widely in photocatalytic applications. The titanium dioxide can be used as a photocatalyst under irradiation of ultraviolet light. The photocatalytic applications of titanium dioxide are limited by (i) wide band gap (ii) fast recombination of positive holes and electrons and (iii) activation under ultraviolet light only. Various modifications in the structure of titanium dioxide have been suggested to overcome the cited limitations. The doping of other elements in the structure of titanium dioxide and the formation of heterojunctions between titanium dioxide and other semiconductor metal oxides have been reported as effective attempts to enhance the photocatalytic performance of titanium dioxide.

Conflict of interest

The authors declare no conflict of interest.

Author details

Atta ul Haq[1], Muhammad Saeed[1*], Samreen Gul Khan[1] and Muhammad Ibrahim[2]

1 Department of Chemistry, Government College University Faisalabad, Pakistan

2 Department of Applied Chemistry, Government College University Faisalabad, Pakistan

*Address all correspondence to: msaeed@gcuf.edu.pk

References

[1] Zeng G, Chen M, Zeng Z. Risks of neonicotinoid pesticides. Science. 2013; 340:1403.

[2] Santos-Ebinuma VC, Roberto IC, Simas Teixeira MF, et al. Improving of red colorants production by a new Penicillium purpurogenum strain in submerged culture and the effect of different parameters in their stability. Biotechnology Progress. 2013;29:778–785.

[3] Anandhan M, Prabaharan T. Environmental Impacts of Natural Dyeing Process Using Pomegranate Peel Extract as a Dye. International Journal of Applied Engineering Research. 2018;13: 7765–7771. http://www.ripublication.com.

[4] Kaur S, Rani S, Mahajan RK, et al. Synthesis and adsorption properties of mesoporous material for the removal of dye safranin: Kinetics, equilibrium, and thermodynamics. Journal of Industrial and Engineering Chemistry. 2015;22: 19–27.

[5] Ajmal A, Majeed I, Malik RN, et al. Principles and mechanisms of photocatalytic dye degradation on TiO 2 based photocatalysts: A comparative overview. RSC Advances. 2014;4: 37003–37026.

[6] Wang M, Chamberland N, Breau L, et al. An organic redox electrolyte to rival triiodide/iodide in dye-sensitized solar cells. Nature Chemistry. 2010;2: 385–389.

[7] Lekang OI, Marie Bomo A, Svendsen I. Biological lamella sedimentation used for wastewater treatment. Aquacultural Engineering. 2001;24:115–127.

[8] Ali I. Water treatment by adsorption columns: Evaluation at ground level. Separation and Purification Reviews. 2014;43:175–205.

[9] Oe T. Performance of membrane filtration system used for water treatment. Desalination. 1996;106:107–113.

[10] Joss A, Zabczynski S, Göbel A, et al. Biological degradation of pharmaceuticals in municipal wastewater treatment: Proposing a classification scheme. Water Research. 2006;40:1686–1696.

[11] Gogate PR, Pandit AB. A review of imperative technologies for wastewater treatment I: Oxidation technologies at ambient conditions. Advances in Environmental Research. 2004;8:501–551.

[12] Oturan MA, Aaron JJ. Advanced oxidation processes in water/wastewater treatment: Principles and applications. A review. Critical Reviews in Environmental Science and Technology. 2014;44:2577–2641.

[13] Pelaez M, Nolan NT, Pillai SC, et al. A review on the visible light active titanium dioxide photocatalysts for environmental applications. Applied Catalysis B: Environmental. 2012;125: 331–349.

[14] Mills A, Le Hunte S. An overview of semiconductor photocatalysis. Journal of Photochemistry and Photobiology A: Chemistry. 1997;108:1–35.

[15] Carp O, Huisman CL, Reller A. Photoinduced reactivity of titanium dioxide. Progress in Solid State Chemistry. 2004;32:33–177.

[16] Umar IG, Halim AA. Heterogeneous photocatalytic degradation of organic contaminants over titanium dioxide: A review of fundamentals, progress and problems. Journal of Photochemistry and Photobiology C: Photochemistry Reviews. 2007;9:1–12.

[17] Ni M, Leung MKH, Leung DYC, et al. A review and recent developments in photocatalytic water-splitting using TiO_2 for hydrogen production. Renewable and Sustainable Energy Reviews. 2007;11:401–425.

[18] Banerjee S, Dionysiou DD, Pillai SC. Self-cleaning applications of TiO_2 by photo-induced hydrophilicity and photocatalysis. Applied Catalysis B: Environmental. 2015;176–177:396–428.

[19] Kapteijn F, Rodriguez-Mirasol J, Moulijn JA. Heterogeneous catalytic decomposition of nitrous oxide. Applied Catalysis B: Environmental. 1996;9: 25–64.

[20] Carey JH, Lawrence J, Tosine HM. Photodechlorination of PCB's in the presence of titanium dioxide in aqueous suspensions. Bulletin of Environmental Contamination and Toxicology. 1976;16: 697–701.

[21] Matsunaga T. Photoelectrochemical sterilization of microbial cells by semiconductor powders. FEMS Microbiology Letters. 1985;29:211–214.

[22] Akira F, Kenichi H. Electrochemical Photolysis of Water at a Semiconductor Electrode. Nature. 1972;238:37–38.

[23] Di Paola A, Bellardita M, Palmisano L. Brookite, the least known TiO_2 photocatalyst. Catalysts. 2013;3: 36–73.

[24] Kandiel TA, Robben L, Alkaim A, et al. Brookite versus anatase TiO2 photocatalysts: Phase transformations and photocatalytic activities. Photochemical and Photobiological Sciences. 2013;12:602–609.

[25] Hurum DC, Agrios AG, Gray KA, et al. Explaining the enhanced photocatalytic activity of Degussa P25 mixed-phase TiO_2 using EPR. Journal of Physical Chemistry B. 2003;107:4545–4549.

[26] Kawahara T, Konishi Y, Tada H, et al. A patterned TiO_2(anatase)/ TiO_2(rutile) bilayer-type photocatalyst: Effect of the anatase/rutile junction on the photocatalytic activity. Angewandte Chemie. 2002;41:2811–2813.

[27] Mills A, O'Rourke C, Moore K. Powder semiconductor photocatalysis in aqueous solution: An overview of kinetics-based reaction mechanisms. Journal of Photochemistry and Photobiology A: Chemistry. 2015;310: 66–105.

[28] Liu B, Zhao X, Terashima C, et al. Thermodynamic and kinetic analysis of heterogeneous photocatalysis for semiconductor systems. Physical Chemistry Chemical Physics. 2014;16: 8751–8760.

[29] Ângelo J. Development and Characterization of Titania-based Photocatalysts and their Incorporation in Paint Coatings. 2016;1–172.

[30] Montoya JF, Peral J, Salvador P. Comprehensive kinetic and mechanistic analysis of TiO_2 photocatalytic reactions according to the direct-indirect model: (I) theoretical approach. Journal of Physical Chemistry C. 2014;118:14266–14275.

[31] Adriana Zaleska. Doped-TiO_2 a review. Recent Patents on Engineering. 2008;2:8.

[32] Ohtani B, Ogawa Y, Nishimoto SI. Photocatalytic activity of amorphous-anatase mixture of titanium(IV) oxide particles suspended in aqueous solutions. Journal of Physical Chemistry B. 1997;101:3746–3752.

[33] Ohtani B, Nishimoto SI. Effect of surface adsorptions of aliphatic alcohols and silver ion on the photocatalytic activity of TiO_2 suspended in aqueous solutions. Journal of Physical Chemistry. 1993;97:920–926.

[34] Li Z, Jia Z, Li W, et al. Synthesis of Ag/AgCl Nanoparticles Immobilized on CoFe2O4 Fibers and Their Photocatalytic Degradation for Methyl Orange. Xiyou Jinshu Cailiao Yu Gongcheng/Rare Metal Materials and Engineering. 2017;46:3669–3674.

[35] Li W, Ma Z, Bai G, et al. Dopamine-assisted one-step fabrication of Ag@AgCl nanophotocatalyst with tunable morphology, composition and improved photocatalytic performance. Applied Catalysis B: Environmental. 2015;174–175:43–48.

[36] Lu L, Li J, Yu J, et al. A hierarchically porous $MgFe_2O_4/\gamma$-Fe_2O_3 magnetic microspheres for efficient removals of dye and pharmaceutical from water. Chemical Engineering Journal. 2016;283:524–534.

[37] Saeed M, Muneer M, Khosa MKK, et al. Azadirachta indica leaves extract assisted green synthesis of Ag-TiO_2 for degradation of Methylene blue and Rhodamine B dyes in aqueous medium. Green Processing and Synthesis. 2019;8: 659–666.

[38] Chen G, Si X, Yu J, et al. Doping nano-Co_3O_4 surface with bigger nanosized Ag and its photocatalytic properties for visible light photodegradation of organic dyes. Applied Surface Science. 2015;330: 191–199.

[39] Komaraiah D, Radha E, Sivakumar J, et al. Photoluminescence and photocatalytic activity of spin coated Ag^+ doped anatase TiO_2 thin films. Optical Materials. 2020;108.

[40] Komaraiah D, Radha E, Kalarikkal N, et al. Structural, optical and photoluminescence studies of sol-gel synthesized pure and iron doped TiO_2 photocatalysts. Ceramics International. 2019;45:25060–25068.

[41] Choi J, Park H, Hoffmann MR. Effects of single metal-ion doping on the visible-light photoreactivity of TiO_2. Journal of Physical Chemistry C. 2010; 114:783–792.

[42] Bharti B, Kumar S, Lee HN, et al. Formation of oxygen vacancies and Ti^{3+} state in TiO_2 thin film and enhanced optical properties by air plasma treatment. Scientific Reports. 2016;6.

[43] Wu C, Gao Z, Gao S, et al. Ti^{3+} self-doped TiO_2 photoelectrodes for photoelectrochemical water splitting and photoelectrocatalytic pollutant degradation. Journal of Energy Chemistry. 2016;25:726–733.

[44] Varma RS, Thorat N, Fernandes R, et al. Dependence of photocatalysis on charge carrier separation in Ag-doped and decorated TiO_2 nanocomposites. Catalysis Science and Technology. 2016; 6:8428–8440.

[45] Banerjee S, Pillai SC, Falaras P, et al. New insights into the mechanism of visible light photocatalysis. Journal of Physical Chemistry Letters. 2014;5: 2543–2554.

[46] Li X, Liu P, Mao Y, et al. Preparation of homogeneous nitrogen-doped mesoporous TiO2 spheres with enhanced visible-light photocatalysis. Applied Catalysis B: Environmental. 2015;164:352–359.

[47] Di Valentin C, Pacchioni G, Selloni A, et al. Characterization of paramagnetic species in N-doped TiO_2 powders by EPR spectroscopy and DFT calculations. Journal of Physical Chemistry B. 2005;109:11414–11419.

[48] Ren W, Ai Z, Jia F, et al. Low temperature preparation and visible light photocatalytic activity of mesoporous carbon-doped crystalline TiO_2. Applied Catalysis B: Environmental. 2007;69:138–144.

[49] Ma X, Dai Y, Guo M, et al. Relative photooxidation and photoreduction

activities of the {100}, {101}, and {001} Surfaces of Anatase TiO_2. Langmuir. 2013;29:13647–13654.

[50] Magalhães P, Andrade L, Nunes OC, et al. Titanium dioxide photocatalysis: Fundamentals and application on photoinactivation. Reviews on Advanced Materials Science. 2017;51:91–129.

[51] Rauf MA, Meetani MA, Hisaindee S. An overview on the photocatalytic degradation of azo dyes in the presence of TiO2 doped with selective transition metals. Desalination. 2011;276:13–27.

[52] Tong T, Zhang J, Tian B, et al. Preparation of Fe^{3+}-doped TiO_2 catalysts by controlled hydrolysis of titanium alkoxide and study on their photocatalytic activity for methyl orange degradation. Journal of Hazardous Materials. 2008;155:572–579.

[53] Abd-Rabboh HSM, Benaissa M, Hamdy MS, et al. Synthesis of an efficient, and recyclable mesoporous $BiVO_4/TiO_2$ direct Z-scheme heterojunction by sonochemical route for photocatalytic hydrogen production and photodegradation of rhodamine B dye in the visible region. Optical Materials. 2021;114:110761.

[54] Topkaya E, Konyar M, Yatmaz HC, et al. Pure ZnO and composite ZnO/ TiO_2 catalyst plates: A comparative study for the degradation of azo dye, pesticide and antibiotic in aqueous solutions. Journal of Colloid and Interface Science. 2014;430:6–11.

[55] Wang L, Fu X, Han Y, et al. Preparation, characterization, and photocatalytic activity of TiO 2/ZnO nanocomposites. Journal of Nanomaterials. 2013;2013: 321459. https://doi.org/10.1155/2013/321459.

[56] Thennarasu G, Kavithaa S, Sivasamy A. Photocatalytic degradation of Orange G dye under solar light using nanocrystalline semiconductor metal oxide. Environmental Science and Pollution Research. 2012;19:2755–2765.

[57] Saeed M, Ibrahim M, Muneer M, et al. ZnO-TiO_2: Synthesis, Characterization and Evaluation of Photo Catalytic Activity towards Degradation of Methyl Orange. Zeitschrift fur Physikalische Chemie. 2019;235(3): 225–237.

[58] Mousavi M, Ghasemi JB. Novel visible-light-responsive Black-TiO_2/ $CoTiO_3$ Z-scheme heterojunction photocatalyst with efficient photocatalytic performance for the degradation of different organic dyes and tetracycline. Journal of the Taiwan Institute of Chemical Engineers. 2021; 121:168–183.

[59] Hieu VQ, Phung TK, Nguyen TQ, et al. Photocatalytic degradation of methyl orange dye by T_i3C_2–TiO_2 heterojunction under solar light. Chemosphere. 2021;276: 130154.

[60] Gnanasekaran L, Pachaiappan R, Kumar PS, et al. Visible light driven exotic p (CuO) - n (TiO_2) heterojunction for the photodegradation of 4-chlorophenol and antibacterial activity. Environmental Pollution. 2021; 278:117304.

[61] Chen Y, Wang K, Lou L. Photodegradation of dye pollutants on silica gel supported TiO_2 particles under visible light irradiation. Journal of Photochemistry and Photobiology A: Chemistry. 2004;163:281–287.

[62] Wang CC, Lee CK, Lyu M Du, et al. Photocatalytic degradation of C.I. Basic Violet 10 using TiO_2 catalysts supported by Y zeolite: An investigation of the effects of operational parameters. Dyes and Pigments. 2008;76:817–824.

[63] Li F, Sun S, Jiang Y, et al. Photodegradation of an azo dye using immobilized nanoparticles of TiO_2

supported by natural porous mineral. Journal of Hazardous Materials. 2008; 152:1037–1044.

[64] Chen CC, Lu CS, Chung YC, et al. UV light induced photodegradation of malachite green on TiO_2 nanoparticles. Journal of Hazardous Materials. 2007; 141:520–528.

[65] XIAO Y, DANG L, AN L, et al. Photocatalytic Degradation of Rhodamine B and Phenol by TiO_2 Loaded on Mesoporous Graphitic Carbon. Chinese Journal of Catalysis. 2008;29:31–36.

[66] Saquib M, Muneer M. TiO_2/ mediated photocatalytic degradation of a triphenylmethane dye (gentian violet), in aqueous suspensions. Dyes and Pigments. 2003;56:37–49.

[67] Wu CH. Effects of operational parameters on the decolorization of C.I. Reactive Red 198 in UV/TiO_2-based systems. Dyes and Pigments. 2008;77: 31–38.

[68] Belessi V, Romanos G, Boukos N, et al. Removal of Reactive Red 195 from aqueous solutions by adsorption on the surface of TiO_2 nanoparticles. Journal of Hazardous Materials. 2009;170:836–844.

[69] Mascolo G, Comparelli R, Curri ML, et al. Photocatalytic degradation of methyl red by TiO_2: Comparison of the efficiency of immobilized nanoparticles versus conventional suspended catalyst. Journal of Hazardous Materials. 2007; 142:130–137.

[70] Carpio E, Zúñiga P, Ponce S, et al. Photocatalytic degradation of phenol using TiO_2 nanocrystals supported on activated carbon. Journal of Molecular Catalysis A: Chemical. 2005;228:293–298.

[71] Chiou CH, Wu CY, Juang RS. Influence of operating parameters on photocatalytic degradation of phenol in UV/TiO_2 process. Chemical Engineering Journal. 2008;139:322–329.

[72] Silva CG, Faria JL. Effect of key operational parameters on the photocatalytic oxidation of phenol by nanocrystalline sol-gel TiO_2 under UV irradiation. Journal of Molecular Catalysis A: Chemical. 2009;305:147–154.

[73] Ohko Y, Ando I, Niwa C, et al. Degradation of bisphenol A in water by TiO_2 photocatalyst. Environmental Science and Technology. 2001;35:2365–2368.

[74] Wang H, Wang HL, Jiang WF, et al. Photocatalytic degradation of 2,4-dinitrophenol (DNP) by multi-walled carbon nanotubes (MWCNTs)/TiO_2 composite in aqueous solution under solar irradiation. Water Research. 2009; 43:204–210.

[75] Gu L, Chen Z, Sun C, et al. Photocatalytic degradation of 2, 4-dichlorophenol using granular activated carbon supported TiO_2. Desalination. 2010;263:107–112.

[76] Fan C, Xue P, Sun Y. Preparation of Nano-TiO_2 doped with cerium and its photocatalytic activity. Journal of Rare Earths. 2006;24:309–313.

[77] Sun L, Li J, Wang CL, et al. An electrochemical strategy of doping Fe^{3+} into TiO_2 nanotube array films for enhancement in photocatalytic activity. Solar Energy Materials and Solar Cells. 2009;93:1875–1880.

[78] Tian B, Li C, Gu F, et al. Flame sprayed V-doped TiO_2 nanoparticles with enhanced photocatalytic activity under visible light irradiation. Chemical Engineering Journal. 2009;151:220–227.

[79] Ji T, Yang F, Lv Y, et al. Synthesis and visible-light photocatalytic activity of Bi-doped TiO_2 nanobelts. Materials Letters. 2009;63:2044–2046.

Section 3

Environmental Applications of Titanium Dioxide

Chapter 5

Titanium Dioxide Thin Films for Environmental Applications

Wafa Selmi, Nabil Hosni, Jamila Ben Naceur, Hager Maghraoui-Meherzi and Radhouane Chtourou

Abstract

The environmental pollution and the rapid depletion of fossil fuel caused by the rapid increase in industrial production became serious problems for humans. These issues have inspired many researchers to found eco-friendly materials, which can degrade pollutants and produce green energy. Titanium dioxide (TiO_2) thin films are one of the important and promising semiconductor materials for environmental and energy applications because of their unique optical and electronic properties. In this chapter, an overview of the background of TiO_2 structure and the different methods of synthesis TiO_2 thin films were carried out. The photocatalytic water treatment and the water split for H_2 production by TiO_2 thin films were investigated. The strong influence on photocatalytic and water split efficiency of TiO_2 thin films by crystal structure, surface area, crystalline structure, average particle size and porosity were summarized.

Keywords: titanium dioxide, thin films, photocatalytic, water splitting, H_2 production, energy and clean environmental

1. Introduction

The industrial revolution has generated the rapid depletion of fossil fuel and the environmental pollution, which have become the most pressing human problems. Nowadays, urgent need to explore environmentally friendly technologies is indispensable to get clean energy and protect the environment.

Titanium dioxide (TiO_2) have been widely investigated in academic research [1, 2] and extensively involved in industrial applications to their abundancy, durability, no toxicity, the high photoactivity and the photoelechemical proprieties.

TiO_2 thin films were successfully synthesis by physical and chemical techniques such as pulsed laser deposition (PLD), molecular beam epitaxy (MBE), RF magnetron sputtering, electrodeposition, sol–gel, hydrothermal, spin-spraying, spin coating, successive ionic layer adsorption and reaction (SILAR), chemical vapor deposition (CVD) and chemical bath deposition (CBD).

Since the discover in the 70's by Fujishima and Honda [3] many research have investigated the production of hydrogen by TiO_2 photoelectrodes under ultraviolet light. For that, TiO_2 semiconductor photocatalysis is considered as the promising material to address both hydrogen production and pollutant degradation.

2. Structures of TiO_2

Titanium dioxide (TiO_2) is an n-type semiconducting material with very interesting properties, such as chemical stability, nontoxicity, low cost, availability, good mechanical flexibility, conductivity and high photocatalytic activity. TiO_2 has three different polymorphs (**Figure 1**) anatase, rutile and brookite [4]. Crystallographic data of TiO_2 structures are summarized in **Table 1**. The most stable form of TiO_2 is rutile. All three polymorphs can be synthesized by many methods.

Anatase is a metastable mineral form of titanium dioxide (TiO_2) which crystallize in the tetragonal system with $I4_1/amd$ space group. Rutile is the most common natural form of TiO_2 and crystallizes in the same system of anatase with $P4_2/mnm$ space group and brookite has an orthorhombic crystalline structure with Pbca space group. In all forms, titanium (Ti^{4+}) atoms are coordinated by six oxygen (O^{2-}) atoms, forming the octahedral TiO_6 were titanium atom (Ti^{4+}) is in the center and oxygen atoms (O^{2-}) are at corners.

TiO_2 is the most crystalline semiconductor used in photocatalytic process, due to the bandgap energy being relatively wide (Eg = 3.2 eV for anatase; Eg = 3.0 eV

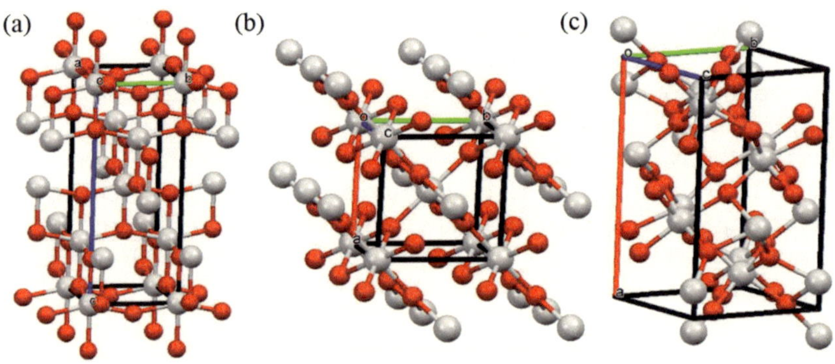

Figure 1.
Titanium dioxide structures (a) anatase, (b) rutile and (c) brookite.

	Anatase	Rutile	Brookite
System	Tetragonal	Tetragonal	Orthorhombic
Space group	$I4_1/amd$	$P4_2/mnm$	Pbca
a (Å)	3.789	4.594	9.184
b (Å)	3.789	4.594	5.447
c (Å)	9.537	2.959	5.145
α (°)	90	90	90
β (°)	90	90	90
γ (°)	90	90	90
V ($Å^3$)	136.93	62.45	257.38
Z	4	2	8
Eg (eV)	3.2	3.0	3.1

Table 1.
Crystallographic data of TiO_2 structures.

for rutile; Eg = 3.1 eV for brookite), the material can only be activated by UV irradiation with $\lambda < 380$ nm.

3. Morphologies of TiO_2

The proprieties of TiO_2 are strongly depends on crystal structure, surface area, crystalline structure, average particle size and porosity. Controlling the morphology of TiO_2 nanomaterials is a highly active area of research. In particular, there is many morphologies of TiO_2 such as nanowires, nanorods, nanotubes and nanosheets.

3.1 Nanowires

Nanowires (**Figure 2**) are one-dimensional nanostructure in cylindrical form. They have lengths in the range of a few micrometers to centimeters and their diameters are in the nanometer range. L. Li et al. [4] have demonstrate that the ITO nanowires coated by TiO_2 have grown by a thermal evaporation method. The high annealing temperatures between 350 and 600°C increase the crystallinity of TiO_2 shell and suppress electron recombination in the core-shell nanostructures. The tin-doped indium oxide (ITO)-TiO_2 core-shell nanostructures are tested as the photoanode for DSSCs. The vertically aligned nanowires are expected to efficiently transport electrons to the substrate where the current is collected. These nanowires have high density, which enables high dye loading and high current density during device operation. The open-circuit voltage decay (OCVD) measurements of (ITO)-TiO_2 core-shell nanostructures show that the electron lifetime increases by quantity of HfO_2 magnitude insertion. ITO-TiO_2 core-shell nanostructures with HfO_2 blocking layers are promising photoanodes for DSSCs.

3.2 Nanorods

Nanorods (**Figure 3**) are one dimensional nanoscale objects. They have a width in the range of 1–100 nm. In the work of J. Ben Naceur et al. [5], the SEM images

Figure 2.
SEM image of TiO_2 nanowires. (a) TEM image of ITO-TiO_2 core-shell nanostructure (b).

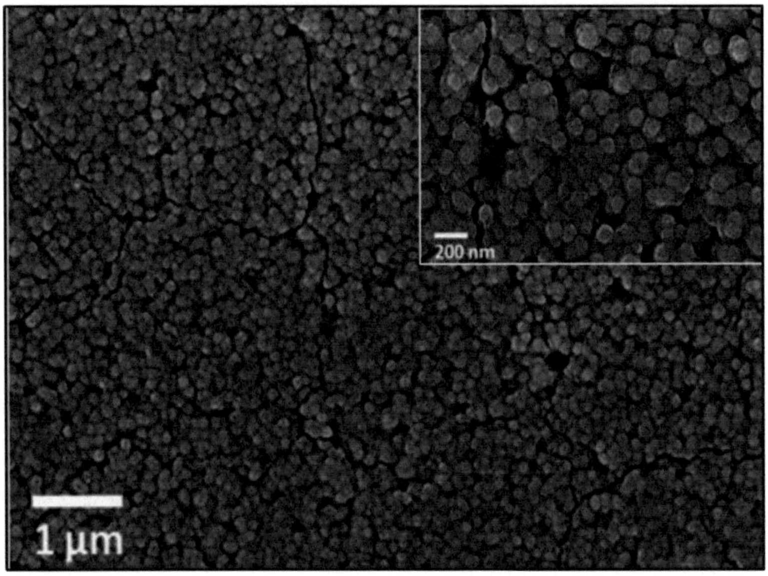

Figure 3.
SEM micrograph of TiO_2 nanorods.

reveal that the entire surfaces of the FTO substrate is uniformly coated by TiO_2 nanorods with an average length and a diameter equal to 1 µm and 60 nm, respectively. Titanium dioxide nanorods arrays (NRAs) photoanodes have been grown by the hydrothermal method on FTO coated glass substrates for different hydrothermal reaction time (5, 10, 15 h). Structural and morphological properties of TiO_2 films confirms the formation of rutile phase with nanorods morphology. The wettability and photoelectrochemical performances of films were investigated. The wettability tests of the sample elaborated at 10 h revealed that this sample is more hydrophilic among all prepared samples for that, it has the best physical properties with a higher photocurrent density equal to 0.22 $mA.cm^{-1}$ at 0.5 V vs. Ag/AgCl.

3.3 Nanotubes

Nanotubes are typically long and thin cylindrical protrusions with sub-micron diameter and lengths in the order of several 100 µm. The SEM images (**Figure 4**), in the work of T-H. Meen et al. [6] show the formation of the TiO_2 nanotubes. To prepared TiO_2 nanotube arrays the electrochemical anodization was used and was tested as photoelectrode of dye-sensitized solar cells. In the SEM analysis, the lengths of TiO_2 nanotube arrays prepared by electrochemical anodization was approximately 10 to 30 µm. After titanium tetrachloride ($TiCl_4$) treatment, the walls of TiO_2 nanotubes were coated with TiO_2 nanoparticles. XRD patterns showed that the oxygen-annealed TiO_2 nanotubes have a better anatase phase. The conversion efficiency with different lengths of TiO_2 nanotube photoelectrodes is 3.21%, 4.35%, and 4.34% with 10, 20, and 30 µm, respectively. The electrochemical impedance spectroscopy analysis, show that the value of Rk (charge transfer resistance related to recombination of electrons) decreases from 26.1 to 17.4 Ω when TiO_2 nanotubes were treated with $TiCl_4$. The treated TiO_2 nanotubes with $TiCl_4$ show that the surface area of nanotubes increase, resulting the increase of dye adsorption and the increase of the conversion efficiency of DSSCs.

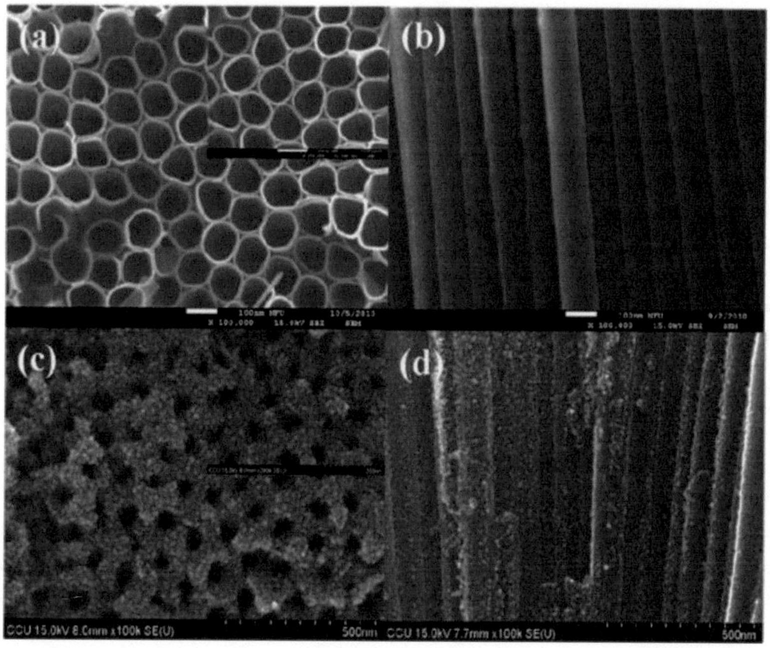

Figure 4.
SEM images of TiO_2 nanotubes (a) top view and (b) side view before $TiCl_4$ treatment, (c) top view and (d) side view after $TiCl_4$ treatment [6].

3.4 Nanosheets

Nanosheet is a two-dimensional nanostructure with thickness in a scale ranging from 1 to 100 nm. As described in work of F. Li et al. [7] the scanning electron microscopy (SEM) image (**Figure 5**) show a single layered 2D morphology of TiO_2 nanosheets. TiO_2 nanosheets are a good carrier of photocatalytic materials and have become attractive materials in the new century because of their high active surface exposure characteristics and special morphology. The preparing TiO_2 nanosheets, was made via hydrothermal calcination method. X-ray powder diffraction (XRD), scanning electron microscopy (SEM), and UV–visible diffuse reflection absorption spectra (DRS) were used to characterize the structure and morphology of the TiO_2 nanosheets. The suitable calcination temperature was 400°C to obtain the TiO_2 nanosheets, with a good hydrogen production rate of 270 μmol/h. The sheet structure of the material was beneficial for improving the photocatalytic water splitting

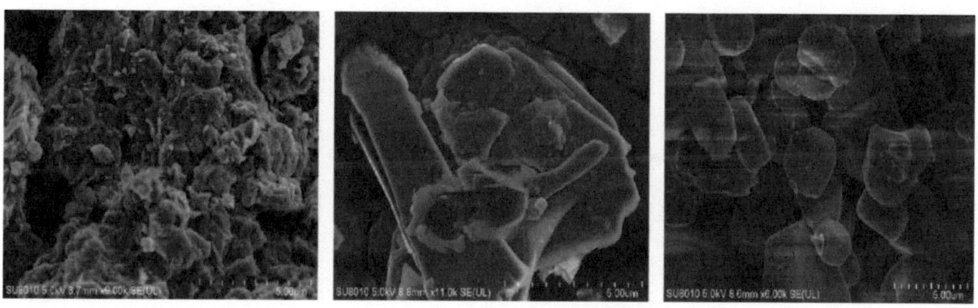

Figure 5.
SEM images of the single-layered 2D mesoporous TiO_2 nanosheets [7].

hydrogen production performance. The research in photocatalytic water splitting of TiO_2 thin films to produce hydrogen are currently a promise topic.

4. Synthesis method of TiO_2

TiO_2 thin films were successfully synthesis by several techniques such as pulsed laser deposition (PLD), molecular beam epitaxy (MBE), RF magnetron sputtering, electrodeposition, sol–gel, hydrothermal, spin-spraying, spin coating, successive ionic layer adsorption and reaction (SILAR), chemical vapor deposition (CVD) and chemical bath deposition (CBD).

4.1 Pulsed laser deposition (PLD) method

The technique of PLD is extremely simple which has been used to deposit high quality films. The technique uses high power laser pulses to melt, evaporate and ionize material from the surface of a target to the substrate (**Figure 6**). *A. Ishii* et al. [8] have elaborated TiO_2 film into glass substrates. For the period of deposition, the pressure of PO_2 in PLD chamber was around 1 to 9 Pa and the temperature was controlled by an infrared lamp heater and a Si plate from room temperature to 600°C for 20 min to make 100–150 nm thick. TiO_2 thin films showed excellent optical properties, with $n = 3.14$ and $k < 0.05$ at $\lambda = 400$ nm.

4.2 Molecular beam epitaxy (MBE) method

The technique of MBE consist to send one or more molecular jets to a substrate with achieve epitaxial growth (**Figure 7**). It makes it possible to grow nanostructured samples of several cm^2 at a rate one atomic monolayer per second. S. Naseem et al. [9] have utilized a Ti electron-beam evaporator and molecular oxygen introduced by a sapphire-sealed leak valve to grown anatase phase TiO_2 doped by Co. The growth of TiO_2 were controlled by a quartz deposition monitor, and under 399.96×10^{-7} Pa of O_2. The 8% cobalt doped TiO_2 film has shown 91% degradation

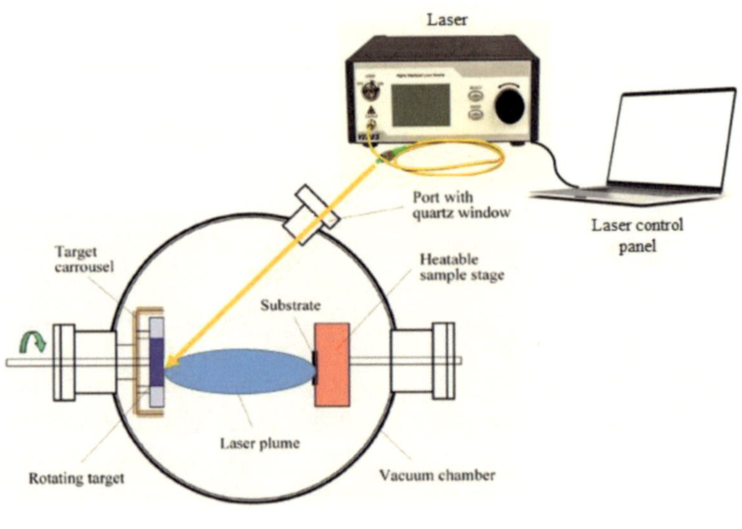

Figure 6.
Pulsed laser deposition (PLD) method.

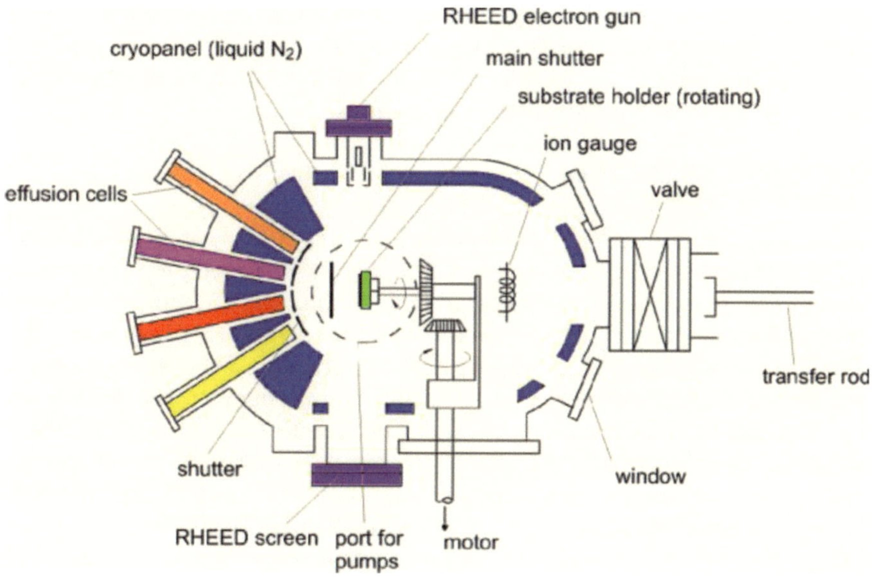

Figure 7.
Molecular beam epitaxy (MBE) method.

with methylene blue and 88% degradation with Azo dye in 70 min under visible light irradiation exhibiting excellent photocatalytic performance.

4.3 Radio frequency magnetron sputtering method

The technique of RF magnetron sputtering (**Figure 8**) has been widely adopted for the high-rate deposition of thin films. This deposition technology involving a gaseous plasma which is generated and confined to a space containing the surface of the target which is eroded by high-energy ions within the plasma, and the liberated atoms travel through the vacuum environment and deposit onto a substrate to form a thin film. J. Singh et al. [10] have prepared TiO_2 thin films deposited on silica glass substrates by RF magnetron sputtering combined with thermal annealing

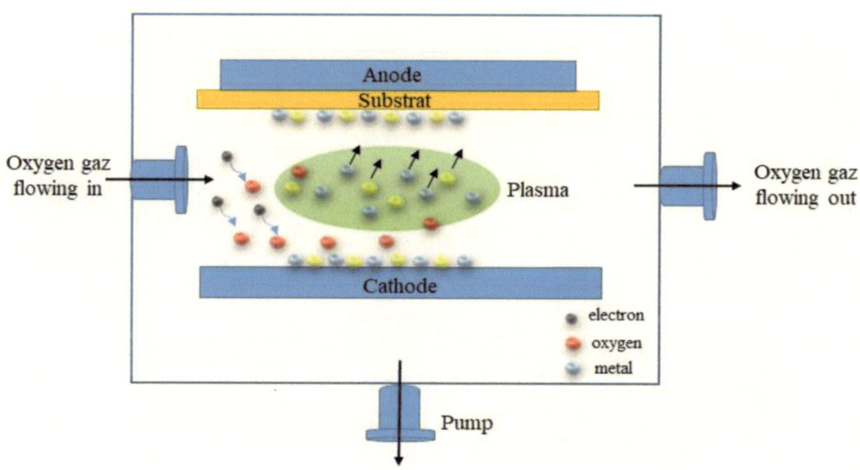

Figure 8.
Radio frequency magnetron sputtering method.

at 2.5×10^{-3} Torr and substrate temperature of 200°C. TiO_2 film exhibits highly enhanced photocatalytic activity leading to complete photocatalytic degradation of 2.1 μM MB in water in only 45 minutes of sun light irradiation, which is very promising for practical photocatalytic applications.

4.4 Electrodeposition method

Electrodeposition (**Figure 9**) is a flexible low-cost method of fabrication of films. The principles of the electrodeposition process are based on principles of electrochemical phenomena associated with the reduction or deposition of electroactive and accompanying species on the cathode surface. H. Wang et al. [11] have elaborated TiO_2 nanotube arrays by the combination of the electrodeposition method and the anodic aluminum oxide (AAO) templating method. A three-electrode potentiostatic system with a saturated calomel electrode (SCE) as a reference electrode and a platinum plate as a counter electrode was used for the electrodeposition method at room temperature. The potential used for the deposition is −0.8, −1.0 V in the electrolyte solution of 0.1 M $TiCl_3$. The pH was maintained at 2.0 by adding a few drops of 2 M Na_2CO_3. After the deposition, the nanotubes in the AAO template were rinsed with double distilled water (DDW), then dried in air at room temperature and finally the samples were heated at 500°C for 2 h under N_2 atmosphere.

4.5 Sol-gel method

The sol–gel technique (**Figure 10**) is a wet low-temperature method that involves the formation of an inorganic colloidal suspension (sol) and gelation of the sol in a continuous liquid phase (gel) to form a three-dimensional network structure. J. Ben Naceur et al. [12] have elaborated TiO_2 on ITO by sol–gel technique. To obtain stable solution, J. Ben Naceur have used 1 mol tetrabutyl-orthotitanate [$Ti(OCH_2CH_2CH_2CH_3)_4$], 1 mol H_2O, 4 mol butanol [CH_3-$(CH_2)_3$-OH], and 3 mol acetic acid [CH_3COOH]. The solution was stirred for 1 h at room temperature, and then a gel film was formed on ITO glass substrate.

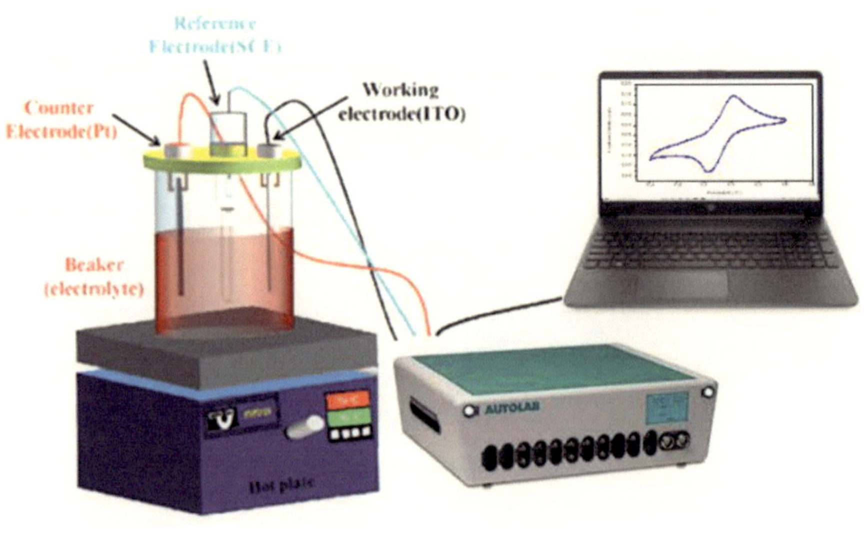

Figure 9.
Schematic of the electrodeposition method.

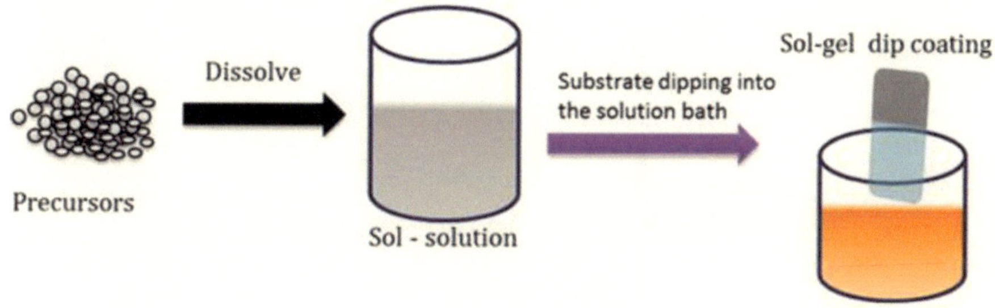

Figure 10.
Schematic of the sol–gel method.

4.6 Hydrothermal method

The hydrothermal technique (**Figure 11**) is a convenient wet way to produce well crystalline materials, with suitably tuned size and shape of particles at temperatures between 180 and 300°C. J. Ben Naceur et al. [13] have prepared TiO_2 nanorods (NRAs) thin film by mixing of hydrochloric acid (HCl), distilled water and Titanium-isopropoxide [$Ti(OCH(CH_3)_2)$]. The mixture was magnetically agitated then transferred to Teflon-coated stainless-steel autoclave. The substrate was placed inside a furnace and kept at 180°C for 5 hours. The hydrothermal method has same advantages such as easy to obtain nanotube morphology, variation in the synthesis method can be implemented to enhance the properties of TiO_2 nanotubes, and it is a feasible method for different applications.

4.7 Spin-spraying method

The spin-spraying technique (**Figure 12**) consists of pulverization of an oxidizing and treatment solutions respectively, onto substrates mounted on a rotating table at a constant temperature. M. O. Abou-Helal et al. [14] have deposited TiO_2 on glass substrates. The solution was prepared by dissolve Titanium (IV) isobutoxide [$Ti((CH_3)_2CHCH_2O)_4$], in a mixture of HNO_3 and methanol solution. The deposition parameters used are 0.1 M of prepared solution, with rate of 1.0 ml/min, at 450–600°C for 5–30 min spray time.

4.8 Spin-coating method

The spin-coating technique (**Figure 13**) is used to deposit uniform thin films onto flat substrates. Usually a small amount of coating material is applied on the

Figure 11.
Schematic of the hydrothermal method.

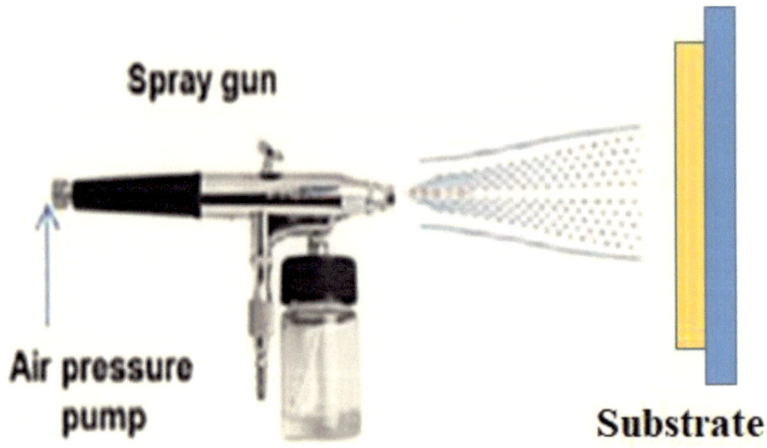

Figure 12.
Schematic of the spin-spraying method.

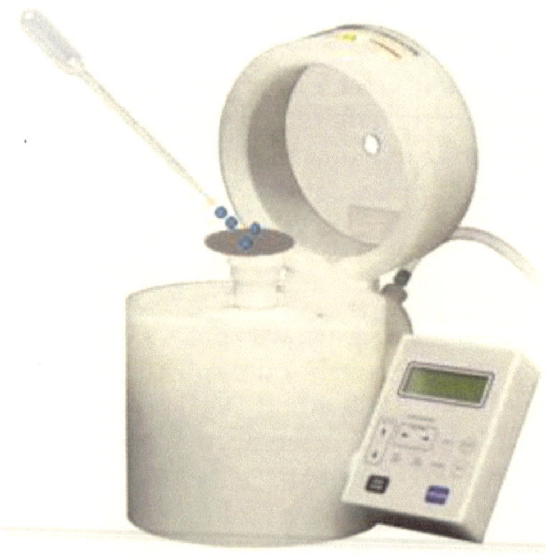

Figure 13.
Schematic of the spin-coating method.

center of the substrate, which is either spinning at low speed or not spinning at all. The substrate is then rotated at speed up to 10,000 rpm to spread the coating material by centrifugal force. F. Joudi et al. [15] have prepared TiO_2 thin film by mixing 2 g of P25 in 10 ml of solution of ethanol and acetylacetone under stiring for 10 min. Then the solution was deposited on substrates by spin coating technique. To remove organic solvents from the samples, the films were dried at 150°C and annealed at 450°C for 2 h.

4.9 Successive ionic layer adsorption and reaction (SILAR) method

The SILAR technique (**Figure 14**) is a simple, less expensive, and less time-consuming method for the deposition of binary semiconducting thin films. It is also applicable in the deposition of large-area thin films. V. L. Patil et al. [16] have succeeded to fabricate nanogranular TiO_2 thin films by SILAR. In the initial step, glass substrate was immersed vertically in first beaker contains 2.5 ml $TiCl_3$ (pH = 1)

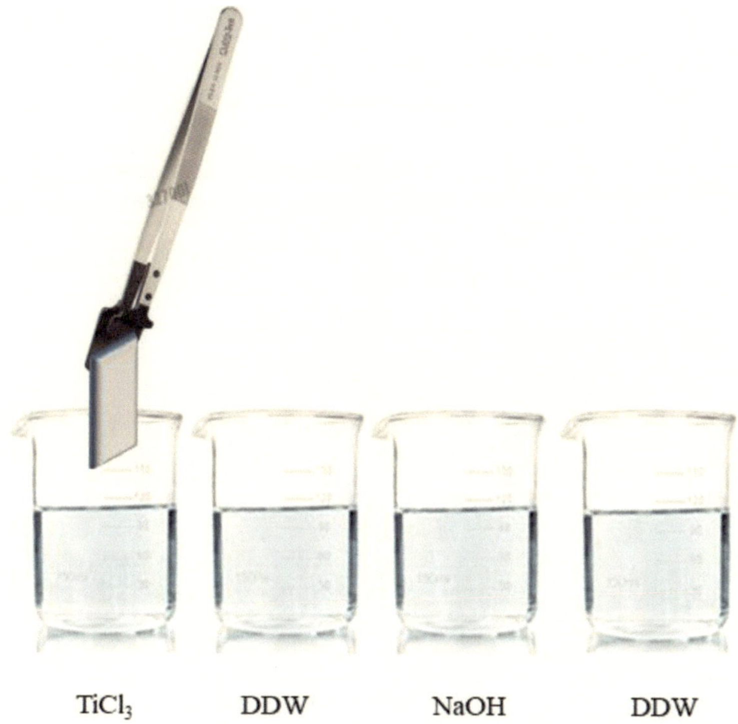

Figure 14.
Schematic of the successive ionic layer adsorption and reaction method.

in 50 ml double distilled water for 10 s, so that a layer of titanium water complex [$Ti(OH)_3$] was adsorbed on the substrate surface. Then the substrate was rinsed using double distilled water (DDW) for 5 s. Then, the substrate was immersed in third beaker contains 0.1 M NaOH solution with pH = 12 for 10 s in which the adsorbed Ti-species with NaOH forms into stable TiO_2 on the substrate. Finally, the TiO_2 and adsorbed hydroxide rinsed in the fourth beaker for 5 s. The nanogranular TiO_2 thin films shows a good gas sensitivity towards NO_2 gas at 50 ppm with good selectivity.

4.10 Chemical vapor deposition (CVD)

The CVD technique (**Figure 15**) is a vacuum deposition method used to produce high quality, high-performance thin films. The substrate is exposed to one or more volatile precursors, which react and/or decompose on the substrate

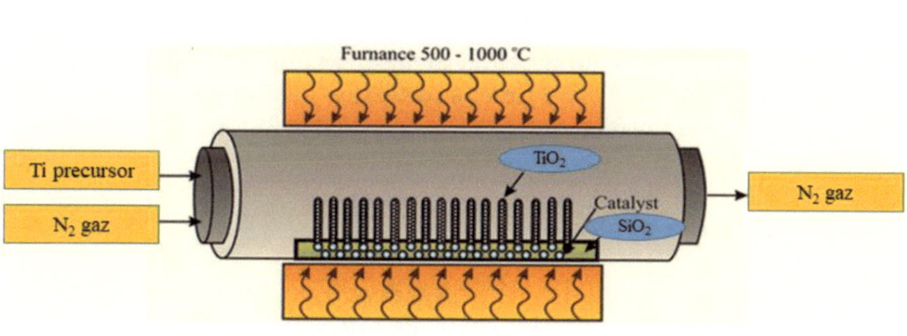

Figure 15.
Schematic of the chemical vapor deposition method.

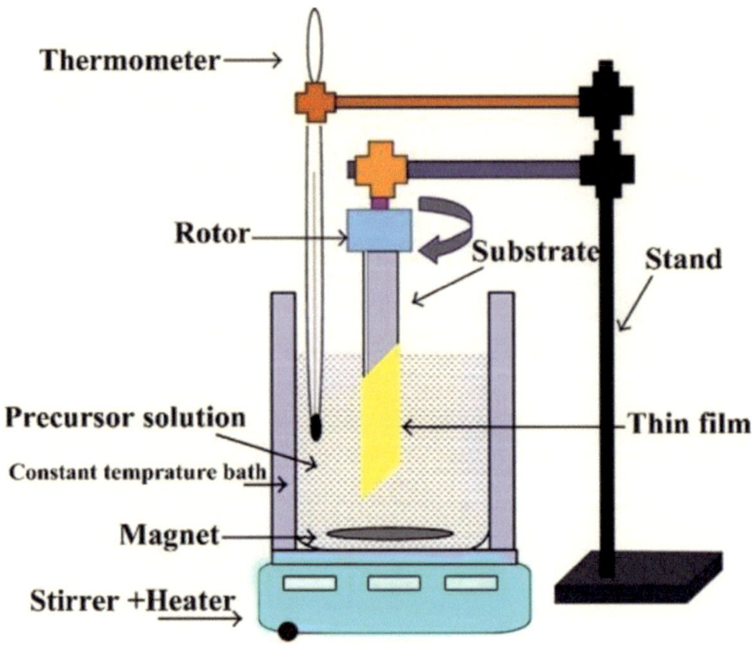

Figure 16.
Schematic of the chemical bath deposition method.

surface to produce the desired deposit. Frequently, volatile by-products are also produced, which are removed by gas flow through the reaction chamber. A. M. Alotaibi et al. [17] have elaborated TiO_2 into SiO_2 substrate by mixing 0.5 g of $[Ti(OCH_2CH_2CH_2CH_3)_4]$ in 20 mL of methanol. The solution was transferred to the reactor and the deposition has been made at 550°C for 40 min. After deposition, the reactor was cooled to room temperature by N_2.

4.11 Chemical bath deposition (CBD)

The CBD technique (**Figure 16**) is one of the cheapest methods to deposit stable, adherent, uniform, and hard thin films by immersing substrate in bath. S.V. Kite et al. [18] have succeed to deposed TiO_2 thin films by simple chemical bath deposition method onto conducting and non-conducting glass substrates. The solution was prepared with ratio of 1: 2: 8 of titanium tetra isopropoxide (TTIP) as a precursor, propan-2-ol (C_3H_8O) and ethanol (EtOH) as solvents respectively. Firstly, the TTIP was mixed with C_3H_8O at constant stirring for 10 min then, EtOH was added. The obtained solution was kept at room temperature with constant stirring to get a clear and homogeneous solution. The substrate was dipped vertically inside the resulting chemical bath for 5 minutes and then substrates pulled out and allowed to dry the step was repeated for 2 to 3 times and finally, the films were rinsed with the distilled water, dried in the oven at 100°C for 2 h then annealed.

5. Water purification by TiO_2 thin films

The conduction band potentials of TiO_2 is near 0 V, which is the reduction potential of H^+ to H_2 and the large bandgap energy of the semiconductors makes them highly promising for photocatalytic and photoelectrochemical water splitting.

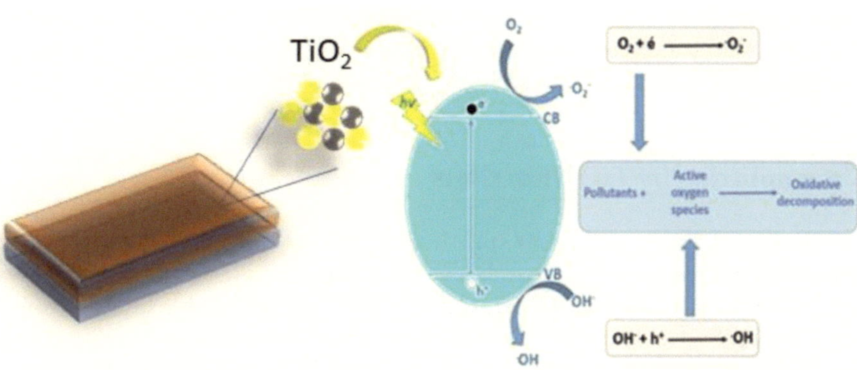

Figure 17.
Schematic of TiO_2 photocatalytic mechanism.

The heterogeneous photocatalysis is widely used for the degradation of water pollutants. The process of photocatalysis can be described as the interaction of incident light with the photocatalyst, causing the formation of electrons and holes (**Figure 17**). The electrons and holes migrate to the surface of the semiconductor without recombining involved the oxydo-reduction reactions. In addition, the holes can also form hydroxyl radicals (·OH) with strong oxidizing properties and the photoexcited electrons can produce superoxide radicals ($\cdot O^{2-}$) and (·OH). These free radicals and e^-/h^+ pairs are highly reactive to decomposed organic substance, environmental pollutants or harmful microorganisms.

The mentioned reactions can be expressed as follow:

$$h^+ + H_2O \rightarrow \cdot OH + H^+ \quad (1)$$

$$e^- + O_2 \rightarrow \cdot O^{2-} \quad (2)$$

$$h^+ + OH^- \rightarrow \cdot OH \quad (3)$$

$$Dye + \cdot OH + \cdot O^{2-} \rightarrow CO_2 + H_2O \quad (4)$$

During the last two decades, researchers have widely studied the purification of water by TiO_2 thin films for their excellent optical and catalytic properties. TiO_2 thin films can act as oxidation and reduction catalysts for organic and inorganic contaminants.

Q. Zhu et al. [19] reported that the synthesis $P/Ag/Ag_2O/Ag_3PO_4/TiO_2$ composite films have degraded 99.9% of rhodamine B (Rh B) after 60 min under simulated solar. J. Singh et al. [20] demonstrated that the synthesized nanocomposite thin films Ag-TiO_2 exhibit a higher photocatalytic activity for the degradation of methylene blue (MB) in water under sun light. S. Yan et al. [21] have determined the photocatalytic performance of MS@TiO_2@PPy through the degradation of soluble organic dye rhodamine B (Rh B) in water under simulated solar irradiation illumination. After being irradiated by simulated sunlight, the degradation efficiency of RhB solution reached up to 90% after 90 min. D. Tekin et al. [22] have reported the efficient application of the ZnO/TiO_2 thin film prepared by the sol–gel method to degrade Orange G dye which show a good performance. K.B. Chaudhari et al. [23] have fabricated TiO_2 film onto glass substrate using a simplified chemical bath deposition method at moderately low temperature. The photocatalytic activity of the TiO_2 thin film was tested by photodegradation of congo red dye solution which proved enhanced photocatalytic activity under visible light.

6. Water splitting by TiO_2 thin films

The first investigation of water splitting in 1972 by Fujishima and Honda, has attracted a large attention of researchers to developed photoelectrochemical water splitting (PEC) system for the generation of clean and renewable energy from solar light and water.

The hydrogen is produced from water using sunlight and semiconductors materials, which use light energy to directly dissociate water molecules into hydrogen and oxygen. Over the past few years, there is great interest on the evolution of hydrogen by PEC water splitting using various semiconducting oxides such as TiO_2, which is considered as suitable photoanode for water splitting. In the PEC water splitting systems (**Figure 18**), the incident photons, with the energy level of the corresponding material band gap, induce the generation of electrons and holes in the conduction band (CB) and valence band (VB), respectively. In a photoanode, holes are driven to photoelectrode's surface to perform the oxygen evolution, meanwhile electrons are collected by the back contact and close the circuit performing the proton reduction reaction in the counter electrode.

The water splitting reaction can be expressed as below:

$$2H_2O \rightarrow 2H_2(g) + O_2(g) \tag{5}$$

A. Sreedhar et al. [24] have demonstrated that the TiO_2 films grown at a higher ion-beam-energy equal to 110 eV have a photocurrent density of 1.2210^{-5} A/cm^2 under UV light and the TiO_2 films decorated by AgNPs under visible illumination

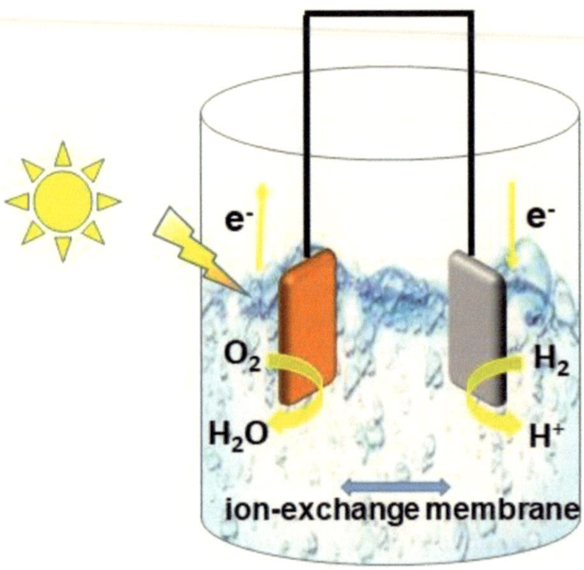

Figure 18.
Schematic of TiO_2 water split mechanism.

have superior photocurrents equal to 6.5310^{-5} A/cm^2, which are a promising photoanodes.

In another work of A. Sreedhar et al. [25] have illustrated that the ZnO/TiO_2 films fabricated at 120 nm yielded with a dominant photocurrent density 4.710^{-6} A/cm^2 is higher ten times than that of ZnO/TiO_2–60 nm films (4.4910^{-7} A/cm^2) and proposed that the film thickness parameter could be a promising pathway to promote the visible-light-driven PEC water splitting activity. C. W. Kim et al. [26] have reported that the TiO_2 photoanode fabricated through hydrothermal method produce 18% of incident photon-to-current conversion efficiency at 0.65 V Ag/AgCl potential under AM 1.5 G illumination and the hydrogen production reached to 0.07 mmol cm^{-2} for 12 hrs. The TiO_2 films with modified morphology synthesis by M. Ibadurrohman et al. [27] have photoconversion efficiency of 1.15% at −0.50 V vs. HgO/Hg, which is an exceptional PEC response and can be considered as promising photoanode.

7. Conclusion

Titanium dioxide TiO_2 is considered as a promising semiconductor for water treatment and H_2 production due to their physical, structural and optical properties under UV and sun light. In this chapter, the structure, the morphology and the synthetic methods of TiO_2 thin films were summarized. The recent researches have confirmed the effect of morphologies of the elaborate nanostructured TiO_2 thin films on proprieties. The photo activity of TiO_2 thin films has been evaluated for the removal of dyes, pharmaceuticals, PCCPs, pesticides, and industrial additives. TiO_2 thin films have the possibility to produce hydrogen from solar energy and water with significant efficiencies. TiO_2 thin films show a good performance to degrade pollutants and can be considered as a promising photoanode for water splitting.

Acknowledgements

This project is carried out under the MOBIDOC scheme, through the PromESsE project managed by the ANPR and the PRF program EnviNanoMat. Grant ref.: PRF2019-D4P2 funded by The Ministry of Higher Education and Scientific Research.

Conflict of interest

We state that the article is original and all authors are aware of its content and approve its submission. This article has not been published previously, and it is not under consideration for publication elsewhere. I confirm that there is no conflict of interest exists.

Author details

Wafa Selmi[1]*, Nabil Hosni[2], Jamila Ben Naceur[1], Hager Maghraoui-Meherzi[2] and Radhouane Chtourou[1]

1 Laboratory of Nanomaterials and Renewable Energy Systems, Research and Technology Center of Energy, Borj-Cedria Science and Technology Park, Hammam-Lif, Tunisia

2 Laboratory of Analytical Chemistry and Electrochemistry, Faculty of Science, University of Tunis El Manar, Tunis, Tunisia

*Address all correspondence to: selmiwafa88@gmail.com

References

[1] Kang X, Liu S, Dai Z, He Y, Song X, Tan Z. Titanium Dioxide: From Engineering to Applications. Catalysts. 2019; 9: 191-196. DOI:10.3390/catal9020191

[2] Guana X, Dub J, Mengc X, Suna Y, Sunc B, Hu Q. Application of titanium dioxide in arsenic removal from water: A review. Journal of Hazardous Materials. 2012; 216: 1-16. DOI:10.1016/j.jhazmat.2012.02.069

[3] Fujishima A, Honda K. Electrochemical Photolysis of Water at a Semiconductor Electrode. Nature 1972; 238: 37-38. DOI:10.1038/238037a0

[4] Li. L, Xu. C, Zhao. Y, Ziegler. K. J, Tin-Doped Indium Oxide-Titania Core-Shell Nanostructures for Dye-Sensitized Solar Cells, Advances in Condensed Matter Physics, 2014; 903294: 1-6. DOI:10.1155/2014/903294

[5] Ben Naceur. J, Jrad. F, Souiwa. K, Ben Rhouma. F, Chtourou. R, Hydrothermal reaction time effect in wettability and photoelectrochemical properties of TiO_2 nanorods arrays films. Optik 2021; 239: 166794. DOI:10.1016/j.ijleo.2021.166794

[6] Meen T-H, Jhuo Y-T, Chao S-M, Lin N-Y, Ji L-W, Tsai J-K, Wu T-C, Chen W-R, Water W, Huang C-J. Effect of TiO_2 nanotubes with $TiCl_4$ treatment on the photoelectrode of dye-sensitized solar cells. Nanoscale Research Letters. 2012; 7: 579-589. DOI:10.1186/1556-276X-7-579

[7] Li. F, Huang. Y, Peng. H, Cao. Y, Niu. Y, Preparation and Photocatalytic Water Splitting Hydrogen Production of Titanium Dioxide Nanosheets. Hindawi International Journal of Photoenergy. 2020; 3617312: 1-6 DOI:10.1155/2020/3617312

[8] Ishii A, Nakamura Y, Oikawa I, Kamegawa A, Takamura H. Low-temperature preparation of high-n TiO_2 thin film onglass by pulsed laser deposition. Applied Surface Science. 2015; 347: 528-534. DOI:10.1016/j.apsusc.2015.04.125

[9] Swaleha N S, Igor V, Pinchuk I-V, Luo Y-K, Kawakami R-K, Khana S, Husain S, Khan W. Epitaxial growth of cobalt doped TiO_2 thin films on $LaAlO_3$ (100) substrate by molecular beam epitaxy and their opto-magnetic based applications. Applied Surface Science. 2019; 493: 691-702. DOI:10.1016/j.apsusc.2019.07.017

[10] Singh J, Saif A. Khan S-A, Shah J, Kotnala R-K, Mohapatra S. Nanostructured TiO_2 thin films prepared by RF magnetron sputtering for photocatalytic applications. Applied Surface Science. 2017; 422: 953-961. DOI:10.1016/j.apsusc.2017.06.068

[11] Wang H, Song Y, Liu W, Yao S, Zhang W. Template synthesis and characterization of TiO_2 nanotube arrays by the electrodeposition method. Materials Letters. 2013; 93: 319-321. DOI:10.1016/j.matlet.2012.11.056

[12] Ben Naceur J, Gaidi M, Bousbih F, Mechiakha R, Chtourou R. Annealing effects on microstructural and optical properties of Nanostructured-TiO_2 thin films prepared by sol–gel technique. Current Applied Physics. 2012; 12: 422-428. DOI:10.1016/j.cap.2011.07.041

[13] Ben Naceur J, Ouertani R, Chakhari W, Chtourou R. Photo-electrochemical properties of Sb_2S_3/TiO_2 heterostructures integrally synthesis by hydrothermal method. Journal of Materials Science: Materials in Electronics. 2019; 30: 5631-5639. DOI:10.1007/s10854-019-00856-6

[14] Abou-Helala M-O, Seeber W-T. Preparation of TiO_2 thin films by spray pyrolysis to be used as a photocatalyst.

Applied Surface Science. 2002; 195: 53-62. DOI:10.1016/S0169-4332 (02)00533-0

[15] Joudi F, Ben Naceur J, Ouertani R, Chtourou R. A novel strategy to produce compact and adherent thin films of SnO_2/TiO_2 composites suitable for water splitting and pollutant degradation. Journal of Materials Science: Materials in Electronics. 2019; 30: 167-179

[16] Patil V L, Vanalakar S A, Shendage S S, Patil S P, Kamble A S, Tarwal N L, Sharma K K, Kim J H, Patil P S. Fabrication of nanogranular TiO_2 thin films by SILAR technique: Application for NO_2 gas sensor. Inorganic and Nano-Metal Chemistry. 2019; 49: 1-7 DOI:10.1080/24701556.2019.1599948

[17] Abdullah A M, Sathasivam S, Williamson B A D, Kafizas A, Sotelo-Vazquez C, Taylor A, Scanlon D O, Parkin I P. Chemical Vapor Deposition of Photocatalytically Active Pure Brookite TiO_2 Thin Films. Chemistry of Materials. 2018; 30 (4): 1353-1361. DOI:10.1021/acs.chemmater.7b04944

[18] Kite S V, Sathe D J, Patil S S, Bhosale P N, Garadkar K M. Nanostructured TiO_2 thin films by chemical bath deposition method for high photoelectrochemical performance. Materials Research Express. 2019; 6: 026411-026421. DOI:10.1088/2053-1591/aaed81

[19] Zhu Q, Hu X, Stanislaus M S, Zhang N, Xiao R, Liu N, Yang Y. A novel P/Ag/Ag2O/Ag3PO4/TiO2 composite film for water purification and antibacterial application under solar light irradiation. Science of the Total Environment. 2017; 577: 236-244. DOI:10.1016/j.scitotenv.2016.10.170

[20] Singh J, Sahu K, Pandey A, Kumar M, Ghosh T, Satpati B, Som T, Varma S, Avasthi D K, Mohapatr S. Atom beam sputtered Ag-TiO2 plasmonic nanocomposite thin films for photocatalytic applications. Applied Surface Science. 2017; 411: 347-354. DOI:10.1016/j.apsusc.2017.03.152

[21] Yan S, Li Y, Xie F, Wu J, Jia X, Yang J, Song H, Zhang Z. Environmentally Safe and Porous MS@TiO_2@PPy Monoliths with Superior Visible-Light Photocatalytic Properties for Rapid Oil–Water Separation and Water Purification. ACS Sustainable Chemistry and Engineering. 2020; 8 (13): 5347-5359. DOI:10.1021/acssuschemeng.0c00360

[22] Tekin D, Kiziltas H, Ungan H. Kinetic evaluation of ZnO/TiO_2 thin film photocatalyst in photocatalytic degradation of Orange G. Journal of Molecular Liquids. 2020; 306: 12905-12913. DOI:10.1016/j.molliq.2020.112905

[23] Chaudhari K B, Rane Y N, Shende D A, Gosavi N M, Gosavi S R. Effect of annealing on the photocatalytic activity of chemically prepared TiO_2 thin films under visible light. Optik – International Journal for Light and Electron Optics. 2019; 193: 163006-163016. DOI:10.1016/j.ijleo.2019.163006

[24] Sreedhar A, Sreekanth T V M, Hyuk J, Kwon J, Yi J, Sohn YY, Gwag JS. Ag nanoparticles decorated ion-beam-assisted TiO_2 thin films for tuning the water splitting activity from UV to visible light harvesting. 2017; 43 (15): 12814-12821. DOI:10.1016/j.ceramint.2017.06.171

[25] Sreedhar A, Jung H, Kwona J H, Yi J, Sohn Y, Gwag J S. Novel composite ZnO/TiO_2 thin film photoanodes for enhanced visible-light-driven photoelectrochemical water splitting activity. Journal of Electroanalytical Chemistry. 2017; 804: 92-98. DOI:10.1016/j.jelechem.2017.09.045

[26] Kim C W, Yeo S J, Hui-Ming Cheng, Kang Y S. Selectively Exposed Crystal Facet-Engineered TiO_2 Thin Film

Photoanode for the Higher Performance of Photoelectrochemical Water Splitting Reaction. Energy Environ. Sci. 2015; 8: 3646-3653. DOI:10.1039/C5EE02300A

[27] Ibadurrohman M, Hellgardt K. Morphological Modification of TiO_2 Thin Films as Highly Efficient Photoanodes for Photoelectrochemical Water Splitting. ACS Appl. Mater. Interfaces 2015; 7(17): 9088-9097. DOI:10.1021/acsami.5b00853

Chapter 6

Titanium Dioxide – A Missing Photo-Responsive Material for Solar-Driven Oil Spill Remediation

Haruna Adamu

Abstract

TiO_2 nanoparticles have been extensively investigated for environmental applications, particularly in the photocatalytic decomposition of organic pollutants using solar energy. The TiO_2-derived photocatalysts attract attention because of their photocatalytic efficiency and activity under a wide range of environmental conditions in response to superior structural and electronic properties. Consequently, TiO_2 compares with other common semiconductors used for environmental photocatalytic applications, TiO_2 is widely being considered close to an ideal semiconductor for photocatalysis. However, despite the impressive photocatalytic and material properties of titanium dioxide, TiO_2 has not to this point been incorporated within commercial hub of oil spill remediation products. Therefore, this chapter covers the description of inevitable technical details required for unveiling the full potential of solar-driven photooxidation potency of TiO_2, which have been the major challenges that halt its translation to commercial use in oil spill remediation. This at the end would underpin and make TiO_2-derived materials a substitute ready to be commercially accepted as a promising method for remediation of oil-polluted aquatic and soil environments.

Keywords: Photo-remediation, oil-spill, solar-radiation, pollution

1. Introduction

The aquatic and terrestrial environments are undergoing constant compositional change due to the continuous introduction of chemicals initiating pollution problems, which considered as part of the dominant threats to living systems surviving on the earth. Crude oil is a focal commodity upon which the economy of the world relies on and thus, its production is the largest and most profitable business in the world. This has, however, created burden and disturbances in ecosystems with attendant environmental quality imbalances. From its development phase to a production phase, many disasters occur in oil industries. An Oil spill is the most important type of disaster which usually occurs and causes a lot of environmental distress. Oil spill mishaps often happen during drilling, production, transportation, transfer, and storage [1]. Besides, the extensive utilisation of crude oil products and the discharges of oily wastewater have also caused increasingly serious oil spills pollution in the harbour and riverine areas as well as other water bodies [2, 3]. In effect, oil spills not only cause loss of energy source but also have long-term damaging impacts on the ecological environment upon which our society relies [3–6]. And so, the negative impacts of oil spills to aquatic and terrestrial ecosystems can be significantly tremendous and unimaginable [7].

It is, therefore, in recent years, the problems of oil spills worldwide have attracted constant concern because of their ecological damage and environmental pollution. Oil spills in an aquatic environment is generally much more damaging since can spread to a distance of hundreds of miles in a thin oil slick layer covering the water surface. This eventually causes the chemical components and elements of the spilled oil to impact negatively on marine life, birds, photosynthesis in plants and as a result, disrupts the normal ecosystem services and structural food chain [1, 3, 4]. Similarly, oil spills pollution could also potentially impose disastrous effects on land [8]. The damage is hard to measure and contain since it involves complex ecosystems. Moreover, low-density spilled oils have an insufficient viscosity to pull together and can therefore speedily spread and damage unimaginable portions of land. On the other hand, high-density spilled oils are too viscous to be dispersed sufficiently well in the soil environment and thus, cause agglomeration that can give rise to stronger adhesive forces of attraction between oil and soil constituents. In either of the two scenarios, it may take land years to recover, during which spilled oils are able to destroy soils, its ecosystems, and biodiversity [9]. This is because oil contamination on land reduces plants' and the soil's ability to pull water from the ground and hold oxygen for plants' growth and micro-creatures survival, respectively. Thus, existing vegetation and fauna-diversity are prone to suffocation due to oil saturation and acts as a barrier, preventing water and oxygen getting to flora and micro-fauna, correspondingly [10]. Accordingly, transporting oil from production sources to consumption locations entails risks, most notably, the risk of accidental oil spills, which causes severe damage to ecosystems and loss to human society [11]. In addition, oil spill is a serious environmental problem not only because of its ability to pollute large areas with associated consequences, but also the longest period of management that usually leads to a heavy financial burden to industries and socio-economic afflictions to society in the immediate vicinity of the affected areas [12]. This is quite challenging because the consequences are not conditional upon the particular geographic, ecological and societal settings in which the disaster occurs, rather viewed as a global problem since crude oil is obviously traded inter-regionally and continently [3, 11]. As such, the damaging impact and compositional alteration of the environment due to oil spills is one of the major concerns of today's world. For example, the tropical Gulf of Mexico oil spill reminds the world again of the importance of oil spill clean-up and environmental remediation [3]. Therefore, an efficient, economical and environmental friendly remedial action is urgently needed as a solution to oil spills pollution problems for the extermination of threats to plants, animals and human life on the gulf coasts and terrestrial environment.

Current remediation techniques for oil spills are typically classified as physical, biological, and chemical. These are the three primary remediation technologies that have widely been applied for addressing or decontamination of oil spills floating and/or dispersing in water and soil environments [13, 14]. The physical method has been considered as one of the most resourceful and inexpensive strategies for oil spills management [15], which is used to remove oil slicks from affected areas by functional materials. However, the process mainly involves the transfer of spilled oil from one environmental phase to another where disposal of oil-soaked agglomerates could also be another source or cause of environmental pollution. On the other hand, the biological method would have been the most attractive option, the hydrophobicity of weathered oil contributes to its low bioavailability to microbial actions, which increases the time for biodegradation and natural attenuation [16].

Although it can degrade oil without any recontamination [17, 18], but high-cost and a long period of action limit its practical application in emergency oil spill incidents that demand an economical and efficient approach [19]. The chemical method mostly involves the use of surfactants, dispersants, and solidifiers. Amongst the three, surfactants only break up oil into tiny droplets directing to help natural oil-eating microbes further break down the hydrocarbons. In contrast, dispersant perhaps do more harm than good. Dispersants hide the oil spills problem. It is used to accelerate the dispersion of the volume of oil into the water column, to reduce the visibility of oil pollution and of the potential impact on the biodiversity of the affected environment [20]. For the solidifiers, are mainly applied to immobilise oil to curtail further spread from concentrated and chunks of floating spilled oil on water or infiltrate into the soil when it occurs on the land surface. Unfortunately, solidified oil always requires to be removed and otherwise, the natural attenuation process of dispersion and volatilisation/evaporation will be inhibited leaving residues of solidified oil to be persistent due to slow weathering processes [13, 21].

It is believed that an oil spill spreads quickly and escalates rapidly and therefore, high speed of action is crucial. For this reason, the real short-time removal process of spilled oil from the environment, including water-bound systems, is imperatively needed for environmental sustainability. Although the application of TiO_2 in the clean-up of oil spills is a chemical method of environmental pollution remediation, in recent years, TiO_2 amongst the metal oxide semiconductors, has been considered as the most widely and well-studied material for the degradation of recalcitrant organic pollutants including spilled oil [22–24]. This is directly connected to its high photocatalytic efficiency, physicochemical stability, high photonic efficiency, and an absence of biological toxicity in bulk form. It also blends under a wide range of environmental conditions for its activity, including stability in acidic and basic aqueous media and activity under ambient temperatures, and most importantly widely available at low cost [25]. Despite these impressive photocatalytic and material properties, TiO_2 solar-driven remediation, as an *in-situ* self-remediation technique and a sustainable solution due to availability of the material and abundance of solar radiation, has not been fully developed. Moreover, it has not been, to this point, adequately incorporated within commercial oil spills remediation products. Thus, the question here that requires a wide spectrum of discursive clarification is that 'will TiO_2 sunlight-driven photocatalytic remediation ever be fit for oil spills pollution tragedy in water and soil environments'? Or, is TiO_2 a missing material-based novel technique for solar-driven oil spill remediation?

Therefore, in this chapter, the properties of TiO_2 that whether or not make it fit to be considered as an ideal material for *in-situ* solar-driven photocatalytic remediation of oil spills, particularly in regions with high sunlight exposition and intensity, as well as the challenges that greatly restrict its application and the ability to translate and incorporate TiO_2-containing materials to commercial use in oil spills remediation are discussed. This is aimed at providing research directions that can be skewed to work on facts rather than an impression in the design and development of TiO_2-containing materials primarily for the solar-driven photocatalytic remediation of oil spills for environmental sustainability.

2. TiO_2 photocatalytic applications in environmental remediation

The potentials of applying photocatalysis in environmental management technology, particularly on pollution remediation, with prime focus on TiO_2 have been proven with spectacular results and reported in a number of research reviewed articles [12, 26–28]. The science of remediation involves removal, separation, containment and destruction of pollutants or contaminants from host environmental media such as surface-and-ground water and soil. With respect to destruction, unlike the other remediation options that not only transfer pollutants from one medium to another, destruction converts contaminants to innocuous products, such as CO_2 and H_2O. In view of that, the photocatalytic destruction of contaminants in the environment, especially the application of irradiated TiO_2-containing materials for the remediation of contaminants from environmental media has been applied successfully for a wide variety of compound [29, 30] such as alkanes, aliphatic alcohols, aliphatic carboxylic acids, alkenes, phenols, aromatic carboxylic acids, dyes, PCB's, simple aromatics, halogenated alkanes and alkenes, surfactants, and pesticides as well as for the reduction of heavy metals (Cr, U, As, Pb, Hg, Cd) from aqueous environments to soil surfaces [31–33]. In many cases, complete mineralisation of organic compounds has been reported [34, 35]. It is in records that in the overall field of semiconductor photocatalysis, both in fundamental research and practical environmental applications, TiO_2 has so far been shown to be the most promising material used for both circumstances because it is highly photo-reactive, cheap, non-toxic, chemically and biologically inert, and photo-stable [26, 27]. The major advantages of TiO_2 photocatalysis are that its process is not an energy-intensive pollution management method and is photo-responsive to renewable and pollution-free solar energy. Also, unlike the conventional environmental pollution treatment methods, TiO_2 photocatalysis does not transform parent pollutants to more refractory types, instead converts pollutants to innocuous products, such as CO_2 and H_2O. Besides, the reaction conditions are mild accompanied with modest reaction time and can be applied in gaseous, aqueous and solid phase pollution remediation techniques [30, 36–43]. Therefore, TiO_2 photocatalysis has the advantage of not only minimising pollution remediation project costs, but also resulted in the remediation reactions with the desired products in the most environmentally harmonious and safe ecologically.

The TiO_2 photocatalytic remediation processes exploits the high reactivity of oxygen superoxide ($O_2^{\bullet-}$) and hydroxyl ($OH^{\bullet}$) radicals as the oxidation driving forces resulting in the formation of benign by-products (i.e., H_2O and CO_2) of photo-mineralisation of toxic organic pollutants. Similarly, if a suitable scavenger or surface defect state is available to trap photogenerated electrons and holes and recombination is halted, a reductive pathway by a conduction-band electron(s) is also initiated. However, in very small TiO_2 nanoparticle suspensions both species are present on the surface. Therefore, simultaneous consideration of both the oxidative and the reductive pathways is required. In the remediation of organic pollutants of which spilled oil inclusive, TiO_2 utilises the oxidising power of the holes either directly or indirectly. On the other hand, to prevent a build-up of electrons, oxygen in the reaction environment serves as an electron acceptor that reacts with the electrons. The oxygen used in the process is atmospheric oxygen, and therefore, in general, there is no need for additional oxidising agents. Accordingly, TiO_2 has taken a highly prominent position of usage in solving environmental pollution problems both in aquatic and terrestrial environments. Similarly, it has taken centre stage in the campaign for solar-driven photocatalytic remediation of oil spills, as TiO_2 demonstrated the capability to develop such a green system that utilises renewal energy source and converts organic contaminants to innocuous products, such as CO_2 and H_2O that are environment friendly (**Figure 1**). From this perspective, as the most promising solar responsive semiconductor, TiO_2-based materials are therefore expected to play a major role to curb serious environmental and pollution challenges through the utilisation of renewable solar energy. Therefore, oxidation of organic compounds by TiO_2-based materials is of considerable interest

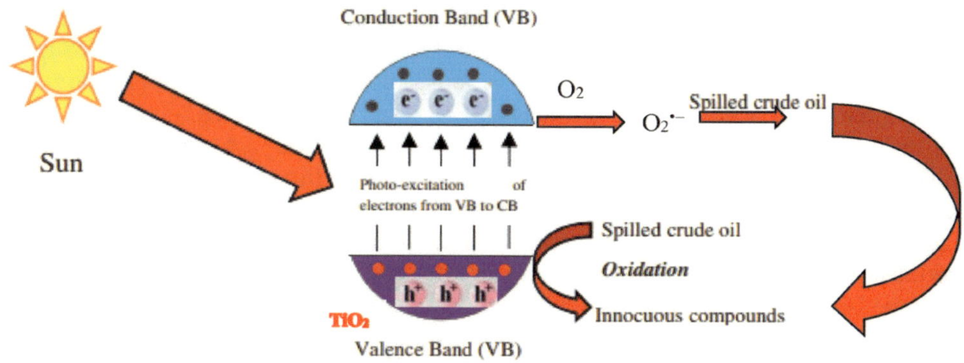

Figure 1.
Photoremediation of spilled oil into innocuous compound by TiO_2 photocatalysis. (Sourced from Nyankson et al. [44] oil photocat itself) but modified.

for environmental applications, particularly for the control and eventual destruction or elimination of hazardous wastes such as spilled oils in aquatic and/or terrestrial environmental compartments. Accordingly, the oxidation process of TiO_2 is indiscriminate and therefore, leading to the mineralisation of almost all-organic pollutants in surface waters and soils.

3. Will TiO_2 photocatalytic remediation ever Be enough? The oil spills pollution tragedy in water and soil environments

The use of titanium dioxide (TiO_2) nanoparticles in oil spill remediation has not fully taken a positional value as a solution to oil pollution problem. The magnitude of being not considered for use in oil spill remediation remains on the impression that TiO_2 in powder form has a strong tendency to form much larger-sized aggregates or cluster and in most cases such agglomerations accounts for its reduced catalytic activity [45]. Another issue of concern involves separation and recovery of suspensions containing nanoparticles along with microparticles after use. These are some of the commercialisation-related challenges that impeding the acceptance of TiO_2 nanoparticles in oil spill remediation. This is enough to inspire scientists and engineers, particularly surface scientists and engineers around the world to become involved with fabrication and evaluation of TiO_2-containing materials for oil spill remediation, so as to make their functionality fully incorporated and suit with aquatic and terrestrial environmental requirements that will subsequently be translated to their commercial use in oil spill remediation. Currently, there has been limited information and engagements on the issue. In this context, because of impressive photocatalytic and material properties of TiO_2, fabrication of TiO_2-containing materials to commercial use in oil spill remediation need to be considered and given the desired attention due to the fact that oil spill pollution is one of the most disastrous infractions of environmental ethics.

Although TiO_2 has been demonstrated to have potential application in the fields of oil spill remediation with a number of advantages, however, the success of its application within the context of solar-driven technology is dependent upon definite factors encompasses of both material properties and environmental dynamic forces. On the material properties of TiO_2 nanoparticles, the issue of dispersion that leads to inevitable secondary pollution and low reusability; effect of aggregation; zero response to visible range of solar spectrum, which constituted its low efficiency of the remediation process under visible light irradiation that greatly restricts its applications in surface waters and soils; wettability of a TiO_2 surface that depends upon the topography and the chemical composition of the surface; the problem of oil coagulation in aqueous environment and oil adsorption over TiO_2 largely depends on engineered surface chemistry of TiO_2-containing material; and zero self-photochargeability. For TiO_2 to be incorporated within environmentally acceptable bulk material that can be translated to commercial use in oil spill remediation, intensity of light and the amount of oxygen available in the pollution-troubled environment; turbulence of water and wind current, which disturbs oil-material binding forces in surface water and soil environments must be considered. Therefore, if all these factors can be fully considered during the architectural and surface engineering design of TiO_2-containing photocatalyst for oil spill remediation in surface waters and soils, TiO_2-containing material would be considered as the most suitable candidate for oil spill remediation due to its impressive photocatalytic and material properties, as well as environmental suitability and compatibility. However, for discovery and initiation of a new brand of TiO_2-containing photocatalyst for oil spill remediation under solar irradiation, the following need to be fully explored for the practical utility of this technique in commercial scale.

3.1 The problem of dispersion and aggregation of TiO_2 nanoparticles

On the issue of whether TiO_2 nanoparticles remains dispersed or forms much larger-sized aggregates or clusters that affect its photocatalytic activity in soil and surface water environments has already been documented and is strongly influenced by the ionic strength and pH of the aqueous suspensions in which TiO_2 nanoparticles is placed [46]. This clustering of TiO_2 nanoparticles is consistent with the principles of colloidal chemistry of other metal oxide nanoparticles, which rate of formation of nanoparticle aggregates in aqueous suspensions is a function of ionic strength and of the nature of the electrolyte in a moderately acid to circumneutral pH range typical of soil and surface water conditions [46–48]. It is true that clustering of TiO_2 nanoparticles has important repercussions for its practical uses in soils and surface waters. However, such problem can be overcome during architectural fabrication of nano-engineered surface of TiO_2-containing material suitable for oil spill remediation by making sure that all the environmental parameters considered fall within ranges likely to be encountered in nature, specifically in situations where TiO_2 nanoparticles enters into contact with surface waters and soils. In another words, the wettability of the flat surface of TiO_2-containing material can be engineered to assume the natural hydrophobicity of butterfly wings or lotus leaves, which forms solid-water or solid-soil solution static wetting mechanism that can enhance distribution by creation of evasion for the influence of ionic strength and pH of aqueous suspensions. With the intention to artificially make TiO_2-containing photocatalyst as hydrophobic surfaces by introducing environmentally acceptable bulk material such as organo-clayed material to create roughness and reduced surface energy, the relationships between the water contact angle on a rough surface (θ_{rough}) and that on a flat surface (θ_{flat}) for homogeneous and heterogeneous wettings can be described by the Wenzel and the Cassie–Baxter equations, respectively. These two equations are shown as follows:

Wenzel's equation:

$$\mathrm{Cos}\,\theta_{rough} = r\,\mathrm{Cos}\,\theta_{flat} \tag{1}$$

Cassie–Baxter's equation:

$$\text{Cos}\,\theta_{\text{rough}} = \varphi_S \text{Cos}\,\theta_{\text{flat}} - (1 - \varphi_S) \tag{2}$$

where r is the roughness factor, defined as the ratio of the actual surface area to the geometrical one, and φ_S is the area fraction of the solid surface that comes into contact with water. Both the theories pointed out that a rough surface is essential for enhancing hydrophobicity and they are commonly used to explain the wetting behaviour on rough hydrophobic surfaces.

In addition, the wettability behaviour that occurs at the interface of the solid, air, water and oil can also be described on the value of the contact angle alone, where surface properties are usually categorised as hydrophilic, hydrophobic and superhydrophobic. If the water contact angle (θ) is less than 90°, the surface is described as hydrophilic, if θ is between 90° and 150° then hydrophobic and if θ is above 150°, the surface is described as superhydrophobic. The water contact angle (CA) θ is usually used to measure the wettability of a flat surface of the nanocomposites, which will depend on the solid–vapour, solid–liquid, and liquid–vapour surface tensions, and can be expressed by Young's equation:

$$\text{Cos}\,\theta = \frac{\gamma_{SV} - \gamma_{SL}}{\gamma_{LV}} \tag{3}$$

where γ_{SV}, γ_{SL} and γ_{LV} are the interfacial tensions between solid and vapour, solid and liquid, and liquid and vapour, respectively, as shown in **Figure 2**.

Such material modifications may be of particular importance in saline environments where high ionic strength could promote coagulation and precipitation of dispersed TiO_2. In a positive note, for example, TiO_2 photocatalysts modified with hydrophobic coatings remain dispersed within organic target pollutants and do not lose photo-degradation efficiency after salt addition [49]. Hence, this result can help to delineate such potential limitation for *in-situ* application of TiO_2 nanoparticles in oil spill remediation, and can also provide insight to guide future material development effort in relation to controlling the buoyancy, hydrophobicity and other desired surface properties of TiO_2-containing photocatalyst to enable oxidative radicals generation in proximity to floatable hydrophobic compounds such as oil spills.

Another difficulty that perhaps has not given TiO_2 edge in oil spill remediation involves separation and recovery of suspensions containing nanoparticles and/or microparticles after use, which leading to inevitable secondary pollution and low reusability. To address this problem, immobilisation of TiO_2 on a support material would render nanoparticles recovery unnecessary. Besides, immobilisation of TiO_2 unto a support is thus an advantage because it allows reusability of material in a number recycles. Therefore, for oil spill photo-remediation in surface waters, fabrication of a floatable photocatalytic material by immobilising TiO_2 on

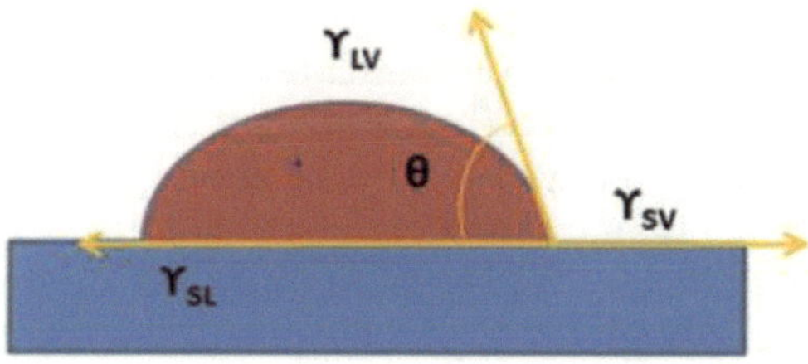

Figure 2.
The interfacial tensions between solid and vapour, solid and liquid, and liquid and vapour.

environmentally acceptable bulk material can serve as a relevant brand of solution to the problem. For example, a floatable photocatalytic material by immobilising TiO_2 on expanded perlite, a siliceous rock of volcanic origin was fabricated and reported [50]. Similarly, TiO_2 nanotube films was successfully anchored on the surfaces of levees for use in degradation of low-level oil spills, e.g., oil spills in harbours [51]. A floatable TiO_2-containing photocatalyst can not only maximise the illumination/light utilisation process, especially in a system with solar irradiation, but could also maximise addition of oxygen (oxygenation) to the photocatalytic system by the proximity with the air/water interface, especially for non-stirred reactions [51]. Accordingly, proximity of TiO_2-containing material with the air/water interface can result in high concentrations of surface oxygen molecules to act as the primary electron acceptor, which can therefore trap electrons resulting in prevention of recombination of the electron–hole pairs and increasing the rate of electron scavenging by O_2 resulting in the formation of an increased yield of superoxide radicals ($O_2^{\bullet-}$) that can directly or indirectly contribute to the degradation and mineralisation of spilled oils. This can increase the rate of oxidation of spilled oil during photo-remediation, as floatable TiO_2-containing photocatalyst can have the ability to interact with floating oil. The potential benefit of engineering of the surface of TiO_2-containing photocatalyst that is sufficiently buoyant to maintain close proximity with floating oil and highly hydrophobic to act at the oil–water interface is to facilitate interaction with the powerful oxidative radicals generated in the process. Therefore, such proximity can be an advantage against low quantum efficiency due to the low rate of electron transfer to oxygen resulting in a high recombination of the photogenerated electron–hole pairs.

3.2 The problem of visible-light response of TiO_2

The use of TiO_2 in remediation of oil spills is sufficient enough because of its impressive photocatalytic and material properties, but one of the major technical challenges that restrict its large-scale application and its use in oil spill remediation technologies is that TiO_2 has a relatively wide bandgap (~3.2 eV, which falls in the UV range of the solar spectrum), and therefore it is unable to harness visible light thus ruling out sunlight as the energy source of its photo activation [52–55]. Because TiO_2 nanoparticles are only responsive to UV light of the solar spectrum, it means could only convert or utilise less than 5% of the total solar radiation (**Figure 3**), which is quite a small proportion of the solar spectrum. On the other hand, the solar spectrum that is mainly composed of about 95% visible light does not have enough energy to excite TiO_2 to be photocatalytically active (**Figure 4**). In view of that, one of the major challenges that hold the throat of scientists, government and entrepreneurial organisations is the development of material(s) using "clean and renewable" energy applications based on the sounding calls and/or demands of Green and Sustainable Science, to relieve the environmental burden due to pollution. In the quest to improve the photocatalysis ability of TiO_2 by responding to visible light or solar, several attempts have been made and shown that visible light responsive modified-TiO_2 based materials for environmental applications are sufficiently promising. Significant progresses have been made in the synthesis of novel materials and nano-structures of TiO_2 meant for efficient processes for the degradation of pollutants, particularly organics. As such, photocatalysis of TiO_2 can be considered as a well understood field of study; yet, immense challenges and opportunities exist in realising this technology for large scale practical applications in the decontamination of the environment, particular in relation to oil spill remediation [57]. Fortunately, fabrication of photo-active TiO_2 in a wide range of solar spectrum plus coating turns its engineered surface into a new smart and environmentally resilient

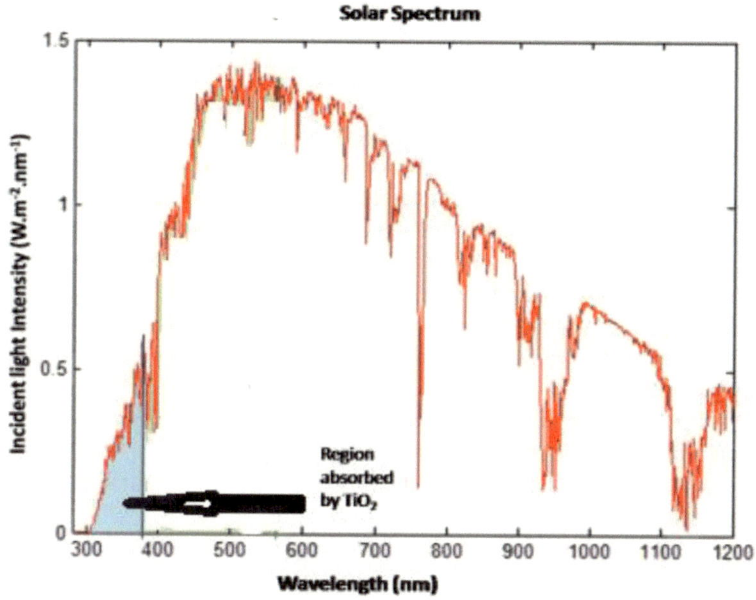

Figure 3.
Absorption region of TiO_2 in solar spectrum.

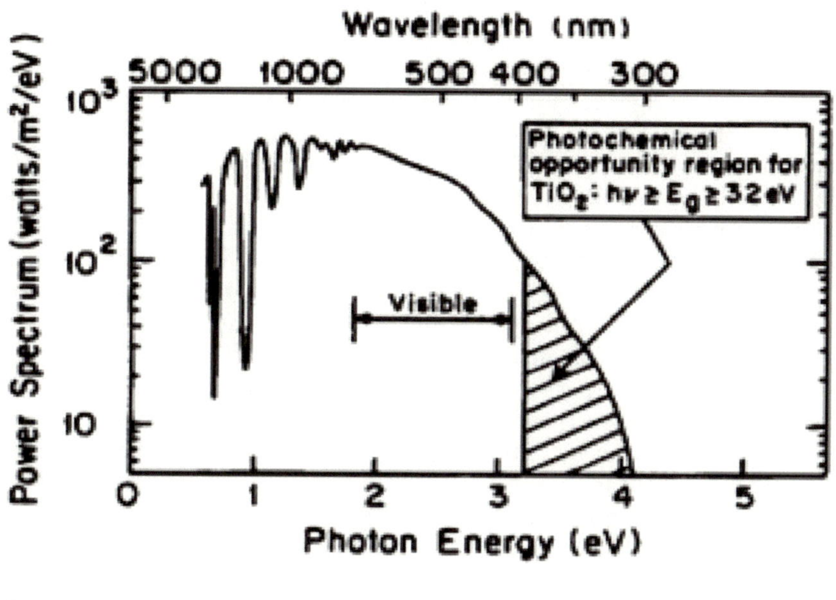

Figure 4.
Absorption of solar spectrum against bang gap of TiO_2. Sourced from Linsebigler et al. [56].

material that once exposed to solar light will be able to function well for the designated purpose.

Undoubtedly, TiO_2 is an efficient photocatalyst in the UV region, which corresponding to an absorption threshold of 390 nm. This restrains its utilisation in the visible range (400–800 nm) for practical applications using solar radiation as the light source. Therefore, for it to be use for oil spill clean-up whose prominent superiority is *in-situ* remediation under visible light irradiation, the surface of the material must be re-engineered. The technological application of TiO_2

photocatalysis in oil spills remediation processes require the development of TiO_2-containing materials that are efficiently responsive to sunlight, since sunlight is the only free source of photons that can yield the desirable clean-up of huge volume of oil spills in a bearable cost. On the account that greater part of the solar radiation that reaches Earth is comprised of visible light couple with a minute fraction of ultraviolet radiation, several improvements have been made to overcome the limitation of solar spectrum to initiate photoexcitation over TiO_2 and optical responsiveness to the visible light region after modification has been reported feasibly [58–61]. Accordingly, it becomes apparent that with such ground-breaking discoveries, TiO_2 is a suitable and an excellent candidate for oil spills remediation and that in addition can pave ways and increase interest for its incorporation within commercial oil spill remediation products in coming years, because of this unique and superior optical property.

Another issue of concern that also limits all day(s) of full application of TiO_2 in oil spills remediation is one major drawback associated to all the traditional photocatalysts like TiO_2 that they can only work under illumination. Surface chemistry and engineering has provided a solution to this limitation, where visible-light-driven energy storage photo-responsive TiO_2-containing photocatalysts have been developed and have been widely used in photocatalysis in dark in recent years [62–72]. Upon advancement, energy storage photocatalysts that are full-sunlight-driven made up of UV–visible-NIR with possession of long-lasting energy storage ability have also been advanced technologically. The materials exhibit a strong absorption at full-sunlight spectrum (300–1,000 nm) that cut-across UV–visible-NIR with a super-long energy storage time. In a system like this, the material system is composed of two kinds of composite materials [65, 71], namely light harvesting material and energy storage material. The light harvesting material is the material capable of absorbing light to generate electron–hole pairs while the energy storage material is the material in charge of trapping and saving the electrons or holes transferred from light harvesting centres during illumination, and releasing them in dark. In the architectural design of such new materials, hydrogen-treated (because hydrogen treatment can extend the light absorption threshold of TiO_2 to NIR) [73] and bulk surface modified-TiO_2 functions as the light harvesting material and also serves as a candidate in charge of the electrons or holes generation simultaneously, while a co-catalyst such Na_xMoO_3 is mainly made to display self-photochargeability effect by trapping and saving electrons [74]. The extraordinary full-spectrum absorption effect and long persistent energy storage ability make such material a potential solar-energy storage and an effective photocatalyst in practice, as such the material has dual functions by harnessing solar energy to excite electrons, store electrons and when light is over in a time there is no sunlight can still do the remediation reaction by allowing the stored electrons to go back to the photocatalyst and initiate the generation of oxygen superoxide radicals ($O_2^{\bullet-}$) for the degradation and mineralisation of pollutants under treatment. The oil molecules adsorbed over the surface or in the pores of TiO_2-containing photocatalyst can be directly oxidised by the $O_2^{\bullet-}$ during the night operational process. The possible reaction pathways could be presented as shown below in Eqs. (4) and (5). The participation of crucial active species of $O_2^{\bullet-}$ in the photocatalytic remediation of diesel oil was detected under visible light illumination [75]. Accordingly, the non-stopped generation and the intensity of $O_2^{\bullet-}$ species in the remediation environment, it simultaneously expands the photocatalytic capacity of TiO_2-containing photocatalyst with the increase of time. It means that long-time illumination can further enhance the photocatalytic remediation effect both in the day time and night. Therefore, the drawback of TiO_2-containing photocatalysts that they can only function under illumination has been overcome and the long persistent energy storage ability of the photocatalysts allows not only be used during daytime, but also be used during the night. Consequently, TiO_2-containing photocatalyst endowed with this new optical and electronic properties still presents TiO_2 as a missing material for its potential application using solar energy utilisation for oil spills remediation in all day(s) operational process.

$$e- + O_2 \rightarrow O_2^{\bullet -} \quad (4)$$

$$\text{Spilled oil} + O_2^{\bullet -} \rightarrow \text{Degradation products } (CO_2 + H_2O) \quad (5)$$

In addition to above, to further enhance the solar-driven activity of TiO_2, up-conversion luminescence agent was coupled with TiO_2 to transform the unused near-infrared (NIR) sunlight tail into UV–vis radiation available for photoreaction activation and the result demonstrated promising contribution suggesting that it can be used for treating surface water and soil pollution problems using solar light [74]. This is an alternative approach of enhancing solar absorption ability of TiO_2 and the process is of considerable interest for photocatalytic processes because it produces UV–visible range from infrared light sources through multi-photon and energy transfer mechanisms. In the up-conversion photonic processes, materials such as rare-earth (RE) doped materials appear as one of the most promising candidates for efficient up-conversion luminescence that assist in the long-wavelength light harvesting of solar irradiation [76–78]. In fact, this technological advancement has already been applied in agricultural production by improving the sunlight conversion efficiency of the photosynthetic process. For that reason, when applied in TiO_2 photocatalysis, transforming the incoming infrared light into UV–visible light provides extra photons for absorption by TiO_2 and therefore, the process cannot only optimise TiO_2 photocatalytic remediation process, but also the incident radiation can lead to an endless range of possibilities. Interestingly, amongst the possibilities is that the process improves the photocatalytic activity of TiO_2 even when solar radiation intensity is low. Although the solar irradiance to be received by a body of water or soil in a particular location depends on its position in the Earth. The locations on the equator of the Earth receive solar radiation at a higher intensity (irradiance) than the norther and southern hemispheres (**Figure 5**). This means that more solar radiation reaches the surface at these altitudes. In other words, all locations receives visible light in the same wavelengths, but the brightness and intensity are very different. However, with a system of TiO_2

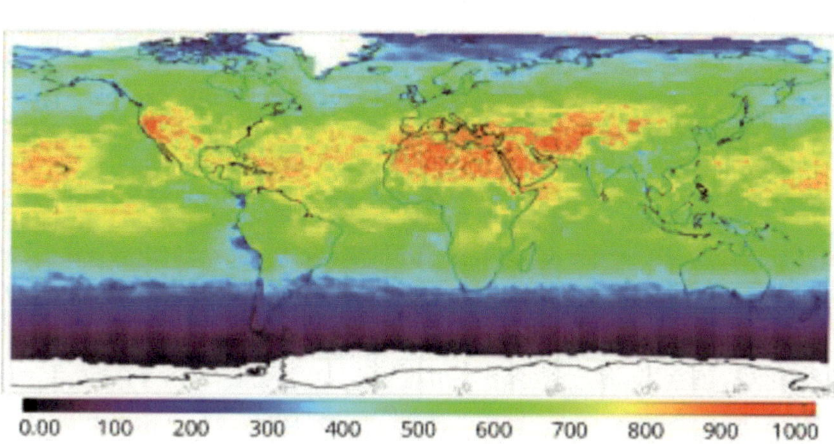

Figure 5.
Surface solar irradiance receives by different locations of the earth. Sourced from Wang et al. [79].

comprised of up-conversion luminescence agent, the problem of solar intensity in different locations of the Earth would no longer be an issue of challenge. This means that the perceived disadvantage of the location of the North and South poles with smaller solar exposition than the equator when it comes to application of photoremediation is now a false impression. Besides, the process also decreases the irradiation time needed for decontamination by solar light. As a result, when such technological process is applied in oil spills remediation, the rare-earth doped materials amalgamated with TiO_2 would facilitate increased solar absorption and higher energy conversion efficiency. This serves as a clear testimony that oil spills remediation can be driven by sunlight using TiO_2-containing photocatalyst, making the remediation process a zero-cost of energy and resulting in considerable economic savings.

3.3 The problem of oil coagulation in aqueous environment and adsorption of oil droplets over TiO_2

Because of surface tension of oil droplets on the surface of water, it hardly makes oil droplets lay flatter instead ball up. This could cancel the solar irradiation to get into micro-crevices of the oil droplets even better, which in turn can result in promptness to sluggish photoremediation process. Besides, ball up formation as the result of coagulation of oil in aqueous environment could create a kind of blanket that can entrap colloidal particles of TiO_2-containing photocatalyst and cut bridge for harvesting of light meant to activate the photoremediation process. Therefore, to apply TiO_2 nanocomposite in the remediation of oil spills remediation in surface waters as well as soils and make it of practical significance and attractive for large-scale environmental applications, the TiO_2-containing material need to be made with surface property that at same time bring about attraction of lay flattered oil droplets and adsorption to the surface of buoyant TiO_2 nanocomposite, for effective solar-driven remediation of spilled oils. Such problem can be eliminated with an organic-based material that has a reduced surface energy, low density, high porosity and adsorption ability as well as good elasticity. For example, organobentonite was sufficiently enough to disturb the surface tension of oil droplets on the surface of water [18], which in other words, can make oil droplets lay flatter instead to ball up. Consequently, TiO_2-containing nanocomposite that is architecturally constructed with superhydrophobicity and superoleophilicity can selectively and smartly facilitate controllable separation of oil from oil/water mixtures and subsequently photoremediate adsorbed oil over the material.

The architectural design of porosity of buoyant TiO_2 nanocomposite cannot only increase the specific surface area but also provide a number of adsorption sites and paths for oils to be in retention in the material. The pore design can primarily decrease the density of the material and enable it easy to enclose oils for rapid degradation and mineralisation. For example, graphene nanosheet can be anchored into the in the architectural framework of TiO_2 nanocomposite to generate micro-pore, meso-pore and macro-pore structures, as it was reported that it provided abundant adsorption sites for oils and organic [80–83]. For recycling and avoidance of damage, when C-C bonds are architecturally ross-linked in the framework of TiO_2 nanocomposite the carbon based architecture of the material can perfectly remain in its original composition and framework. Hence, a combination of adsorption with photochemical remediation of spilled oils through the utilisation of solar energy over TiO_2-based structural framework is another elegant route that can directly decompose spilled oil into inorganics without any further procedural treatment and thus, could be highly promising for practical applications in both surface waters and soils.

3.4 The oil-TiO_2 nanocomposite binding strength in turbulent flow

Turbulent motions due to strong tide and wind in surface water and soil environments can disturb oil-TiO_2 nanocomposite binding strength and stability. These turbulent motions can produce an external breaking force that destroys binding between two bodies. The binding strength can be related to the inter-particle bonds between aggregate components which involves surface interaction between oil and TiO_2 nanocomposite. Therefore, an oil-TiO_2 nanocomposite binding force will be broken when the shear force applied to their surface of contact is larger than the bonding strength within the cohesion. As such, it is important to architecturally design TiO_2-containing nanocomposite with capacity to resist shear force in a situation of turbulent condition be it in water or soil environment. This will play an important role in determining performance and general acceptance of TiO_2-containing nanocomposite in real oil spill remediation application. The strength of surface interaction between oil and TiO_2-containing nanocomposite will be controlled by two counteracting forces under a given turbulent condition, namely, the oil-TiO_2-containing nanocomposite binding force and the turbulent breaking force of the fluid-material surface contact. The binding force is related to the material's morphological characteristics, and the breakage of binding is to be governed by turbulence kinetic parameters [84]. In this part, the oil-material's binding force and the turbulent breaking force of fluid can be deduced from morphological characteristics of material and the force of attraction between oil and the material under application. However, a method for quantitative evaluating the strength of oil-material's binding force can also be developed based on the binding and the breaking forces of the surface contact between oil and TiO_2-containing nanocomposite. For easy comprehension, under a given turbulent condition, it will be deduced that the critical condition of the breakage of surface interaction between oil and TiO_2-containing nanocomposite is considered to be the binding force equal to the breaking force, which can be written in the following form.

$$B_f = F \tag{6}$$

When TiO_2-containing nanocomposite is architecturally designed with given pore size, an increase in effective adsorption that bring spilled oils much closer to the material can facilitate an increase in oil density that would be larger enough to be stored in the macropores of the nanocomposites. This in effect can result in enhancement of adherence of the oils to the TiO_2-containing nanacomposites, which is beneficial for the improvement in oil retention capacity of the material that can allow adsorbed oil to resist turbulent motion either cause by tide or wind in water or soil environment. In addition, in response on the problem of turbulent motion particularly in surface water environment, TiO_2-containing nanocomposite can be engineered to a strong magnetic response to an external magnetic field according to the magnetization curve. In effect, it is expected that an architecturally designed TiO_2-containing nanocomposite can be easily controlled by an external magnetic field, and then oils can be made to be strongly attracted to the magnetic component of the nanocomposite with which the adsorbed oils are to be retained structural stable. This can provide high contacting rate between spilled oils and the material as well as additional kinetic energy that could enhance the overall degradation and mineralisation rate of spilled oils in surface waters and soils. Hence, removal of spilled oils from the surface of ocean and soil environments can also be achieved in large scale through strategic TiO_2 photoremediation process.

4. Conclusion

The problem of oil spill accidents unto aquatic and terrestrial environments remains one of the series of severe environmental and ecological damages on mother Earth planet, which when not properly managed causes long-term great distortion of ecological equilibrium that consumes lots of financial and bio-diversified resources. To address this kind of environmental issue with vigour considering the problem of complexity of oil spill strongly interested to be managed within the shortest possible treatment time through use of renewable and cost-free energy source for just to maintain ecological equilibrium, development of solar-driven oil spill remediating material with highly desired self-multifaceted features and functions is unwaveringly needed.

Amongst the remediation technology which would completely remove spilled oils from surface waters and soils with cost-free energy, TiO_2 photocatalysis has a proven potential to treat "difficult-to-remove" spilled oils inexpensively and thus, is expected to play an important role in large-scale oil spill remediation challenges. Despite the substantial progress made in TiO_2 photocatalysis, considerable opportunities and commercialisation-related challenges still remain in oil spill remediation using TiO_2. This clearly demonstrates that gaps exist between material research and application studies for practical application of TiO_2-containing nanocomposite in oil spill remediation. However, as the complexities and hindrances involved in oil spill remediation using TiO_2 photocatalysis can be modelled to overcome the limitations, such provides the ground basis for designing better TiO_2-containing nanocomposite for utilisation of full spectrum of solar radiation that is adequate to meet the demands of large-scale commercial use of TiO_2 in oil spill remediation. It is expected that this fundamental understanding of remedies to overcome TiO_2 limitations in oil spill remediation dispels the fear of whether or not modified-TiO_2 can perform well and therefore, it needs to be considered for incorporation within commercial oil spill remediation products over the coming years.

List of nomenclature

C-C	carbo-carbon bond
CO_2	carbon dioxide
H_2O	water
NIR	natural infrared
$OH^•$	hydroxyl radical
$O_2^{•-}$	super peroxide oxygen radical
pH	potential of hydrogen (in a scale used to specify acidity or basicity of an aqueous solution)
RE	rare earth
TiO_2	titanium dioxide
UV	ultraviolet

Author details

Haruna Adamu
Department of Environmental Management Technology, Abubakar Tafawa Balewa University, Bauchi, Nigeria

*Address all correspondence to: hadamu2@atbu.edu.ng

References

[1] Fingas MF. The Basics of Oil Spill Clean-up. Lewis Publishers, London, 2001.

[2] Zhang X, Wang Z, Liu X, Hu X, Liang X, Hu Y. Degradation of diesel pollutants in Huangpu-Yangtze River estuary wetland using plant-microbe systems. International Biodeterioration and Biodegradation. 2013;**76**:71-75.

[3] Schnoor JL. The gulf oil spill. Environmental Science and Technology. 2010; 4833.

[4] Ceylan D, Dogu S, Karacik B, Yakan SD, Okay OS, Okay O. Evaluation of butyl rubber as sorbent material for the removal of oil and polycyclic aromatic hydrocarbons from seawater. Environmental Science and Technology. 2009;**43**(10):3846-3852.

[5] Al-Majed AA, Adebayo AR, Hossain ME. A sustainable approach to controlling oil spills. Journal of Environmental Management. 2012;**113**: 213-227.

[6] D'Auria M, Emanuele L, Racioppi R, Velluzzi V. Photochemical degradation of crude oil: Comparison between direct irradiation, photocatalysis, and photocatalysis on zeolite. Journal of Hazardous Materials. 2009;**164**(1):32-38.

[7] García-Martínez MJ, Da Riva I, Canoira L, Llamas JF, Alcántara R, Gallego JL. Photodegradation of polycyclic aromatic hydrocarbons in fossil fuels catalysed by supported TiO_2. Applied Catalysis B: Environmental. 2006;**67**(3-4):279-289.

[8] King SM, Leaf PA, Olson AC, Ray PZ, Tarr MA. Photolytic and photocatalytic degradation of surface oil from the Deepwater Horizon spill. Chemosphere. 2014;**95**:415-422.

[9] Emam EA, Aboul-Gheit NA. Photocatalytic degradation of oil-emulsion in water/seawater using titanium dioxide. Energy Sources, Part A: Recovery, Utilization, and Environmental Effects. 2014;**36**(10): 1123-1133.

[10] Fan J, Zhao L, Yu J, Liu G. The effect of calcination temperature on the microstructure and photocatalytic activity of TiO_2-based composite nanotubes prepared by an in situ template dissolution method. Nanoscale. 2012;**4**(20):6597-6603.

[11] Liu S, Sun X, Li JG, Li X, Xiu Z, Yang H, Xue X. Fluorine-and iron-modified hierarchical anatase microsphere photocatalyst for water cleaning: facile wet chemical synthesis and wavelength-sensitive photocatalytic reactivity. Langmuir. 2010;**26**(6): 4546-4553.

[12] Hoffmann MR, Martin ST, Choi W, Bahnemann DW. Environmental applications of semiconductor photocatalysis. Chemical Reviews. 1995; **95**(1):69-96.

[13] Chen X, Mao SS. Titanium dioxide nanomaterials: synthesis, properties, modifications, and applications. Chemical Reviews. 2007;**107**(7): 2891-2959.

[14] Herrmann JM. Heterogeneous photocatalysis: fundamentals and applications to the removal of various types of aqueous pollutants. Catalysis Today. 1999;**53**(1):115-129.

[15] Kobwittaya K, Sirivithayapakorn S. Photocatalytic reduction of nitrate over TiO_2 and Ag-modified TiO_2. Journal of Saudi Chemical Society.2014;**18**(4): 291–298.

[16] Le TN, Ton NQ, Tran VM, Dang Nam N, Vu TH. TiO_2 nanotubes with different Ag loading to enhance

visible-light photocatalytic activity. Journal of Nanomaterials. 2017;2017.

[17] Yang Y, Liu E, Dai H, Kang L, Wu H, Fan J, Hu X, Liu H. Photocatalytic activity of Ag–TiO_2-graphene ternary nanocomposites and application in hydrogen evolution by water splitting. International Journal of Hydrogen Energy. 2014;**39**(15): 7664-7671.

[18] Gitipour S, Bowers MT, Huff W, Bodocsi A. The efficiency of modified bentonite clays for removal of aromatic organics from oily liquid wastes. Spill Science and Technology Bulletin. 1997; **4**(3):155-164.

[19] Doerffer JW. Mechanical Response Technology to an Oil Spill in Oil Spill Response in the Marine Environment. Pergramon Press, Oxford, 1992, pp.133–220.

[20] Brame JA, Hong SW, Lee J, Lee SH, Alvarez PJ. Photocatalytic pre-treatment with food-grade TiO2 increases the bioavailability and bioremediation potential of weathered oil from the Deepwater Horizon oil spill in the Gulf of Mexico. Chemosphere. 2013;**90**(8): 2315-2319.

[21] Rocha ORS, Duarte MMMB, Dantas RF, Duarte MML, Silva VL. Solar photo-Fenton treatment of petroleum extraction wastewater. Desalination and Water Treatment. 2013;**51**(28-30): 5785-5791.

[22] Zhou Q, Zhong YH, Chen X, Liu JH, Huang XJ, Wu YC. Adsorption and photocatalysis removal of fulvic acid by TiO 2–graphene composites. Journal of Materials Science. 2014;**49**(3): 1066-1075.

[23] Wen Y, Ding H, Shan Y. Preparation and visible light photocatalytic activity of Ag/TiO_2/graphene nanocomposite. Nanoscale. 2011;**3**(10):4411-4417.

[24] Lee C, Wei X, Kysar JW, Hone J. Measurement of the elastic properties and intrinsic strength of monolayer graphene. Science. 2008;**321**(5887): 385-388.

[25] Barthlott W, Neinhuis C. Purity of the sacred lotus, or escape from contamination in biological surfaces. Planta. 1997;**202**(1):1-8.

[26] Schneider J, Matsuoka M, Takeuchi M, Zhang J, Horiuchi Y, Anpo M, Bahnemann DW. Understanding TiO_2 photocatalysis: mechanisms and materials. Chemical Reviews. 2014;**114**(19):9919-9986.

[27] Friedmann D, Mendive C, Bahnemann D. TiO_2 for water treatment: parameters affecting the kinetics and mechanisms of photocatalysis. Applied Catalysis B: Environmental. 2010;**99**(3-4): 398-406.

[28] Nakata K, Fujishima A. TiO2 photocatalysis: Design and applications. Journal of Photochemistry and Photobiology C: Photochemistry Reviews. 2012;**13**(3):169-189.

[29] Adamu H, Dubey P, Anderson JA. Probing the role of thermally reduced graphene oxide in enhancing performance of TiO2 in photocatalytic phenol removal from aqueous environments. Chemical Engineering Journal. 2016;**284**:380-388.

[30] Kabra K, Chaudhary R, Sawhney RL. Treatment of hazardous organic and inorganic compounds through aqueous-phase photocatalysis: A review. Industrial and Engineering Chemistry Research. 2004;**43**(24): 7683-7696.

[31] Litter MI. Last advances on TiO_2-photocatalytic removal of chromium, uranium and arsenic. Current Opinion in Green and Sustainable Chemistry. 2017;**6**:150-158.

[32] Djellabi R, Ghorab FM, Nouacer S, Smara A, Khireddine O. Cr(VI) photocatalytic reduction under sunlight followed by Cr (III) extraction from TiO_2 surface. Materials Letters. 2016; **176**:106-119.

[33] Ku Y, Jung IL. Photocatalytic reduction of Cr(VI) in aqueous solutions by UV irradiation with the presence of titanium dioxide. Water research. 2001;**35**(1):135-142.

[34] Ghasemi B, Anvaripour B, Jorfi S, Jaafarzadeh N. Enhanced photocatalytic degradation and mineralization of furfural using UVC/TiO_2/GAC composite in aqueous solution. International Journal of Photoenergy. 2016;2016.

[35] Bailón-García E, Elmouwahidi A, Álvarez MA, Carrasco-Marín F, Pérez-Cadenas AF, Maldonado-Hódar FJ. New carbon xerogel-TiO_2 composites with high performance as visible-light photocatalysts for dye mineralization. Applied Catalysis B: Environmental. 2017;**201**:29-40.

[36] Ahmad R, Ahmad Z, Khan AU, Mastoi NR, Aslam M, Kim J. Photocatalytic systems as an advanced environmental remediation: Recent developments, limitations and new avenues for applications. Journal of Environmental Chemical Engineering. 2016;**4**(4):4143-4164.

[37] Suárez S, Arconada N, Castro Y, Coronado JM, Portela R, Durán A, Sánchez B. Photocatalytic degradation of TCE in dry and wet air conditions with TiO_2 porous thin films. Applied Catalysis B: Environmental. 2011;**108**: 14-21.

[38] Zhang L, Li P, Gong Z, Li X. Photocatalytic degradation of polycyclic aromatic hydrocarbons on soil surfaces using TiO2 under UV light. Journal of Hazardous Materials. 2008;**158**(2-3): 478-484.

[39] Abramović B, Šojić D, Anderluh V. Visible-light-induced Photocatalytic Degradation of Herbicide Mecoprop in Aqueous Suspension of TiO_2. Acta Chimica Slovenica. 2007 1;**54**(3): 558–564.

[40] Choi H, Stathatos E, Dionysiou DD. Photocatalytic TiO_2 films and membranes for the development of efficient wastewater treatment and reuse systems. Desalination. 2007;**202** (1-3):199-206.

[41] Akarsu M, ASİLTÜRK M, Sayilkan F, Kiraz N, Arpaç E, Sayilkan H. A novel approach to the hydrothermal synthesis of anatase titania nanoparticles and the photocatalytic degradation of rhodamine B. Turkish Journal of Chemistry. 2006;30(3):333-343.

[42] Johnson BF. Nanoparticles in catalysis. Topics in Catalysis. 2003; **24**(1):147-159.

[43] Higarashi MM, Jardim WF. Remediation of pesticide contaminated soil using TiO_2 mediated by solar light. Catalysis Today. 2002;**76**(2-4): 201-207.

[44] Nyankson E, Rodene D, Gupta RB. Advancements in crude oil spill remediation research after the Deepwater Horizon oil spill. Water, Air, & Soil Pollution. 2016;**227**(1):29.

[45] Xuzhuang Y, Yang D, Huaiyong Z, Jiangwen L, Martins WN, Frost R, Daniel L, Yuenian S. Mesoporous structure with size controllable anatase attached on silicate layers for efficient photocatalysis. The Journal of Physical Chemistry C. 2009;**113**(19):8243-8248.

[46] Dunphy Guzman KA, Finnegan MP, Banfield JF. Influence of surface potential on aggregation and transport of titania nanoparticles. Environmental Science and Technology. 2006;**40**(24):7688-7693.

[47] Gilbert B, Lu G, Kim CS. Stable cluster formation in aqueous suspensions of iron oxyhydroxide nanoparticles. Journal of colloid and interface science. 2007;**313**(1):152-159.

[48] Ridley MK, Hackley VA, Machesky ML. Characterization and surface-reactivity of nanocrystalline anatase in aqueous solutions. Langmuir. 2006;**22**(26):10972-10982.

[49] Sunada F, Heller A. Effects of water, salt water, and silicone overcoating of the TiO_2 photocatalyst on the rates and products of photocatalytic oxidation of liquid 3-octanol and 3-octanone. Environmental Scienceand Technology. 1998;**32**(2):282-286.

[50] Rodríguez-González V., Hinojosa-Reyes M. Waste-Porous-Based Materials as Supports of TiO_2 Photocatalytic Coatings for Environmental Applications. In: Martínez L., Kharissova O., Kharisov B. (eds) Handbook of Ecomaterials. Springer, Cham. 2019, pp. 1751-1775.

[51] Hsu YY, Hsiung TL, Wang HP, Fukushima Y, Wei YL, Chang JE. Photocatalytic degradation of spill oils on TiO2 nanotube thin films. Marine pollution bulletin. 2008;**57**(6-12): 873-876.

[52] Romero-Gómez P, Rico V, Espinós JP, González-Elipe AR, Palgrave RG, Egdell RG. Nitridation of nanocrystalline TiO_2 thin films by treatment with ammonia. Thin Solid Films. 2011;**519**(11):3587-3595.

[53] Wang P, Zhou T, Wang R, Lim TT. Carbon-sensitized and nitrogen-doped TiO_2 for photocatalytic degradation of sulfanilamide under visible-light irradiation. Water Research. 2011; **45**(16):5015-5026.

[54] Nawi MA, Jawad AH, Sabar S, Ngah WW. Immobilized bilayer TiO_2/ chitosan system for the removal of phenol under irradiation by a 45 watt compact fluorescent lamp. Desalination. 2011;**280**(1-3):288-296.

[55] Bai A, Liang W, Zheng G, Xue J. Preparation and enhanced daylight-induced photo-catalytic activity of transparent C-Doped TiO 2 thin films. Journal of Wuhan University of Technology-Material Science Edition. 2010;**25**(5):738-742.

[56] Linsebigler AL, Lu G, Yates Jr JT. Photocatalysis on TiO_2 surfaces: principles, mechanisms, and selected results. Chemical reviews. 1995;**95**(3): 735-758.

[57] Narayan R. Titania: a material-based approach to oil spill remediation? Materials today. 2010;**13**(9):58-59.

[58] Chakrabarti S. Solar photocatalysis for environmental remediation. The Energy and Resources Institute (TERI). 2017.

[59] Mahlambi MM, Ngila CJ, Mamba BB. Recent developments in environmental photocatalytic degradation of organic pollutants: the case of titanium dioxide nanoparticles—a review. Journal of Nanomaterials. 2015.

[60] Daghrir R, Drogui P, Robert D. Modified TiO_2 for environmental photocatalytic applications: a review. Industrial and Engineering Chemistry Research. 2013;**52**(10):3581-3599.

[61] Colina-Márquez J, Machuca-Martínez F, Puma GL. Radiation absorption and optimization of solar photocatalytic reactors for environmental applications. Environmental Science and Technology. 2010;**44**(13):5112-5120.

[62] Tatsuma T, Saitoh S, Ngaotrakanwiwat P, Ohko Y, Fujishima A. Energy storage of TiO_2–WO_3 photocatalysis systems in the gas

phase. Langmuir. 2002:**18**(21): 7777-7779.

[63] Takahashi Y, Tatsuma T. Visible light-induced photocatalysts with reductive energy storage abilities. Electrochemistry Communications. 2008;**10**(9):1404-1407.

[64] Ngaotrakanwiwat P, Tatsuma T. Optimization of energy storage TiO_2–WO_3 photocatalysts and further modification with phosphotungstic acid. Journal of Electroanalytical Chemistry. 2004;**573**(2):263-269.

[65] Cao L, Yuan J, Chen M, Shangguan W. Photocatalytic energy storage ability of TiO_2-WO_3 composite prepared by wet-chemical technique. Journal of Environmental Sciences. 2010;**22**(3):454-459.

[66] Li YJ, Cao TP, Shao CL, Wang CH. Preparation and energy stored photocatalytic properties of WO_3/TiO_2 composite fibers. Chemical Journal of Chinese Universities-Chinese. 2012;**33**: 1552-1558.

[67] Ngaotrakanwiwat P, Tatsuma T, Saitoh S, Ohko Y, Fujishima A. Charge–discharge behavior of TiO_2–WO_3 photocatalysis systems with energy storage ability. Physical Chemistry Chemical Physics. 2003;**5**(15): 3234-3237.

[68] Takahashi Y, Ngaotrakanwiwat P, Tatsuma T. Energy storage TiO_2–MoO_3 photocatalysts. Electrochimica acta. 2004;**49**(12):2025-2029.

[69] Ohko Y, Saitoh S, Tatsuma T, Fujishima A. $SrTiO_3$-WO_3 photocatalysis systems with an energy storage ability. Electrochemistry. 2002; **70**(6):460-462.

[70] Ngaotrakanwiwat P, Meeyoo V. TiO_2–V_2O_5 nanocomposites as alternative energy storage substances for photocatalysts. Journal of nanoscience and nanotechnology. 2012; **12**(1):828-833.

[71] Ng C, Ng YH, Iwase A, Amal R. Visible light-induced charge storage, on-demand release and self-photorechargeability of WO_3 film. Physical Chemistry Chemical Physics. 2011;**13**(29):13421-11326.

[72] Park H, Bak A, Jeon TH, Kim S, Choi W. Photo-chargeable and dischargeable TiO_2 and WO_3 heterojunction electrodes. Applied Catalysis B: Environmental. 2012;**115**: 74-80.

[73] Chen X, Liu L, Peter YY, Mao SS. Increasing solar absorption for photocatalysis with black hydrogenated titanium dioxide nanocrystals. Science. 2011;**331**(6018):746-750.

[74] Borges ME, Sierra M, Méndez-Ramos J, Acosta-Mora P, Ruiz-Morales JC, Esparza P. Solar degradation of contaminants in water: TiO_2 solar photocatalysis assisted by up-conversion luminescent materials. Solar Energy Materials and Solar Cells. 2016; **155**:194-201.

[75] Qiu H, Hu J, Zhang R, Gong W, Yu Y, Gao H. The photocatalytic degradation of diesel by solar light-driven floating BiOI/EP composites. Colloids and Surfaces A: Physicochemical and Engineering Aspects. 2019;**583**:123996.

[76] Mendez-Ramos J, Acosta-Mora P, Ruiz-Morales JC, Hernandez T, Borges ME, Esparza P. Turning into the blue: materials for enhancing TiO_2 photocatalysis by up-conversion photonics. RSC Advances. 2013;**3**(45): 23028-23034.

[77] Obregón Alfaro S, Colón Ibáñez G. Evidence of upconversion luminescence contribution to the improved photoactivity of erbium doped TiO_2

systems. Chemical Communications. 2012;**48**:7865-7867.

[78] Wang W, Huang W, Ni Y, Lu C, Xu Z. Different upconversion properties of β-$NaYF_4$: Yb^{3+}, Tm^{3+}/Er^{3+} in affecting the near-infrared-driven photocatalytic activity of high-reactive TiO_2. ACS applied Materials and Interfaces. 2014;6 (1):340-348.

[79] Wang P, Stammes P, der A RV, Pinardi G, Roozendael MV. FRESCO+: an improved O_2 A-band cloud retrieval algorithm for tropospheric trace gas retrievals. Atmospheric Chemistry and Physics. 2008;**8**(21):6565-6576.

[80] Chen B, Ma Q, Tan C, Lim TT, Huang L, Zhang H. Carbon-based sorbents with three-dimensional architectures for water remediation. Small. 2015;**11**(27):3319-3336.

[81] Allahbakhsh A, Bahramian AR. Self-assembled and pyrolyzed carbon aerogels: an overview of their preparation mechanisms, properties and applications. Nanoscale. 2015;**7**(34): 14139-14158.

[82] Gupta S, Tai NH. Carbon materials as oil sorbents: a review on the synthesis and performance. Journal of Materials Chemistry A. 2016;**4**(5):1550-1565.

[83] Xue Z, Cao Y, Liu N, Feng L, Jiang L. Special wettable materials for oil/water separation. Journal of Materials Chemistry A. 2014;**2**(8): 2445-2460.

[84] Rong H, Gao B, Dong M, Zhao Y, Sun S, Yue Q, Li Q. Characterization of size, strength and structure of aluminum-polymer dual-coagulant flocs under different pH and hydraulic conditions. Journal of Hazardous Materials. 2013;**252**:330-337.

Section 4

Miscellaneous Applications of Titanium Dioxide

Chapter 7

Titanium Dioxide and Its Applications in Mechanical, Electrical, Optical, and Biomedical Fields

Rajib Das, Vibhav Ambardekar and Partha Pratim Bandyopadhyay

Abstract

Titanium dioxide (TiO_2), owing to its non-toxicity, chemical stability, and low cost, is one of the most valuable ceramic materials. TiO_2 derived coatings not only act like a ceramic protective shield for the metallic substrate but also provide cathodic protection to the metals against the corrosive solution under Ultraviolet (UV) illumination. Being biocompatible, TiO_2 coatings are widely used as an implant material. The acid treatment of TiO_2 promotes the attachment of cells and bone tissue integration with the implant. In this chapter, the applications of TiO_2 as a corrosion inhibitor and bioactive material are briefly discussed. The semiconducting nature and high refractive index of TiO_2 conferred UV shielding properties, allowing it to absorb or reflect UV rays. Several studies showed that a high ultraviolet protection factor (UPF) was achieved by incorporating TiO_2 in the sunscreens (to protect the human skin) and textile fibers (to minimize its photochemical degradation). The rutile phase of TiO_2 offers high whiteness, and opacity owing to its tendency to scatter light. These properties enable TiO_2 to be used as a pigment a brief review of which is also addressed in this chapter. Since TiO_2 exhibits high hardness and fracture toughness, the wear rate of composite is considerably reduced by adding TiO_2. On interacting with gases like hydrogen at elevated temperatures, the electrical resistance of TiO_2 changes to some different value. The change in resistance can be utilized in detecting various gases that enables TiO_2 to be used as a gas sensor for monitoring different gases. This chapter attempts to provide a comprehensive review of applications of TiO_2 as an anti-corrosion, wear-resistant material in the mechanical field, a UV absorber, pigment in the optical sector, a bioactive material in the biomedical field, and a gas sensor in the electrical domain.

Keywords: Titanium dioxide, properties, applications, corrosion resistance, wear resistance, UV absorber, biomaterials, gas sensors

1. Introduction

Titanium dioxide (TiO_2) is a naturally occurring oxide of titanium. It is also referred to as titanium (IV) oxide or titania. TiO_2 is a cheap and widely available white oxide ceramic having a molecular mass of 79.86 g/mol, a density of 3.9–4.2 g/cm^3, a refractive index in the range of 2.5–2.75, and Mohs hardness of 5.5–7 [1]. It occurs in three crystalline forms: rutile, anatase, and brookite. Both rutile and anatase have a tetragonal structure, whereas brookite has an orthorhombic structure. In industrial applications, only anatase and rutile phases of TiO_2 are used [1]. TiO_2 also serves as a semiconductor, with a band gap of 3.2 eV for anatase and 3.0 eV for rutile. TiO_2 is non-toxic, chemically as well as photo-chemically stable, non-flammable, and biocompatible [2]. TiO_2 is often deposited as thin films or thick film coatings to impart anti-wear and corrosion-resistant properties [3]. It is also used in gas sensing and biomedical applications. Because of its UV absorption ability, TiO_2 has also been used in sunscreens. TiO_2 is also suitable to be used as white pigments. In the past few decades, research activities on nanomaterials have grown rapidly since materials in nano size exhibit completely different properties as compared to their bulk properties. As a result, TiO_2 is one of the most extensively used nano-size materials and is found to be useful in a wide range of applications [4].

2. Applications of titanium dioxide

Owing to its appealing electrical, optical, and mechanical attributes, TiO_2 coating is commonly utilized for gas detecting, wear-resistant, UV shielding, and corrosion-resistant applications. It is also used as a pigment in paints, coatings, cosmetics, plastics, etc. TiO_2 also plays an important role in the fabrication of medical implants. **Figure 1** depicts the applications of TiO_2 in various sectors. These applications of TiO_2 are discussed in the subsequent sections.

2.1 Corrosion resistance

From a technical standpoint, it is crucial to keep metals free of corrosion [5]. One of the most effective techniques to protect metals from corrosion is to apply a protective layer to the metal's surface [6, 7]. Organic coatings are often used in the industry for such purposes [8]. In addition to organic coatings, ceramic coatings have gained popularity in this field due to their superior resistance to oxidation and corrosion in high-temperature or corrosive environments [9–12]. TiO_2 coating is an example of a ceramic coating that is commonly used as a protective layer [13].

There are different ways of depositing TiO_2 films over the metallic surface. Researchers have attempted a variety of methods of depositing TiO_2 on the substrate to investigate its ability in protecting the substrate from corrosion. Masalski et al. [14] proposed plasma assisted chemical vapor deposition (PACVD) as one such approach to obtain TiO_2 films on the 316 steel. The un-coated specimen showed the pitting nucleation (breakdown) at 0.2–0.3 V. However, pitting corrosion was not observed for TiO_2-coated specimen even at 3 V. Furthermore, the current densities of TiO_2-coated specimen were found to be significantly lower than those of the uncoated sample. This demonstrated the efficacy of TiO_2 as a corrosion inhibitor.

Ceramic coatings are often deposited using plasma spraying [15]. However, the metal substrate and bond coat are exposed to corrosion owing to the existence of pores within the coating [16]. Yan et al. [17] tested the corrosion resistance of alumina (Al_2O_3) composite coating in dilute hydrochloric acid (HCl) solution. They reported that the connectivity of pores in the composite coating was lowered by adding 13 wt. % TiO_2 to Al_2O_3. As a result, the composite coating exhibited better resistance to corrosion than the Al_2O_3 coating without any dopants.

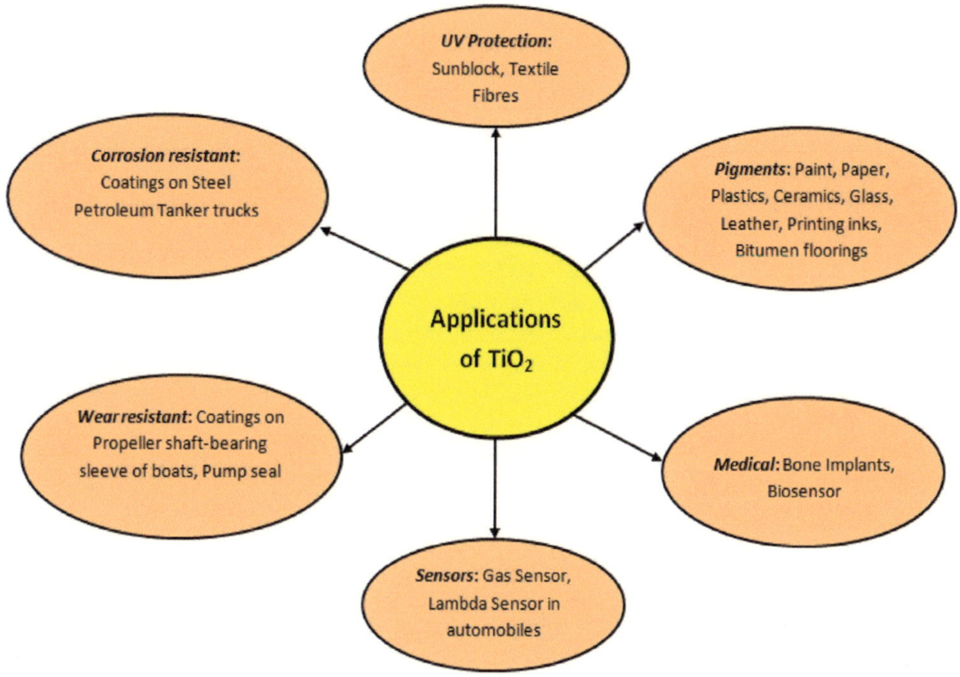

Figure 1.
Some applications of TiO_2 in different domains.

Several researchers tested the potential of TiO_2 thin films under UV light. TiO_2 is an n-type semiconductor, upon exposed to UV illumination, the electrons flow towards the metal through the conduction band of TiO_2. Consequently, the metal's potential will be lower than that required for oxidation. If this occurs, metals can be protected from corrosion. Moreover, TiO_2 film is not decomposed and can act as a non-sacrificial anode. Copper and stainless steel could be cathodically protected using TiO_2 film under UV illumination [18]. Ohko et al. [18] prepared TiO_2 films on 304 stainless steel (SUS 304) using the spray pyrolysis technique. It was subjected to a corrosion test in a sodium chloride (NaCl) solution with a pH higher than 3. When irradiated with UV light of intensity 10 mW/cm^2, they observed that the photopotential of TiO_2 coated specimen was lower than that of uncoated SUS 304. It showed that under illumination, a photo-electrochemical property was exhibited by TiO_2 that rendered cathodic protection to the metals [19].

Titanium coating can shield the aluminum alloy from pitting corrosion. The thermal oxidation of this coating can further improve the anti-corrosion property of the substrate. This is owing to the formation of a dense TiO_2 layer of rutile phase on the surface of the titanium coating during thermal oxidation. This formation of the TiO_2 layer is responsible for the reduction in corrosion rates. The oxidation temperature and time are the two factors that have a direct impact on the degree of improvement [20].

Shen et al. [19] studied the corrosion protection behavior of nano TiO_2 coatings on 316 L stainless steel in both dim and UV illumination conditions. A sol–gel method was used to create TiO_2 nanoparticle coating, which was subsequently exposed to hydrothermal treatment. When tested in 0.5 $molL^{-1}$ NaCl solution in a dark environment, the coating exhibited excellent resistance to corrosion as it acted like a ceramic protective shield on the metal's surface. This was corroborated by the fact that in comparison to uncoated steel, the corrosion current density was decreased by three orders of magnitude and the corrosion resistance was enhanced by more than a hundredfold for TiO_2 nanoparticle coated stainless steel. However, the electrons generated under UV illumination offered cathodic protection to the stainless steel substrate. Mahmoud et al. [21] also reported that the TiO_2 layer deposited on weathering steel displayed higher anti-corrosion properties than the bare steel in NaCl aqueous solution, under UV light.

Shan et al. [22] employed the atomic layer deposition (ALD) technique to deposit a thin TiO_2 film of thickness 50 nm onto stainless steel. X-ray diffraction (XRD) results indicated the amorphous structure of TiO_2. When steel was evaluated for corrosion, the corrosion potential increased from −0.96 eV to −0.63 eV on applying TiO_2 coating. In addition, the corrosion current density was reduced from 7.0×10^{-7} A/cm^2 for uncoated steel to 6.3×10^{-8} A/cm^2 for coated steel. This implied that the TiO_2 film was effective in shielding the stainless steel substrate against the corrosive agents.

Anti-corrosion coatings are usually made of epoxy resins. However, micro-pores are created during their curing process. The corrosive medium can easily penetrate the epoxy coating through these micro-pores, making the substrate highly susceptible to corrosion. To improve the anti-corrosion characteristics of epoxy coatings, Yu et al. [23] developed hybrid modified graphene oxide (GO) sheets by incorporating nano-TiO_2 on the surface of graphene oxide using 3-aminopropyltriethoxysilane. The interlayer gap of the sheets was observed to rise as a result of this. Owing to this greater interlayer spacing, the TiO_2-GO hybrids were easily exfoliated and disseminated in the epoxy coating. The electrochemical impedance spectroscopy (EIS) test revealed that adding merely 2 wt. % of TiO_2-GO hybrid to epoxy resulted in a tremendous improvement in the corrosion resistance. This was attributed to the sheet-like structure of hybrid which behaved as an additional barrier layer, preventing the corrosive liquid medium from accessing the micro-pores. Excellent plugging of micro-pores by this hybrid led to an improved anti-corrosion performance of such coating.

Khalajabadi et al. [24] studied the effect of adding TiO_2 nanopowders on the anticorrosion performance of magnesium/hydroxyapatite (Mg/HA)-based nanocomposite for medical applications. TiO_2 doped nanocomposite was synthesized by the milling-pressing-sintering technique. They reported that the addition of 15 wt. % TiO_2 nanopowders and a drop in HA amount to 5 wt. %, resulted in a decrease in the number of pores and HA agglomeration. Moreover, the wettability of the samples after sintering was reduced owing to the formation of magnesium titanate ($MgTiO_3$) nanoflakes. This obstructed the electrolyte from penetrating the nanocomposite. The corrosion current of the composite without TiO_2 was 285.3 μA/cm^2, which drastically reduced to 4.8 μA/cm^2 for nanocomposite containing TiO_2. Similarly, the polarization resistance of Mg/HA increased dramatically from 0.25 kΩ cm^2 to 11.86 kΩ cm^2 with the incorporation of TiO_2. On doping TiO_2, the corrosion rate of composite coating reduced remarkably from 4.28 mm/yr. to 0.1 mm/yr. These findings suggested that the corrosion resistance of Mg/HA-based nanocomposite could be improved significantly by adding TiO_2 nanopowders. Similarly, the addition of TiO_2 nanoparticles in the nickel-tunsgten (Ni-W) alloy matrix enhanced its anti-corrosion characteristics as compared to Ni-W alloy [25]. Poorraeisi et al. [26] also reported that the incorporation of zirconium oxide-titanium oxide (ZrO_2-TiO_2) in hydroxyapatite coating showed better resistance to corrosion than the coating without ZrO_2-TiO_2 reinforcement.

Epoxy coatings can be applied on steel petroleum tanker trucks to protect them from corrosion as they provide a physical barrier layer and offer excellent chemical stability [27]. Several reports have suggested the use of nanoparticles in epoxy coatings to further enhance its anticorrosion characteristics. Nanoparticles possess excellent surface properties and can block pores, cavities, and channels in the coating. As a result, they serve as a buffer against the electrolyte, preventing the corrosive solution from diffusing into the coating. TiO_2/epoxy nanocomposites significantly elevate the corrosion resistance of epoxy resins. The anti-corrosion behavior of poly-dimethylaminosiloxane (PDMAS)/TiO_2 epoxy hybrid nanocomposite coating and traditional epoxy coating were tested using salt spray accelerated corrosion test by Fadl et al. [27]. The scanning electron microscope (SEM) and energy dispersive X-ray spectroscopy (EDS) analysis as well as the weight loss study confirmed that the hybrid nanocomposite coating was superior to conventional epoxy coating in terms of anti-corrosion properties.

Krishna et al. [28] developed TiO_2 film over commercially pure titanium by thermal oxidation. In this report, when thermal air oxidation was carried out at a temperature less than 400°C for 1 hour, the corrosion resistance of the TiO_2 film increased with the increase in the film thickness. They discovered that the rutile phase of TiO_2 films exhibited better anti-corrosion properties than the amorphous phase.

A thin layer of ceramic coating on aluminum alloy is effective in protecting it from corrosion. To achieve long-term protection from corrosion, Merisalu et al. [29] deposited two different layers of coating on aluminum alloy. Initially, a thin film of nanoporous aluminum oxide base layer was deposited on the substrate using a special anodizing process. The nano-sized pores present in this aluminum oxide layer were then sealed by depositing chemically resistant aluminum oxide/titanium oxide (Al_2O_3/TiO_2) nanolaminates using ALD. The Al_2O_3/TiO_2 nanolaminates not only transformed the base layer into a nanocomposite but also covered the entire surface of the base layer to form the top-most layer of coatings. The samples underwent corrosion tests by immersing them in a salt solution for a longer duration. The tests revealed that the top layer of Al_2O_3/TiO_2 nanolaminates significantly improved the corrosion resistance of coating by acting as an ion barrier. This coating was able to withstand the salt solution for 298 days. Atomic layer deposition of Al_2O_3 and TiO_2 nanolaminates as a corrosion inhibitor was also reported in other papers [30, 31].

Biomedical magnesium alloys such as WE43 Mg alloy must exhibit excellent resistance to corrosion in a physiological environment. To protect it from corrosion by simulated body fluid (SBF), Li et al. [32] electrodeposited nanocrystalline zinc (Zn) coating on WE43 Mg alloy. This coating was further chemically treated to form a Titanium oxide-Zinc phosphate layer. In SBF, this composite coating displayed a much lower corrosion current density of $4.1 \pm 0.8\ \mu A/cm^2$ and a much larger resistance of $4.28 \times 10^3\ \Omega\ cm^2$ than uncoated WE and WE alloy with only Zn coating. WE alloy with Zn coating showed poor corrosion resistance because of the establishment of galvanic couples between the Zn coating and WE43 Mg alloy substrate.

Nanomaterials offer a larger surface area than conventional materials which can considerably influence the anticorrosion performance of nano-coatings. Chen et al. [33] observed that the average corrosion potential of bare titanium alloy (Ti-6Al-4 V) in the presence of NaCl solution was 0.316 V which was reduced to 0.07 V for TiO_2 nanoparticle coated titanium alloy.

2.2 UV protection

Ultraviolet (UV) rays coming from the sun are the radiations having a wavelength in the range of 200–400 nm. UV rays are generally known to have detrimental effects on the skin health of the human body. These UV rays can be categorized into three groups: Ultraviolet C (wavelengths range of 200 to 290 nm) denoted as UVC, Ultraviolet B (wavelengths range of 290–320 nm) denoted as

UVB, and Ultraviolet A (wavelengths range of 320–400 nm) denoted as UVA. UVC is blocked by the atmosphere and cannot reach the earth. UVB causes sunburn, however, it is absorbed by the glass and therefore rooms with glass windows can block the UVB rays from entering the room. UVA, on the other hand, can transmit through the glass and inflict serious skin damage, possibly leading to skin cancer [34]. Sunscreen is often applied to protect the skin from these harmful rays. There are two types of sunblock available. One is an organic sunblock, which absorbs UV rays and converts them into heat, and the other one is inorganic. TiO_2 and ZnO are examples of inorganic sunblock [35]. TiO_2 provides exceptional blocking against UVA and UVB radiation owing to its chemical and physical properties. However, its UV protection mechanisms are still being studied [36]. TiO_2 is a semiconductor, and its UV absorption ability can be understood from the band theory of solids. TiO_2 is capable of absorbing UV rays due to the formation, mobility, and separation of photo-generated electrons and holes [37, 38]. Some researchers believe that because of the high refractive index of TiO_2, the UV rays are reflected and/or scattered, resulting in high UV-shielding properties. Apart from protecting the skin from harmful UV rays, the UV-blocking properties of TiO_2 can also benefit the textile industry by minimizing the photochemical degradation and color fading of textile fibers after prolonged UV exposure.

There are several research publications available that indicate the UV protection of textile substrates by TiO_2 [39–44]. The UV resistance performance is measured by calculating the ultraviolet protection factor (UPF) for UVA and UVB. Higher UPF values in the range of 40–50 or above 50 indicate lower transmission of UV rays. A UPF value of 50 indicates that only 1/50 (or 2%) of UV radiation transmits through the textile material and reaches the skin. UV absorption spectra can also be used to assess UV resistance. With a UPF value of 10, an untreated cotton fabric displayed insufficient UV protection, whereas when it was treated with TiO_2 nanoparticles, the cotton material offered a maximum UPF value of 50 [45].

Engineering polymer such as polyetheretherketone (PEEK) suffers from chemical degradation when exposed to UV radiation. This also results in its discolouration and loss of mechanical properties such as ductility. To boost the UV resistance, Bragaglia et al. [46] incorporated submicron size TiO_2 as fillers in the PEEK matrix to form PEEK-TiO_2 composites. The volume fraction of TiO_2 was varied from 0 to 5%. The UV-thermal aging of samples was conducted for 8 hours at a temperature of 70°C using 351 nm peaked UVA radiation with an intensity of 0.77 W/m^2. Following that, the samples were exposed to the humid condition of 100% RH at 50°C for 4 hours. The cycles were repeated for 30 days. The UV aging test revealed that the composite containing 5 vol.% of TiO_2 was effective in retarding the photodegradation of the PEEK polymer because of the UV-blocking action of TiO_2. It was also reported that the tensile strength and ductility of the PEEK- 5 vol. % TiO_2 composite remain unaltered even after exposure to UV rays.

Li et al. [47] prepared TiO_2 coated polyester (PET) to enhance its UV resistance and anti-aging characteristics. To carry out the UV aging test, the specimens (both coated and uncoated fabrics) were exposed to UV irradiation of intensity 0.89 $W/m^2/nm$ for 60°C for a specific period. The effect of UV irradiation on the breaking strength of the fabrics was also tested. The TiO_2-coated PET displayed excellent UV resistance as its UPF was reported to be 130 in contrast to 34 for uncoated PET. After 100 hours of UV exposure, the strength of uncoated PET was reduced by 44% in warp direction whereas, for coated PET, the strength was reduced by 35.6%.

Torbati et al. [34] compared the UV resistance of base sunscreen cream to one containing 0.5% w/w TiO_2 nanoparticle. The sun protection factor (SPF) against UV radiation was measured for both the creams. The cream containing TiO_2 showed a significantly high SPF rating compared to the base cream, indicating superior UV protection by the TiO_2 doped cream.

Natural rubber (NR) is prone to photo-oxidation when exposed to UV light. The double bonds in NR chains are attacked by the UV rays which lead to changes in the mechanical properties. Seentrakoon et al. [48] prepared nanoparticles of rutile TiO_2 (n-TiO_2) from micron-sized rutile TiO_2 (micro-TiO_2) through ultrasonication. This n-TiO_2 was mixed with natural rubber to explore its UV shielding properties. The UV blocking performance of the n-TiO_2/NR composite was compared with the unfilled NR. The UV resistance of the prepared nanocomposites was tested using accelerated weathering tester equipped with a UVA 340 nm fluorescent lamp of light intensity 0.63 W/m^2/nm. The test was conducted at a temperature of 50°C for 24 hours. The extent of retention of the mechanical properties after UV irradiation was also assessed for all the samples. The unfilled NR displayed a strong carbonyl peak after exposure to UV light. This indicated a high degree of degradation of NR by UV rays. On the other hand, the carbonyl peak intensity was significantly reduced with the addition of n-TiO_2. This demonstrated that an effective UV photodegradation prevention was achieved by the incorporation of n-TiO_2. The n-TiO_2 doped NR composite was reported to be more effective in shielding UV rays than the unfilled NR and NR composite containing micron-sized rutile TiO_2 (micro TiO_2/NR). The mechanical strength testing of these specimens revealed that the percentage retention of tensile strength and elongation at break after exposure to UV rays for NR was 51.2%, and 84%, respectively which climbed to 90.6%, and 92.9%, respectively for n-TiO_2/NR. The characteristics of micro-TiO_2/NR were intermediate between those of unfilled NR and n-TiO_2/NR. The considerable increment in the percentage retention of mechanical properties confirmed that n-TiO_2/NR composite provided better UV protection and prevented the NR from the negative impact of UV. The high UV shielding performance of n-TiO_2 than micro-TiO_2 was attributed to the high surface area per particle size of n-TiO_2 which significantly boosted the UV shielding property.

Reinosa et al. [49] formulated a sunscreen using a combination of nano zinc oxide (ZnO) and micro-TiO_2 composite. They claimed that this product not only provided a greater SPF but was also capable of reducing nanoparticle diffusion into the skin of the human body.

Sun et al. [50] investigated the influence of TiO_2 layer thickness on the solar energy conversion efficiency and illumination stability of the polymer solar cell. TiO_2 layers were formed by the spray pyrolysis technique. By functioning as a UV blocker, the photodegradation of the organic solar cell was decreased with an increase in the TiO_2 layer thickness. The thick TiO_2 layer, on the other hand, restricted the amount of light falling on the solar cell and lowered its performance. Considering both the positive and the negative effects, it was stated that the TiO_2 layer of 100 nm thickness was the optimum thickness for this solar cell.

The morphology and the content of TiO_2 largely influence its UV-shielding properties. The increment in UPF value with the addition of TiO_2 has been documented in numerous studies [41, 43]. TiO_2 is also doped with other UV absorbing materials to enhance the UV blocking properties. Noble metals such as gold and silver are also effective UV absorbers and combining TiO_2 with such noble metals improved the UV protection of cotton fabrics [39, 41, 44].

Liang et al. [42] prepared a composite with natural pigment melanin and TiO_2. Melanin protects the living cell from UV radiation. Wool fabrics were treated with such organic–inorganic composite to impart UV protection characteristics. They found that when untreated wool cloth with a UPF of 25 was treated with pure-melanin, pure-TiO_2, and melanin/TiO_2 composite, the UPF value increased to 40, 83.9, and 112.6, respectively. Similarly, Li et al. [47] reported that the TiO_2-coated polyester fabric with a UPF of 130 rose to 135 when coated with a mixture of TiO_2 and benzotriazole (an organic UV absorber).

2.3 Bioactive materials

Medical implants are structures that provide support or can be a substitute for injured biological tissue. A biocompatible medical implant encourages a healthy relationship between the implant and the surrounding tissue. A medical implant must not release any harmful substances into the body. It must not trigger an inflammatory response as well. Some of the examples of implants include artificial hearts, bone implants, dental structures, etc. Titanium (Ti) and its alloys such as nitinol (TiNi) possess desirable mechanical properties and biocompatibility, making them ideal for use as bone implants [51]. But Ti-based metallic materials cannot form chemical bonds with bone tissues. Hence, the formation of new bone becomes complicated during the initial stages of implantation, resulting in low bioactivity and a reduction in the service time of the implant. Furthermore, the release of dangerous metallic ions from the titanium alloy in the biological environment may result in toxic reactions [52]. Therefore, the surface of the implant needs to be modified to encourage the development and growth of the bone tissue on the implants, as well as to enhance the implant's integration with the bone tissues.

Because of its superior biocompatibility and anti-corrosion properties, TiO_2 coatings on the surface of implants have gotten a lot of attention in the biomedical industry [53]. There are various methods of depositing TiO_2 coating which include laser ablation, dip coating, sol–gel process, heat treatment, electrochemical methods [54], sputtering, thermal spraying, etc. [55]. TiO_2 films fabricated by anodic oxidation process in sulfuric acid under potentiostatic regulation may function as a bioactive coating [54]. This implies that in the presence of body fluids, a layer of calcium phosphate may grow on the surface of the TiO_2 film, allowing the implant to bond with the surrounding bone tissues [54]. Zhao et al. [56] plasma sprayed TiO_2 coatings on Ti alloy substrate using nano TiO_2 powders as feedstocks to explore their bioactivity and cytocompatibility. They reported that the acid treatment of plasma sprayed TiO_2 coating using a high concentration of sulfuric acid promoted the formation of apatite on the surface. The bioactivity of TiO_2 could not be enhanced at a low concentration (0.01 M) of sulfuric acid (H_2SO_4), indicating that the concentration of H_2SO_4 influenced the bioactivity of TiO_2 coatings. The in vitro cell culture test showed that the acid treatment of TiO_2 coatings enhanced cell adhesion. This could be attributed to the formation of many hydroxyl groups (Ti-OH bonds) on the surface of TiO_2 by acid treatment. The OH groups enhanced the attachment and adhesion of cells [56]. Several reports are available which revealed the formation of apatite on TiO_2 powders and sol–gel TiO_2 film [57] in SBF. Incorporating other metallic elements such as copper (Cu) into Ti-based material can accelerate cellular activity and stimulate osteogenesis (bone tissue formation). Antibacterial capabilities of Cu are extremely impressive [58]. He et al. [59] prepared Copper oxide (CuO) doped TiO_2 coatings on Ti-based implant material by using a combination of magnetron sputtering and annealing process. The in vitro cytocompatibility tests revealed that the TiO_2/CuO coating displayed no apparent toxicity and supported osteoblast spreading and proliferation. The composite coating outperformed pure-Ti and TiO_2 coating in terms of corrosion resistance and antibacterial potential against *Staphylococcus aureus* bacterial species.

The possible applications of titanium dioxide nanotubes on Ti metal as bone implants was summarized in a review article by Awad et al. [51]. TiO_2 nanotubes of diameters ranging from 30 to 100 nm were reported to increase cell attachment and osseointegration (bonding of implant with bone tissue) [60]. TiO_2 nanotubes can also be filled with drugs or modified with proteins or hydroxyapatite, making them highly essential for bone implants.

TiO_2 nanotubes offer greater specific surface area, that allows biomolecules to be immobilized and employed in biosensor development [61]. Biosensors are analytical devices that combine a bioreceptor (for example, a catalyst) with a transducer to transform a biological response into electrical signals [62]. Owing to the semiconducting attributes of TiO_2 nanotubes, the rapid transport of electrons takes place from the surface reaction to the Ti substrate. This improves the performance of the biosensor and aids in the diagnosis of diseases [63]. For instance, the immobilization of enzyme fructosyl-amino acid oxidase in TiO_2 nanotubes can detect glycated hemoglobin (HbA1c) in a diabetic patient [64].

The biological performance of TiO_2 is influenced by the surface topography and porosities [57]. Garcia-Lobato et al. [55] deposited TiO_2 coatings on 316 L stainless steel plates using the spraying method. The deposition rate, which is a spraying parameter, was varied. A rough and porous TiO_2 layer was obtained at a high deposition rate. Such a TiO_2 layer assisted the nucleation and growth of hydroxyapatite. According to Zhang et al. [65], a micro/nano-structured TiO_2 coating deposited via induction suspension plasma spraying showed better adsorption of protein than the traditional TiO_2 coating deposited by atmospheric plasma spraying or pure Ti with a smooth surface. The cell culture experiment showed that the micro/nano structured TiO_2 coating also facilitated cell attachment, proliferation, and alkaline phosphatase activity.

Metallic implants such as 316 L stainless steel are susceptible to bacterial infection and corrosion [66]. To shield the stainless steel implants from such infection and corrosion, Zhang et al. [66] developed a titanium oxide-polytetrafluoroethylene (TiO_2-PTFE) nanocomposite coating on a polydopamine coated stainless steel using a sol–gel dip coating process. Under UV radiation, TiO_2 nanoparticles inhibited the growth of bacteria. The TiO_2-PTFE coating exhibited excellent antibacterial and anti-adhesion characteristics against both *Escherichia coli* and *Staphylococcus aureus* bacterial strains. The coating also displayed remarkable corrosion resistance in SBF.

2.4 Pigments

TiO_2 exhibits a high refractive index, whiteness, brightness, high opacity, and non-toxicity which make it suitable to be used as white pigments [67]. The high brightness and opacity of TiO_2 can be ascribed to its tendency to scatter light [68]. TiO_2 pigments are often used in paints, coatings, inks, paper, plastics, cosmetics, etc. [1]. According to Fresnel, the larger the refractive index difference between the pigment and the medium, the more light is reflected from the surface and the opacity is enhanced [69]. The refractive indices of rutile and anatase TiO_2 pigments are 2.73 and 2.55, respectively. The particle size of TiO_2 and its dispersion (interparticle separation) have a significant impact on the degree of scattering of light. It has been shown that efficient light scattering of a particular wavelength occurs when the particle size is approximately half that wavelength. As a result, the optimum pigment size for the maximum scattering of visible light is 0.2–0.3 μm. The agglomeration of TiO_2 particles weakens the hiding power of TiO_2 [1]. The degree of the pigment dispersion affects the opacity, tinting strength, brightness, gloss development, and durability of the TiO_2 film. Therefore, the maximum opacity and other essential optical and physical properties in a coating can be achieved by completely dispersing TiO_2 pigments down to their ultimate particle size [69]. The increase in the particle size of TiO_2 above 1 μm harms the film gloss and the degree of dispersion [1]. In addition to TiO_2, some fine particle minerals, also known as pigment extenders, are used as fillers in paints to enhance the optical properties of TiO_2. These pigment extenders avoid the agglomeration of TiO_2 particles and separate the individual particles of TiO_2 to obtain the optimum inter-particle spacing required for maximum opacity. Some of the examples of minerals include calcium carbonate ($CaCO_3$), silica, kaolin, talc, wollastonite, mica, and so on [1]. Some of the ways of fabricating mineral-TiO_2 composites include the mechano-chemical method, chemical precipitation method, etc. Currently, the mineral-TiO_2 composite pigment is used in coatings, plastics, and papermaking [70].

Zhou et al. [71] synthesized barite/TiO_2 composite particles using the chemical precipitation method. Barite and TiO_2 were joined together with the Ti-O-Ba bond. The pigment properties of the composite, such as hiding power and oil adsorption value were 18.5 g/m^2 and 15.5 g/100 g, respectively which were comparable to the pigment properties of TiO_2. Using the same approach, Chen et al. [72] fabricated $CaCO_3$ based-TiO_2 pigment and observed that the hiding power of the end product (23.82 g/m^2) was similar to that of anatase TiO_2 (22.56 g/m^2). Similarly, the illite/TiO_2 composite pigment offered higher whiteness of 95.73% and hiding power of 97.55% than illite [73]. These results indicated that composite pigments could be used in applications, including architectural paints, whitening additives in paper manufacturing, etc.

Wang et al. [67] studied the pigment properties of a mechano-chemically formulated calcined kaolin/TiO_2 composite. The composite showed the whiteness and hiding power of 95.7%, and 17.12 g/m^2, respectively which was close to the pigment properties of pure anatase TiO_2 which offered a whiteness of 95.8%, and hiding power of 15.14 g/m^2. Similar pigment properties were also observed for brucite/TiO_2 composite, wollastonite/anatase TiO_2 composite, and sericite/anatase TiO_2 composite [70]. However, amorphous silica/anatase TiO_2 composite displayed an even better hiding power of 13.07 g/m^2 as compared to pure anatase TiO_2 [74].

Sun et al. [75] adopted another technique called the self-assembly method to fabricate barite/rutile TiO_2 composite. As compared to pure rutile TiO_2, the composite product possessed identical pigment attributes (hiding strength of 12.08 g/m^2 and oil adsorption value of 14.48 g/100 g). Consequently, these composites could easily substitute pure TiO_2 as additives in paper manufacturing industries.

According to Hou et al. [76], the whiteness of TiO_2/wollastonite composite (96.6%) was somewhat higher than that of anatase TiO_2 (96.2%). Therefore, the TiO_2/wollastonite composites could also be used as pigments in coatings.

The dispersion of TiO_2 particles (5–10 vol. %) in a molten glass phase provides whiteness and opacity. These enamels (or glass phases) are coated on metals or ceramics. A desirable whiteness and appearance in enamels can be achieved by controlling the anatase-to-rutile phase ratio. In paper manufacturing industries, pigment coatings are added to improve the printability, smoothness, brightness, and opacity of the paper. A smooth paper surface is obtained by adding TiO_2 pigments. TiO_2 pigments are also added to the textiles fibers to impart opacity and provide protection against visible and UV light. The content of the pigments in the textile fibers ranges from 0.3 to 1 wt. % [1]. TiO_2 pigments are also utilized in artificial leather, cement products, ceramics, glass, cosmetics, laminating papers, pharmaceuticals, moldings, bitumen floorings, printing inks, rubber, putty, shoe creams, etc. [69].

2.5 Wear resistant

When two solid bodies in contact have relative motion, some material is removed from their surfaces [77]. This phenomenon of material removal from the surface owing to rubbing is known as wear. Many engineering components made of metals or alloys fail or their service life is reduced due to wear [78]. So, efforts should be undertaken to minimize this undesirable phenomenon [79]. One method to combat wear is by modifying the surface properties of the material to impart anti-wear characteristics [80]. For instance, depositing a new hard and wear-resistant material onto the substrate can significantly reduce the wear of the substrate material. TiO_2 possesses high hardness and is known to resist wear [81]. Thermally sprayed TiO_2 coatings are often used as wear-resistant coatings in pump seal, propeller shaft-bearing sleeve, etc. [82]. To reduce wear, researchers have employed different methods to produce TiO_2 coatings.

The rutile TiO_2 phase offers low friction and high wear-reducing abilities [83]. It can be produced by the thermal oxidation of Ti-alloys. Krishna et al. [81] deposited Ti coatings on stainless steel by magnetron sputtering. The Ti coating was later converted to TiO_2 by thermal oxidation at 550°C. As-deposited TiO_2 layer exhibited a much higher hardness of 11 GPa (4 times that of the as-deposited Ti), which was close to the hardness of the bulk rutile TiO_2 phase. This enhanced the load-carrying capacity of the oxidized specimen. Sun et al. [84] also fabricated rutile TiO_2 by thermally oxidizing pre-coated titanium on an aluminum alloy (Al-alloy) substrate. For tribological testing, an alumina counterball was used. Severe adhesive wear with stick–slip propensity and high friction (friction coefficient of 0.5–0.8) were reported for the uncoated Al-alloy. Thermally oxidized coatings showed three orders of magnitude reduction in wear rate with no signs of adhesive wear. The oxidized coating offered less friction (friction coefficient t < 0.25), and it did not fail throughout the test. High hardness was responsible for the excellent wear-resistant of the oxidized coatings. Other researchers have also documented the role of thermally oxidized Ti in combating wear [85].

Dejang et al. [86] fabricated Al_2O_3/TiO_2 composite coating with varying contents of TiO_2 (0–20 wt. %) using plasma spraying process and compared its wear performance with monolithic Al_2O_3 coatings. The hardness of the composite coating was found to be lower than that of Al_2O_3 coating owing to the comparatively lower hardness of TiO_2 compared to Al_2O_3. On the other hand, the fracture toughness was improved by increasing the weight fraction of TiO_2. However, the sliding wear test revealed that the wear rate of Al_2O_3 coating was 1.5 times higher than that of 3 wt. % TiO_2 doped Al_2O_3 composite coatings owing to the presence of TiO_2 splats that increased the fracture toughness and decreased the friction coefficient. Owing to the hydrophilic nature, TiO_2 layer can absorb moisture from the air and potentially lowers the friction coefficient.

Baghery et al. [87] used the electrodeposition method to deposit nickel-titania (Ni-TiO_2) nano composite coating. They discovered that increasing the quantity of TiO_2 nanoparticles increased microhardness and wear resistance. The grain refinement and dispersion strengthening mechanisms were responsible for the increase in hardness. Similar strengthening mechanisms were also reported for TiO_2 sol-strengthened copper-tin-polytetrafluoroethylene (Cu-Sn-PTFE) composite coating by Ying et al. [88]. Baghery et al. [87] further observed that the stable friction coefficient for Ni coating was 1 that was reduced to 0.3 for Ni-8.3 wt. % TiO_2 coating. The TiO_2 nanoparticle reinforcement in the coating minimized the direct interaction between the Ni matrix and the abrasive counterbody. Furthermore, TiO_2 nanoparticles detached from the coating due to abrasive action served as a solid lubricant between the two mating surfaces. These mechanisms reduced the friction coefficient and wear rate for TiO_2 doped coatings. Similarly, Li et al. [89] plasma-sprayed chromium oxide (Cr_2O_3) - TiO_2 composite coatings and found that Cr_2O_3 doped with 16 wt. % TiO_2 coatings exhibited the lowest friction coefficient owing to its minimum surface free energy. The presence of $(Cr_{0.88}Ti_{0.12})_2O_3$ phase raised the microhardness of the composite coatings while lowered their friction coefficient. Babu et al. [90] examined the tribological performance of TiO_2 coated aluminum-silicon carbide (Al-SiC) substrate and uncoated Al-SiC. They reported that the plasma sprayed TiO_2 coating (570 $HV_{0.5}$) had an 8-fold higher hardness than the Al-SiC substrate (70 $HV_{0.5}$). This resulted in the reduction in wear rate from 11 mm^3/m for the uncoated specimen to 6 mm^3/m for the coated specimens. The uncoated sample experienced severe abrasive wear with delamination. The plasma-sprayed sample, on the other hand, showed only minor abrasive wear. Ying et al. [88] investigated the wear performance of electrodeposited Cu-Sn-PTFE coating with TiO_2 sol as a reinforcing agent. At 40 ml/L concentration of TiO_2 sol, TiO_2 nanoparticles were found to be well dispersed which strengthened the Cu-Sn-PTFE matrix. This led to an increment in hardness and wear resistance of the TiO_2 doped composite coating. The wear performance of TiO_2 film deposited on commercially pure Ti using the sol–gel method was evaluated by Comakli et al. [91]. The TiO_2 coated specimen showed a low value of friction coefficient because of the self-lubricating property of TiO_2 films, and higher surface hardness than the uncoated specimen. In another investigation, similar results of TiO_2 films improving tribological performance have been published [92]. Barkallah et al. [93] fabricated aluminum oxide/tricalcium phosphate (Al_2O_3/10 wt.% TCP) bioceramic for an orthopedic implant and found that the wear behavior of the biocoating could be improved by incorporating 5 wt.% TiO_2. The hardness and the fracture toughness of the bioceramic without TiO_2 were 3.56 GPa and 8.734 MPa $m^{1/2}$, respectively. Both hardness and fracture toughness were increased to 8.55 GPa (140% increment) and 13 MPa $m^{1/2}$ (48.8% increment) when 5 wt. % TiO_2 was added. This increment in hardness and fracture toughness could be responsible for the improved wear performance. As a result, by adding TiO_2, the tribological performance of composite can be significantly improved.

2.6 Gas sensing

TiO_2 was widely studied by numerous researchers for gas sensing performance as well [94]. The working principle of the TiO_2 metal oxide gas sensor involves adsorption, desorption reactions relevant to air and test gas of interest [94]. TiO_2 surface at ambient temperature consists of adsorption of ambient oxygen in the form of O_2 [94]. This is termed as physically adsorbed oxygen [94]. TiO_2 surface, at elevated temperatures (150–450°C), consists of electron transfer as a result of chemical interaction of ambient oxygen that ultimately leads to the formation of chemical adsorbed oxygen in different forms namely, O_2^-, O^- [94]. This leads to an increase in the sensor resistance under the influence of air (R_a). During gas sensing, test gas such as hydrogen (H_2) reacts with chemically adsorbed oxygen ions to form an oxidized reaction product and this reaction transports electrons back to the conduction band of the TiO_2 layer [94]. Herein, sensor resistance of TiO_2 under the influence of test gas (denoted as R_g) drops to a certain value depending upon the test gas concentration and the change in the electrical signal from R_a to R_g ultimately determines gas response (R_a/R_g) or (R_a-R_g)/R_a [94].

In line with this principle, TiO_2 has been studied by numerous researchers for H_2 [94], carbon monoxide (CO) [95], ammonia (NH_3) [96], etc. In addition, TiO_2 is one of the popular materials for developing air-fuel ratio sensors [97]. A brief review of TiO_2 sensors in the recent literature is detailed in the following paragraphs:

Hydrogen (H_2): H_2 being colorless, odorless, highly combustible gas needs careful attention during its generation, storage, transportation as well [98]. Therefore, considerable efforts have been made by researchers to develop H_2 sensors using TiO_2 in different forms [94]. Tang et al. deposited TiO_2 anatase film using reactive triode sputtering and reported H_2 sensing response at 370°C. Though this paper reported the possibility of TiO_2 films towards H_2 sensing, gas sensing performance was not investigated in detail [99]. Devi et al. reported the synthesis of mesoporous TiO_2 powders and obtained gas response ($R_a/R_g \sim 4.8$) was found superior to that of commercial TiO_2 powder ($R_a/R_g \sim 2.5$) [100]. This was attributed to the larger surface area for efficient gas sensing reactions [100]. Yoo et al. synthesized a nano-fibrillar TiO_2 sensor for H_2 sensing applications at 400°C [101]. Jun et al. reported the H_2 sensing behavior of TiO_2 films grown using the thermal oxidation route and studied the analogy between gas sensor response and film microstructure. Superior H_2 sensor response (R_a/R_g of 1.2×10^6) at 300°C with a response time of 10 s was attributed to the ease of H_2 penetration into the sensing layer owing to its porous morphology [102]. Moon et al. reported a gas sensor response of 250% for TiO_2 film exposed to 100 ppm H_2 gas at 200°C [103]. Moon et al. synthesized meso-porous TiO_2 film by anodization over Ti substrate and reported gas response ($R_a/R_g \sim 2.5$) at 140°C towards 1000 ppm H_2 [104]. In this report, the enhanced H_2 response could be attributed to the mesoporous microstructure of TiO_2 film. The effect of niobium (Nb) doping with TiO_2 film was reported to yield a useful gas response at room temperature [105]. The plausible reason behind room temperature sensing can be attributed to enhanced oxygen adsorption over TiO_2 nanotubes.

Carbon monoxide (CO) is also a colorless, odorless, toxic gas that needs early detection [106]. Harmful levels of CO could be found owing to gasoline engine exhaust, burning of coal in a boiler room, wooden stove exhaust, cigarette smoke, etc. [107]. Quite a few researchers have studied TiO_2 for CO sensing applications. Devi et al. developed a mesoporous TiO_2 particulate sensor and obtained gas response in the presence of 500 ppm CO was ($R_a/R_g \sim 2.4$) at 450°C [100]. Jun et al. prepared TiO_2 film using the micro-arc oxidation method and reported gas response of ($R_a/R_g \sim 3.10$) towards 30 ppm CO at 350°C [95]. Choi et al. observed gas response of ($R_a/R_g \sim 1.4$) at 600°C towards 500 ppm CO using 7.5 wt. % Al-doped TiO_2 sensor using an auto combustion route [108].

Ammonia (NH_3) is a volatile organic compound and being highly flammable and harmful to the respiratory system needs careful monitoring during its handling [109]. TiO_2 has been explored by different researchers for NH_3 sensing applications. Karunagaran et al. deposited TiO_2 thin film using DC magnetron sputtering [96]. A sensitivity factor of 7000 was noted at a temperature of 300°C towards 500 ppm NH_3 [96]. Gardon et al. deposited TiO_2 layer using atmospheric plasma spraying method in which maximum gas response ($(R_a\text{-}R_g)/R_a$) of 7% was attained at 210°C [110]. Dhivya et al. tested NH_3 sensing performance of DC magnetron sputtered TiO_2 film with a gas response (Ra/Rg ~8000) measured at room temperature [111].

In practice, the TiO_2 functional layer finds application as a lambda sensor in the automotive exhaust system between the exhaust manifold and catalytic converter [112]. Herein, the term lambda (designated as 'λ') refers to the air-fuel equivalence ratio that measures the oxygen content in the exhaust gas being analyzed [113]. The lambda sensor is used to properly adjust the fuel amount that is being supplied to the cylinder of the internal combustion engine [114]. This controls the air-fuel mixture thereby ensuring the proper running of the engine [115]. Also, the lambda sensor is used to ensure that the catalytic converter is functioning in an intended manner [116]. The following paragraph presents a review of successful attempts to make TiO_2 based lambda sensors.

In the year 1987, a group of inventors in Japan developed porous TiO_2 coating as lambda sensor to measure air-fuel equivalence ratio at around 1000°C. TiO_2

(50 μm) film was coated over commercially available Al_2O_3 substrate having a pair of platinum (Pt) electrodes. As a result of the change in λ from 1.2 to 0.7, sensor resistance first increased from 10 to 10^4 Ω, reached a saturation value. Upon the change in λ from 0.7 to 1.2, sensor resistance again decreased from 10^4 Ω to the original value, i.e. 10 Ω. Thus, the proposed sensor was a potential candidate to function as a lambda sensor in real-time applications [117].

Francioso et al. developed TiO_2 thin film through the sol–gel route and tested its potential for lambda measurements in real-time applications [113]. Initially, the variation of sensor resistance at 650°C to different nitrogen/oxygen concentrations corresponding to different λ values was measured. In the next step, the dynamic response of the lambda sensor was also measured for the mixture of nitrogen (N_2), oxygen (O_2), carbon dioxide (CO_2), nitrogen oxide (NO), and methane (CH_4) for different λ values. In both cases, namely, under exposure to nitrogen/oxygen mixture and mixture of said gases, the change in sensor resistance was almost similar. Therefore, this work proved the potential of TiO_2 thin film as a lambda sensor [113]. However, the repeatability and stability of the sensor needed improvement owing to the instability of gold electrodes.

In successive attempts, Francioso et al. deposited sol–gel TiO_2 thin film with Pt electrodes [115]. Experiments were performed at varying temperatures (400–700°C) in the presence of 0.1% of O_2. Since maximum gas response determined by the ratio of electric current in the presence of O_2 to that of N_2 (I_{O2}/I_{N2}) was realized at 650°C, the sensor was tested at different O_2 concentrations in the range of 0.2–0.5%. Sensing tests were carried out for other gases, namely, CH_4, CO_2, oxides of nitrogen (NO_x) to ascertain the suitability of the sensor. Sensor performance was finally compared to commercial lambda sensors that showed close agreement between sensing signals [115].

3. Conclusions

This chapter demonstrated the potential of titanium dioxide (TiO_2) in imparting UV protection, anti-wear, corrosion inhibitor, gas detection properties. In the mechanical sector, TiO_2 can be used as a corrosion and wear-resistant material. TiO_2 coatings protect the substrate from the corrosive medium by serving as a ceramic barrier and also provide cathodic protection to the metals under UV illumination because of the photo-electrochemical property of TiO_2. A tremendous increment in the corrosion resistance of the sample with the application of TiO_2 coatings proved the potential of TiO_2 as a corrosion inhibitor. TiO_2, owing to its high hardness, fracture toughness can be embedded in a composite to improve the tribological performance of functional layers in numerous applications. TiO_2 is also found to be a suitable candidate for bone implants. The bioactivity tests revealed that the acid treatment of TiO_2 enhances the cell attachment and the bonding of the implant with the bone tissues. Owing to the high refractive index, TiO_2 layers are applied in sunscreens to protect the human skin from harmful UV rays. TiO_2 is also incorporated in textiles to reduce its photochemical degradation and therefore its mechanical properties are retained even after exposure to UV light. The whiteness, brightness, and hiding power of TiO_2 pigments are utilized in paints, coatings, papers, textile industries, etc. By adding certain minerals the degree of dispersion of TiO_2 pigments can be improved that further enhances the pigment properties of TiO_2. The change in the electrical resistance of TiO_2 layers was exploited to develop a gas sensor for air quality monitoring and a lambda sensor for monitoring of air-fuel ratio of internal combustion engines. TiO_2 nanoparticles have been found to outperform their bulk counterparts in such applications, which can be attributed to their large surface area to volume ratio.

List of nomenclatures

TiO_2	Titanium oxide
NaCl	Sodium Chloride
Mg	Magnesium
$MgTiO_3$	Magnesium titanate
Ni-W	Nickel-Tungsten
ZrO_2	Zirconium oxide
Al_2O_3	Aluminum oxide/Alumina
Zn	Zinc
ZnO	Zinc Oxide
n-TiO_2	Nano titania
micro-TiO_2	Micron-sized titania
TiNi	Nitinol
HCl	Hydrochloric acid
H_2SO_4	Sulfuric acid
OH	Hydroxyl
Cu	Copper
Ti	Titanium
CuO	Copper oxide
TiO_2-PTFE	Titanium oxide-polytetrafluoroethylene
$CaCO_3$	Calcium carbonate
Cu-Sn-PTFE	Copper-tin-polytetrafluoroethylene
Cr_2O_3	Chromium oxide
Al-SiC	Aluminum-Silicon carbide
H_2	Hydrogen
CO	Carbon monoxide
NH_3	Ammonia
Nb	Niobium
N_2	Nitrogen
O_2	Oxygen
CO_2	Carbon dioxide
NO	Nitrogen oxide
CH_4	Methane
Pt	Platinum
NO_x	Oxides of Nitrogen
R_a	Electrical resistance of TiO_2 under the influence of air
R_g	Electrical resistance of TiO_2 under the influence of test gas
λ	Air-fuel equivalence ratio
I_{O2}	Electrical current of TiO_2 under the influence of oxygen
I_{N2}	Electrical current of TiO_2 under the influence of nitrogen
UV	Ultraviolet
UVA	Ultraviolet A
UVB	Ultraviolet B
UVC	Ultraviolet C
UPF	Ultraviolet Protection Factor
PACVD	Plasma Assisted Chemical Vapor Deposition
ALD	Atomic Layer Deposition
XRD	X-Ray Diffraction
GO	Graphene Oxide
EIS	Electrochemical Impedance Spectroscopy
HA	Hydroxyapatite
PDMAS	Poly-dimethylaminosiloxane

SEM	Scanning Electron Microscope
EDS	Energy-Dispersive X-Ray Spectroscopy
PEEK	Polyetheretherketone
PET	Polyester
SPF	Sun Protection Factor
NR	Natural Rubber
SBF	Simulated Body Fluid
PTFE	Polytetrafluoroethylene
TCP	Tricalcium Phosphate

Author details

Rajib Das, Vibhav Ambardekar* and Partha Pratim Bandyopadhyay
Department of Mechanical Engineering, Indian Institute of Technology, Kharagpur, India

*Address all correspondence to: vibhav6305@gmail.com

References

[1] P. Blanchart, Extraction, Properties and Applications of TiO_2, in: Ind. Chem. Oxides Emerg. Appl., John Wiley & Sons Ltd, Chichester, UK, 2018: pp. 255-309. https://doi.org/10.1002/9781119424079.ch6.

[2] M. Gardon, J.M. Guilemany, Milestones in functional titanium dioxide thermal spray coatings: A review, J. Therm. Spray Technol. 23 (2014) 577-595. https://doi.org/10.1007/s11666-014-0066-5.

[3] M. Robotti, Functional surfaces obtained by thermal spray techniques, University of Barcelona, 2016.

[4] A.J. Haider, Z.N. Jameel, I.H.M. Al-Hussaini, Review on: Titanium Dioxide Applications, Energy Procedia. 157 (2019) 17-29. https://doi.org/10.1016/j.egypro.2018.11.159.

[5] U.R. Evans, Electrochemical Mechanism of Atmospheric Rusting, Nature. 206 (1965) 980-982.

[6] S. Kar, S. Paul, P.P. Bandyopadhyay, Processing and characterisation of plasma sprayed oxides: Microstructure, phases and residual stress, Surf. Coatings Technol. 304 (2016) 364-374. https://doi.org/10.1016/j.surfcoat.2016.07.043.

[7] P. Das, S. Paul, P.P. Bandyopadhyay, Preparation of diamond reinforced metal powders as thermal spray feedstock using ball milling, Surf. Coatings Technol. 289 (2016) 165-171. https://doi.org/10.1016/j.surfcoat.2015.12.022.

[8] A. Pilbáth, L. Nyikos, I. Bertóti, E. Kálmán, Zinc corrosion protection with 1,5-diphosphono-pentane, Corros. Sci. 50 (2008) 3314-3321. https://doi.org/10.1016/j.corsci.2008.09.021.

[9] T.-K. Yeh, Y.-C. Chien, B.-Y. Wang, C.-H. Tsai, Electrochemical characteristics of zirconium oxide treated Type 304 stainless steels of different surface oxide structures in high temperature water, Corros. Sci. 50 (2008) 2327-2337. https://doi.org/10.1016/j.corsci.2008.05.012.

[10] S. Datta, D.K. Pratihar, P.P. Bandyopadhyay, Modeling of input-output relationships for a plasma spray coating process using soft computing tools, Appl. Soft Comput. J. 12 (2012) 3356-3368. https://doi.org/10.1016/j.asoc.2012.07.015.

[11] P.P. Bandyopadhyay, S. Siegmann, Friction and wear behavior of vacuum plasma-sprayed Ti-Zr-Ni quasicrystal coatings, Surf. Coatings Technol. 197 (2005) 1-9. https://doi.org/10.1016/j.surfcoat.2004.11.009.

[12] M. Hadad, P.P. Bandyopadhyay, J. Michler, J. Lesage, Tribological behaviour of thermally sprayed Ti-Cr-Si coatings, Wear. 267 (2009) 1002-1008. https://doi.org/10.1016/j.wear.2009.01.013.

[13] H. Yun, J. Li, H.-B. Chen, C.-J. Lin, A study on the N-, S- and Cl-modified nano-TiO_2 coatings for corrosion protection of stainless steel, Electrochim. Acta. 52 (2007) 6679-6685. https://doi.org/10.1016/j.electacta.2007.04.078.

[14] P. Furman, J. Gluszek, J. Masalski, Titanium dioxide film obtained using the MOCVD method on 316L steel, J. Mater. Sci. Lett. 16 (1997) 471-472. https://doi.org/10.1023/A:1018564326769.

[15] S. Ghosh, S. Das, T.K. Bandyopadhyay, P.P. Bandyopadhyay, A.B. Chattopadhyay, Indentation responses of plasma sprayed ceramic coatings, J. Mater. Sci. 38 (2003) 1565-1572. https://doi.org/10.1023/A:1022997203996.

[16] P.P. Bandyopadhyay, Processing and Characterisation of Plasma Sprayed Ceramic Coatings on Steel Substrate, Dr. Diss. (2000).

[17] D. Yan, J. He, X. Li, Y. Liu, J. Zhang, H. Ding, An investigation of the corrosion behavior of Al_2O_3-based ceramic composite coatings in dilute HCI solution, Surf. Coatings Technol. 141 (2001) 1-6. https://doi.org/10.1016/S0257-8972(01)01170-7.

[18] Y. Ohko, S. Saitoh, T. Tatsuma, A. Fujishima, Photoelectrochemical Anticorrosion and Self-Cleaning Effects of a TiO_2 Coating for Type 304 Stainless Steel, J. Electrochem. Soc. 148 (2001) B24. https://doi.org/10.1149/1.1339030.

[19] G.X. Shen, Y.C. Chen, C.J. Lin, Corrosion protection of 316 L stainless steel by a TiO_2 nanoparticle coating prepared by sol-gel method, Thin Solid Films. 489 (2005) 130-136. https://doi.org/10.1016/j.tsf.2005.05.016.

[20] Y. Sun, Thermally oxidised titanium coating on aluminium alloy for enhanced corrosion resistance, Mater. Lett. 58 (2004) 2635-2639. https://doi.org/10.1016/j.matlet.2004.04.001.

[21] M.G. Mahmoud, R. Wang, M. Kato, K. Nakasa, Influence of ultraviolet light irradiation on corrosion behavior of weathering steel with and without TiO_2-coating in 3mass% NaCl solution, Scr. Mater. 53 (2005) 1303-1308. https://doi.org/10.1016/j.scriptamat.2005.07.039.

[22] C.X. Shan, X. Hou, K.L. Choy, Corrosion resistance of TiO_2 films grown on stainless steel by atomic layer deposition, Surf. Coatings Technol. 202 (2008) 2399-2402. https://doi.org/10.1016/j.surfcoat.2007.08.066.

[23] Z. Yu, H. Di, Y. Ma, Y. He, L. Liang, L. Lv, X. Ran, Y. Pan, Z. Luo, Preparation of graphene oxide modified by titanium dioxide to enhance the anti-corrosion performance of epoxy coatings, Surf. Coatings Technol. 276 (2015) 471-478. https://doi.org/10.1016/j.surfcoat.2015.06.027.

[24] S.Z. Khalajabadi, N. Ahmad, A. Yahya, M.A.M. Yajid, A. Samavati, S. Asadi, A. Arafat, M.R. Abdul Kadir, The role of TiO_2 on the microstructure, biocorrosion and mechanical properties of Mg/HA-based nanocomposites for potential application in bone repair, Ceram. Int. 42 (2016) 18223-18237. https://doi.org/10.1016/j.ceramint.2016.08.146.

[25] S. Shajahan, A. Basu, Corrosion, oxidation and wear study of electro-co-deposited ZrO_2-TiO_2 reinforced Ni-W coatings, Surf. Coatings Technol. 393 (2020). https://doi.org/10.1016/j.surfcoat.2020.125729.

[26] M. Poorraeisi, A. Afshar, The study of electrodeposition of hydroxyapatite-ZrO_2-TiO_2 nanocomposite coatings on 316 stainless steel, Surf. Coatings Technol. 339 (2018) 199-207. https://doi.org/10.1016/j.surfcoat.2018.02.030.

[27] A.M. Fadl, M.I. Abdou, M.A. Hamza, S.A. Sadeek, Corrosion-inhibiting, self-healing, mechanical-resistant, chemically and UV stable PDMAS/TiO_2 epoxy hybrid nanocomposite coating for steel petroleum tanker trucks, Prog. Org. Coatings. 146 (2020) 105715. https://doi.org/10.1016/j.porgcoat.2020.105715.

[28] N.G. Krishna, R.P. George, J. Philip, Anomalous enhancement of corrosion resistance and antibacterial property of commercially pure Titanium (CP-Ti) with nanoscale rutile TiO_2 film, Corros. Sci. 172 (2020) 108678. https://doi.org/10.1016/j.corsci.2020.108678.

[29] M. Merisalu, L. Aarik, J. Kozlova, H. Mändar, A. Tarre, V. Sammelselg, Effective corrosion protection of aluminum alloy AA2024-T3 with novel thin nanostructured oxide coating, Surf.

Coatings Technol. 411 (2021). https://doi.org/10.1016/j.surfcoat.2021.126993.

[30] L. Paussa, L. Guzman, E. Marin, N. Isomaki, L. Fedrizzi, Protection of silver surfaces against tarnishing by means of alumina/TiO_2-nanolayers, Surf. Coatings Technol. 206 (2011) 976-980. https://doi.org/10.1016/j.surfcoat.2011.03.101.

[31] E. Marin, L. Guzman, A. Lanzutti, W. Ensinger, L. Fedrizzi, Multilayer Al_2O_3/TiO_2 Atomic Layer Deposition coatings for the corrosion protection of stainless steel, Thin Solid Films. 522 (2012) 283-288. https://doi.org/10.1016/j.tsf.2012.08.023.

[32] J. Li, J. Li, Q. Li, H. Zhou, G. Wang, X. Peng, W. Jin, Z. Yu, P.K. Chu, W. Li, TiO_2-zinc phosphate/nanocrystalline zinc composite coatings for corrosion protection of biomedical WE43 magnesium alloy, Surf. Coatings Technol. 410 (2021). https://doi.org/10.1016/j.surfcoat.2021.126940.

[33] Q. Chen, G.D. McEwen, N. Zaveri, R. Karpagavalli, A. Zhou, Corrosion Resistance of Ti6Al4V with Nanostructured TiO_2 Coatings, First Edit, Elsevier Inc., 2012. https://doi.org/10.1016/B978-1-4557-7862-1.00009-2.

[34] T.V. Torbati, V. Javanbakht, Fabrication of TiO_2/Zn_2TiO_4/Ag nanocomposite for synergic effects of UV radiation protection and antibacterial activity in sunscreen, Colloids Surfaces B Biointerfaces. 187 (2020) 110652. https://doi.org/10.1016/j.colsurfb.2019.110652.

[35] N. Sabzevari, S. Qiblawi, S.A. Norton, D. Fivenson, Sunscreens: UV filters to protect us: Part 1: Changing regulations and choices for optimal sun protection, Int. J. Women's Dermatology. 7 (2021) 28-44. https://doi.org/10.1016/j.ijwd.2020.05.017.

[36] H. Yang, S. Zhu, N. Pan, Studying the mechanisms of titanium dioxide as ultraviolet-blocking additive for films and fabrics by an improved scheme, J. Appl. Polym. Sci. 92 (2004) 3201-3210. https://doi.org/10.1002/app.20327.

[37] Y. Li, W. Xie, X. Hu, G. Shen, X. Zhou, Y. Xiang, X. Zhao, P. Fang, Comparison of Dye Photodegradation and its Coupling with Light-to-Electricity Conversion over TiO_2 and ZnO, Langmuir. 26 (2010) 591-597. https://doi.org/10.1021/la902117c.

[38] X. Wang, Z. Li, J. Shi, Y. Yu, One-Dimensional Titanium Dioxide Nanomaterials: Nanowires, Nanorods, and Nanobelts, Chem. Rev. 114 (2014) 9346-9384. https://doi.org/10.1021/cr400633s.

[39] A. Mishra, B.S. Butola, Deposition of Ag doped TiO_2 on cotton fabric for wash durable UV protective and antibacterial properties at very low silver concentration, Cellulose. 24 (2017) 3555-3571. https://doi.org/10.1007/s10570-017-1352-4.

[40] D. Cheng, M. He, J. Ran, G. Cai, J. Wu, X. Wang, In situ reduction of TiO_2 nanoparticles on cotton fabrics through polydopamine templates for photocatalysis and UV protection, Cellulose. 25 (2018) 1413-1424. https://doi.org/10.1007/s10570-017-1606-1.

[41] S. Li, T. Zhu, J. Huang, Q. Guo, G. Chen, Y. Lai, Durable antibacterial and UV-protective Ag/TiO_2@fabrics for sustainable biomedical application, Int. J. Nanomedicine. Volume 12 (2017) 2593-2606. https://doi.org/10.2147/IJN.S132035.

[42] Y. Liang, E. Pakdel, M. Zhang, L. Sun, X. Wang, Photoprotective properties of alpaca fiber melanin reinforced by rutile TiO_2 nanoparticles: A study on wool fabric, Polym. Degrad. Stab. 160 (2019) 80-88. https://doi.org/10.1016/j.polymdegradstab.2018.12.006.

[43] R.D. Kale, T. Potdar, P. Kane, R. Singh, Nanocomposite polyester fabric based on graphene/titanium dioxide for conducting and UV protection functionality, Graphene Technol. 3 (2018) 35-46. https://doi.org/10.1007/s41127-018-0021-1.

[44] M. Gorjanc, M. Šala, Durable antibacterial and UV protective properties of cellulose fabric functionalized with Ag/TiO_2 nanocomposite during dyeing with reactive dyes, Cellulose. 23 (2016) 2199-2209. https://doi.org/10.1007/s10570-016-0945-7.

[45] M. Radeti, Journal of Photochemistry and Photobiology C : Photochemistry Reviews Functionalization of textile materials with TiO_2 nanoparticles Maja Radeti c, 16 (2013) 62-76.

[46] M. Bragaglia, V. Cherubini, F. Nanni, PEEK -TiO_2 composites with enhanced UV resistance, Compos. Sci. Technol. 199 (2020) 108365. https://doi.org/10.1016/j.compscitech.2020.108365.

[47] C. Li, Z. Li, X. Ren, Preparation and characterization of polyester fabrics coated with TiO_2/Benzotriazole for UV protection, Colloids Surfaces A Physicochem. Eng. Asp. 577 (2019) 695-701. https://doi.org/10.1016/j.colsurfa.2019.06.030.

[48] B. Seentrakoon, B. Junhasavasdikul, W. Chavasiri, Enhanced UV-protection and antibacterial properties of natural rubber/rutile-TiO_2 nanocomposites, Polym. Degrad. Stab. 98 (2013) 566-578. https://doi.org/10.1016/j.polymdegradstab. 2012.11.018.

[49] J.J. Reinosa, C.M.Á. Docio, V.Z. Ramírez, J.F.F. Lozano, Hierarchical nano ZnO-micro TiO_2 composites: High UV protection yield lowering photodegradation in sunscreens, Ceram. Int. 44 (2018) 2827-2834. https://doi.org/10.1016/j.ceramint.2017.11.028.

[50] H. Sun, J. Weickert, H.C. Hesse, L. Schmidt-Mende, UV light protection through TiO_2 blocking layers for inverted organic solar cells, Sol. Energy Mater. Sol. Cells. 95 (2011) 3450-3454. https://doi.org/10.1016/j.solmat.2011.08.004.

[51] N.K. Awad, S.L. Edwards, Y.S. Morsi, A review of TiO_2 NTs on Ti metal: Electrochemical synthesis, functionalization and potential use as bone implants, Mater. Sci. Eng. C. 76 (2017) 1401-1412. https://doi.org/10.1016/j.msec.2017.02.150.

[52] V. Raj, M.S. Mumjitha, Fabrication of biopolymers reinforced TNT/HA coatings on Ti: Evaluation of its Corrosion resistance and Biocompatibility, Electrochim. Acta. 153 (2015) 1-11. https://doi.org/10.1016/j.electacta.2014.10.055.

[53] A. Balamurugan, S. Kannan, S. Rajeswari, Evaluation of TiO_2 coatings obtained using the sol–gel technique on surgical grade type 316L stainless steel in simulated body fluid, Mater. Lett. 59 (2005) 3138-3143. https://doi.org/10.1016/j.matlet.2005.05.036.

[54] E. Santos, N.K. Kuromoto, G.A. Soares, Mechanical properties of TiO_2 films used as biomaterials, Mater. Chem. Phys. 102 (2007) 92-97. https://doi.org/10.1016/j.matchemphys. 2006.11.010.

[55] M.A. Garcia-Lobato, A.I. Mtz-Enriquez, C.R. Garcia, M. Velazquez-Manzanares, F. Avalos-Belmontes, R. Ramos-Gonzalez, L.A. Garcia-Cerda, Corrosion resistance and in vitro bioactivity of dense and porous TiO_2 coatings deposited on 316L SS by spraying method, Appl. Surf. Sci. 484 (2019) 975-980. https://doi.org/10.1016/j.apsusc.2019.04.108.

[56] X. Zhao, X. Liu, J. You, Z. Chen, C. Ding, Bioactivity and cytocompatibility of plasma-sprayed TiO_2 coating treated

by sulfuric acid treatment, Surf. Coatings Technol. 202 (2008) 3221-3226. https://doi.org/10.1016/j.surfcoat.2007.11.026.

[57] T. Peltola, M. Jokinen, H. Rahiala, M. Patsi, J. Heikkila, I. Kangasniemi, A. Yli-Urpo, Effect of aging time of sol on structure andin vitro calcium phosphate formation of sol-gel-derived TiO_2 films, J. Biomed. Mater. Res. 51 (2000) 200-208. https://doi.org/10.1002/(SICI)1097-4636(200008)51:2<200::AID-JBM8>3.0.CO;2-Z.

[58] N. Matsumoto, K. Sato, K. Yoshida, K. Hashimoto, Y. Toda, Preparation and characterization of β-tricalcium phosphate co-doped with monovalent and divalent antibacterial metal ions, Acta Biomater. 5 (2009) 3157-3164. https://doi.org/10.1016/j.actbio.2009.04.010.

[59] X. He, G. Zhang, X. Wang, R. Hang, X. Huang, L. Qin, B. Tang, X. Zhang, Biocompatibility, corrosion resistance and antibacterial activity of TiO_2/CuO coating on titanium, Ceram. Int. 43 (2017) 16185-16195. https://doi.org/10.1016/j.ceramint.2017.08.196.

[60] C.-M. Han, H.-E. Kim, Y.-H. Koh, Creation of hierarchical micro/nano-porous TiO_2 surface layer onto Ti implants for improved biocompatibility, Surf. Coatings Technol. 251 (2014) 226-231. https://doi.org/10.1016/j.surfcoat.2014.04.030.

[61] A.K.M. Kafi, G. Wu, A. Chen, A novel hydrogen peroxide biosensor based on the immobilization of horseradish peroxidase onto Au-modified titanium dioxide nanotube arrays, Biosens. Bioelectron. 24 (2008) 566-571. https://doi.org/10.1016/j.bios.2008.06.004.

[62] W.F. Oliveira, I.R.S. Arruda, G.M.M. Silva, G. Machado, L.C.B.B. Coelho, M.T.S. Correia, Functionalization of titanium dioxide nanotubes with biomolecules for biomedical applications, Mater. Sci. Eng. C. 81 (2017) 597-606. https://doi.org/10.1016/j.msec.2017.08.017.

[63] Z. Zhang, Y. Xie, Z. Liu, F. Rong, Y. Wang, D. Fu, Covalently immobilized biosensor based on gold nanoparticles modified TiO_2 nanotube arrays, J. Electroanal. Chem. 650 (2011) 241-247. https://doi.org/10.1016/j.jelechem.2010.10.016.

[64] Q. Zhao, S. Tang, C. Fang, Y.-F. Tu, TiO_2 nanotubes decorated with gold nanoparticles for electrochemiluminescent biosensing of glycosylated hemoglobin, Anal. Chim. Acta. 936 (2016) 83-90. https://doi.org/10.1016/j.aca.2016.07.015.

[65] W. Zhang, J. Gu, C. Zhang, Y. Xie, X. Zheng, Preparation of TiO_2 coating by induction suspension plasma spraying for biomedical application, Surf. Coatings Technol. 358 (2019) 511-520. https://doi.org/10.1016/j.surfcoat.2018.11.047.

[66] S. Zhang, X. Liang, G.M. Gadd, Q. Zhao, Advanced titanium dioxide-polytetrafluorethylene (TiO_2-PTFE) nanocomposite coatings on stainless steel surfaces with antibacterial and anti-corrosion properties, Appl. Surf. Sci. 490 (2019) 231-241. https://doi.org/10.1016/j.apsusc.2019.06.070.

[67] Z. Lu, M. Ren, H. Yin, A. Wang, C. Ge, Y. Zhang, L. Yu, T. Jiang, Preparation of nanosized anatase TiO_2-coated kaolin composites and their pigmentary properties, Powder Technol. 196 (2009) 122-125. https://doi.org/10.1016/j.powtec.2009.07.006.

[68] I. Ali, M. Suhail, Z.A. Alothman, A. Alwarthan, Recent advances in syntheses, properties and applications of TiO_2 nanostructures, RSC Adv. 8 (2018) 30125-30147. https://doi.org/10.1039/c8ra06517a.

[69] Oil and Colour Chemists' Association Australia, Titanium Dioxide Pigments, in: Surf. Coatings, Springer Netherlands, Dordrecht, 1983: pp. 305-312. https://doi.org/10.1007/978-94-011-6940-0_26.

[70] Y. Liang, H. Ding, Mineral-TiO_2 composites:Preparation and application in papermaking, paints and plastics, J. Alloys Compd. 844 (2020) 156139. https://doi.org/10.1016/j.jallcom.2020.156139.

[71] H. Zhou, M. Wang, H. Ding, G. Du, Preparation and Characterization of Barite/TiO_2 Composite Particles, Adv. Mater. Sci. Eng. 2015 (2015) 1-8. https://doi.org/10.1155/2015/878594.

[72] Y. Chen, H. Yu, L. Yi, Y. Liu, L. Cao, K. Cao, Y. Liu, W. Zhao, T. Qi, Preparation of ground calcium carbonate-based TiO_2 pigment by a two-step coating method, Powder Technol. 325 (2018) 568-575. https://doi.org/10.1016/j.powtec.2017.11.040.

[73] X. Zhao, J. Li, Y. Liu, Y. Zhang, J. Qu, T. Qi, Preparation and mechanism of TiO_2-coated illite composite pigments, Dye. Pigment. 108 (2014) 84-92. https://doi.org/10.1016/j.dyepig.2014.04.022.

[74] X.F. Hou, H. Ding, Y.X. Zheng, M.M. Wang, Preparation and characterisation of amorphous silica/anatase composite through mechanochemical method, Mater. Res. Innov. 17 (2013) 234-239. https://doi.org/10.1179/1432891713Z.000000000222.

[75] S. Sun, H. Ding, H. Zhou, Preparation of TiO_2-coated barite composite pigments by the hydrophobic aggregation method and their structure and properties, Sci. Rep. 7 (2017) 10083. https://doi.org/10.1038/s41598-017-10620-7.

[76] X. Hou, Y. Zhang, H. Ding, P.K. Chu, Environmentally friendly wollastonite@TiO_2 composite particles prepared by a mechano-chemical method, Particuology. 40 (2018) 105-112. https://doi.org/10.1016/j.partic.2017.10.010.

[77] S. Hazra, P.P. Bandyopadhyay, Scratch induced failure of plasma sprayed alumina based coatings, Mater. Des. 35 (2012) 243-250. https://doi.org/10.1016/j.matdes.2011.09.014.

[78] S. Kar, P.P. Bandyopadhyay, S. Paul, Precision superabrasive grinding of plasma sprayed ceramic coatings, Ceram. Int. 42 (2016) 19302-19319. https://doi.org/10.1016/j.ceramint.2016.09.100.

[79] P.P. Bandyopadhyay, S. Das, S. Madhusudan, A.B. Chattopadhyay, Wear and thermal fatigue characteristics of plasma-sprayed alumina coatings, J. Mater. Sci. Lett. 18 (1999) 727-729. https://doi.org/10.1023/A:1006608715370.

[80] D. Mohanty, S. Kar, S. Paul, P.P. Bandyopadhyay, Carbon nanotube reinforced HVOF sprayed WC-Co coating, Mater. Des. 156 (2018) 340-350. https://doi.org/10.1016/j.matdes.2018.06.054.

[81] D.S.R. Krishna, Y. Sun, Thermally oxidised rutile-TiO_2 coating on stainless steel for tribological properties and corrosion resistance enhancement, Appl. Surf. Sci. 252 (2005) 1107-1116. https://doi.org/10.1016/j.apsusc.2005.02.046.

[82] R.S. Lima, B.R. Marple, Thermal spray coatings engineered from nanostructured ceramic agglomerated powders for structural, thermal barrier and biomedical applications: A review, J. Therm. Spray Technol. 16 (2007) 40-63. https://doi.org/10.1007/s11666-006-9010-7.

[83] M. Woydt, A. Skopp, I. Dörfel, K. Witke, Wear engineering oxides/

antiwear oxides, Wear. 218 (1998) 84-95. https://doi.org/10.1080/10402009908982185.

[84] Y. Sun, Tribological rutile-TiO_2 coating on aluminium alloy, Appl. Surf. Sci. 233 (2004) 328-335. https://doi.org/10.1016/j.apsusc.2004.03.241.

[85] H. Dong, T. Bell, Enhanced wear resistance of titanium surfaces by a new thermal oxidation treatment, Wear. 238 (2000) 131-137. https://doi.org/10.1016/S0043-1648(99)00359-2.

[86] N. Dejang, A. Watcharapasorn, S. Wirojupatump, P. Niranatlumpong, S. Jiansirisomboon, Fabrication and properties of plasma-sprayed Al_2O_3/TiO_2 composite coatings: A role of nano-sized TiO_2 addition, Surf. Coatings Technol. 204 (2010) 1651-1657. https://doi.org/10.1016/j.surfcoat.2009.10.052.

[87] P. Baghery, M. Farzam, A.B. Mousavi, M. Hosseini, Ni-TiO_2 nanocomposite coating with high resistance to corrosion and wear, Surf. Coatings Technol. 204 (2010) 3804-3810. https://doi.org/10.1016/j.surfcoat.2010.04.061.

[88] L.X. Ying, K. Wu, D.Y. Li, C.X. Wu, Z. Fu, TiO_2 Sol strengthened Cu-Sn-PTFE composite coatings with high homogeneity and superior resistance to wear, Wear. 426-427 (2019) 258-264. https://doi.org/10.1016/j.wear.2019.01.031.

[89] N. nan Li, G. lu Li, H. dou Wang, J. jie Kang, T. shun Dong, H. jun Wang, Influence of TiO_2 content on the mechanical and tribological properties of Cr_2O_3-based coating, Mater. Des. 88 (2015) 906-914. https://doi.org/10.1016/j.matdes.2015.09.085.

[90] P. Dinesh Babu, B. Prasannakumar, P. Marimuthu, R.K. Mishra, T. Ram Prabhu, Microstructure, wear and mechanical properties of plasma sprayed TiO_2 coating on Al–SiC metal matrix composite, Arch. Civ. Mech. Eng. 19 (2019) 756-767. https://doi.org/10.1016/j.acme.2019.03.001.

[91] O. Çomaklı, M. Yazıcı, H. Kovacı, T. Yetim, A.F. Yetim, A. Çelik, Tribological and electrochemical properties of TiO_2 films produced on Cp-Ti by sol-gel and SILAR in bio-simulated environment, Surf. Coatings Technol. 352 (2018) 513-521. https://doi.org/10.1016/j.surfcoat.2018.08.056.

[92] P.A. Dearnley, K.L. Dahm, H. Çimenoğlu, The corrosion-wear behaviour of thermally oxidised CP-Ti and Ti-6Al-4V, Wear. 256 (2004) 469-479. https://doi.org/10.1016/S0043-1648(03)00557-X.

[93] R. Barkallah, R. Taktak, N. Guermazi, K. Elleuch, J. bouaziz, Mechanical properties and wear behaviour of alumina/tricalcium phosphate/TiO2 ceramics as coating for orthopedic implant, Eng. Fract. Mech. 241 (2021) 107399. https://doi.org/10.1016/j.engfracmech.2020.107399.

[94] Z. Li, Z.J. Yao, A.A. Haidry, T. Plecenik, L.J. Xie, L.C. Sun, Q. Fatima, Resistive-type hydrogen gas sensor based on TiO_2: A review, Int. J. Hydrogen Energy. 43 (2018) 21114-21132. https://doi.org/10.1016/j.ijhydene.2018.09.051.

[95] Y.K. Jun, H.S. Kim, J.H. Lee, S.H. Hong, CO sensing performance in micro-arc oxidized TiO_2 films for air quality control, Sensors Actuators, B Chem. 120 (2006) 69-73. https://doi.org/10.1016/j.snb.2006.01.045.

[96] B. Karunagaran, P. Uthirakumar, S.J. Chung, S. Velumani, E.K. Suh, TiO_2 thin film gas sensor for monitoring ammonia, Mater. Charact. 58 (2007) 680-684. https://doi.org/10.1016/j.matchar.2006.11.007.

[97] N. Yamazoe, G. Sakai, K. Shimanoe, Oxide semiconductor gas sensors, Catal.

Surv. from Asia. 7 (2003) 63-75. https://doi.org/10.1023/A:1023436725457.

[98] V. Aroutiounian, Metal oxide hydrogen, oxygen, and carbon monoxide sensors for hydrogen setups and cells, Int. J. Hydrogen Energy. 32 (2007) 1145-1158. https://doi.org/10.1016/j.ijhydene.2007.01.004.

[99] H. Tang, K. Prasad, R. Sanjinés, F. Lévy, TiO_2 anatase thin films as gas sensors, "Sensors Actuators, B Chem. 26 (1995) 71-75. https://doi.org/10.1016/0925-4005(94)01559-Z.

[100] G.S. Devi, T. Hyodo, Y. Shimizu, M. Egashira, Synthesis of mesoporous TiO_2-based powders and their gas-sensing properties, Sensors Actuators, B Chem. 87 (2002) 122-129. https://doi.org/10.1016/S0925-4005(02)00228-9.

[101] S. Yoo, S.A. Akbar, K.H. Sandhage, Nanocarving of TiO_2 (TiO_2): A novel approach for fabricating chemical sensing platform, Ceram. Int. 30 (2004) 1121-1126. https://doi.org/10.1016/j.ceramint.2003.12.085.

[102] Y.K. Jun, H.S. Kim, J.H. Lee, S.H. Hong, High H_2 sensing behavior of TiO_2 films formed by thermal oxidation, Sensors Actuators, B Chem. 107 (2005) 264-270. https://doi.org/10.1016/j.snb.2004.10.010.

[103] J. Moon, M. Kemell, J. Kukkola, R. Punkkinen, H.P. Hedman, A. Suominen, E. Mäkilä, M. Tenho, A. Tuominen, H. Kim, Gas sensor using anodic TiO_2 thin film for monitoring hydrogen, Procedia Eng. 47 (2012) 791-794. https://doi.org/10.1016/j.proeng.2012.09.266.

[104] J. Moon, H.P. Hedman, M. Kemell, A. Suominen, E. Mäkilä, H. Kim, A. Tuominen, R. Punkkinen, A study of monitoring hydrogen using mesoporous TiO_2 synthesized by anodization, Sensors Actuators, B Chem. 189 (2013) 246-250. https://doi.org/10.1016/j.snb.2013.05.070.

[105] H. Liu, D. Ding, C. Ning, Z. Li, Wide-range hydrogen sensing with Nb-doped TiO_2 nanotubes, Nanotechnology. 23 (2012) 0-6. https://doi.org/10.1088/0957-4484/23/1/015502.

[106] D. Gorman, A. Drewry, Y.L. Huang, C. Sames, The clinical toxicology of carbon monoxide, Toxicology. 187 (2003) 25-38. https://doi.org/10.1016/S0300-483X(03)00005-2.

[107] J.A. Raub, M. Mathieu-Nolf, N.B. Hampson, S.R. Thom, Carbon monoxide poisoning - A public health perspective, Toxicology. 145 (2000) 1-14. https://doi.org/10.1016/S0300-483X(99)00217-6.

[108] Y.J. Choi, Z. Seeley, A. Bandyopadhyay, S. Bose, S.A. Akbar, Aluminum-doped TiO_2 nano-powders for gas sensors, Sensors Actuators, B Chem. 124 (2007) 111-117. https://doi.org/10.1016/j.snb.2006.12.005.

[109] N. Van Toan, C.M. Hung, N. Van Duy, N.D. Hoa, D.T.T. Le, N. Van Hieu, Bilayer SnO_2–WO_3 nanofilms for enhanced NH_3 gas sensing performance, Mater. Sci. Eng. B Solid-State Mater. Adv. Technol. 224 (2017) 163-170. https://doi.org/10.1016/j.mseb.2017.08.004.

[110] M. Gardon, O. Monereo, S. Dosta, G. Vescio, A. Cirera, J.M. Guilemany, New procedures for building-up the active layer of gas sensors on flexible polymers, Surf. Coatings Technol. 235 (2013) 848-852. https://doi.org/10.1016/j.surfcoat.2013.09.011.

[111] P. Dhivya, A.K. Prasad, M. Sridharan, Nanostructured TiO_2 films: Enhanced NH3 detection at room temperature, Ceram. Int. 40 (2014) 409-415. https://doi.org/10.1016/j.ceramint.2013.06.016.

[112] U. Kirner, K.D. Schierbaum, W. Göpel, B. Leibold, N. Nicoloso, W.

Weppner, D. Fischer, W.F. Chu, Low and high temperature TiO_2 oxygen sensors, Sensors Actuators B. Chem. 1 (1990) 103-107. https://doi.org/10.1016/0925-4005(90)80181-X.

[113] L. Francioso, D.S. Presicce, A.M. Taurino, R. Rella, P. Siciliano, A. Ficarella, Automotive application of sol-gel TiO_2 thin film-based sensor for lambda measurement, Sensors Actuators, B Chem. 95 (2003) 66-72. https://doi.org/10.1016/S0925-4005(03)00405-2.

[114] K. Moriya, T. Sako, Oxygen sensor monitoring a deterioration of a three-way catalyst in natural gas fueled engines, Sensors Actuators, B Chem. 73 (2001) 142-151. https://doi.org/10.1016/S0925-4005(00)00683-3.

[115] L. Francioso, D.S. Presicce, M. Epifani, P. Siciliano, A. Ficarella, Response evaluation of TiO_2 sensor to flue gas on spark ignition engine and in controlled environment, Sensors Actuators, B Chem. 107 (2005) 563-571. https://doi.org/10.1016/j.snb.2004.11.017.

[116] J.H. Lee, Review on zirconia air-fuel ratio sensors for automotive applications, J. Mater. Sci. 38 (2003) 4247-4257. https://doi.org/10.1023/A:1026366628297.

[117] K. Naomasa Sunano, Akashi Naotatsu, Asahi Katsuta, Toshio Yoshida, Gas Sensor And Method Of Producing The Same, 4713646, 1987.

Chapter 8

TiO_2 Based Nanomaterials and Their Application as Anode for Rechargeable Lithium-Ion Batteries

Nabil El Halya, Karim Elouardi, Abdelwahed Chari, Abdeslam El Bouari, Jones Alami and Mouad Dahbi

Abstract

Titanium dioxide- (TiO_2-) based nanomaterials have been widely adopted as active materials for photocatalysis, sensors, solar cells, and for energy storage and conversion devices, especially rechargeable lithium-ion batteries (LIBs), due to their excellent structural and cycling stability, high discharge voltage plateau (more than 1.7 V versus Li^+/Li), high safety, environmental friendliness, and low cost. However, due to their relatively low theoretical capacity and electrical conductivity, their use in practical applications, i.e. anode materials for LIBs, is limited. Several strategies have been developed to improve the conductivity, the capacity, the cycling stability, and the rate capability of TiO_2-based materials such as designing different nanostructures (1D, 2D, and 3D), Coating or combining TiO_2 with carbonaceous materials, and selective doping with mono and heteroatoms. This chapter is devoted to the development of a simple and cost-efficient strategies for the preparation of TiO_2 nanoparticles as anode material for lithium ion batteries (LIBs). These strategies consist of using the Sol–Gel method, with a sodium alginate biopolymer as a templating agent and studying the influence of calcination temperature and phosphorus doping on the structural, the morphological and the textural properties of TiO_2 material. Moreover, the synthetized materials were tested electrochemically as anode material for lithium ion battery. TiO_2 electrodes calcined at 300°C and 450°C have delivered a reversible capacity of 266 mAh g^{-1}, 275 mAh g^{-1} with coulombic efficiencies of 70%, 75% during the first cycle under C/10 current rate, respectively. Besides, the phosphorus doped TiO_2 electrodes were presented excellent lithium storage properties compared to the non-doped electrodes which can be attributed to the beneficial role of phosphorus doping to inhibit the growth of TiO_2 nanoparticles during the synthesis process and provide a high electronic conductivity.

Keywords: Lithium-ion Battery, TiO_2 based materials, Gelation of biopolymers, Sodium alginate, conductivity, Phosphorus doping

1. Introduction

In recent years, lithium-ion batteries (LIBs) have been established as efficient electrochemical energy storage devices and have become the best choice for electric vehicles (EVs) and mobile phones due to their long cycle life, low self-discharge rate, high working voltage, high power and energy density [1, 2]. Developing and using LIBs can significantly reduce pollution of combustion gas by replacing traditional transportation powered by gasoline with environmentally friendly electric vehicles. Following their success in the transport sector, batteries have recently been considered for grid applications, contributing this to the improvement of the energy efficiency of solar, wind, tidal and other clean energy technologies. LIBs are therefore considered to be an essential element in the building an energy-sustainable economy [3, 4].

Figure 1 present the working principle of LIB; both anodes and cathodes could possess a host structure for Li^+ ions to ensure a good insertion/ disinsertion of these ions during the charge and discharge. The electrolyte is the polypropylene/ polyethylene which contains lithium salts (i.e., $LiPF_6$) in alkyl organic carbonates. The separator, usually Celgard or Whatman, must allow the diffusion of Li^+ ions between the cathode and the anode during the charging and discharging process [5].

The development of large and efficient batteries operating at high potentials necessitates the use of elements that give low an anode intercalation potential. Today, Li^+ is considered to give the best performances and is therefore widely used. In addition to improving the electrochemical characteristics of anodes, researchers are also concerned with the cost and the environmental impact of the materials under development. In general, an ideal anode material must possess the following characteristics [6, 7]:

- High specific surface area offering more lithium insertion channels,

- Good cycling stability and low volume change during Li ion insertion/ desertion process.

- Large pore size for fast Li^+ ion diffusion and good rate capability,

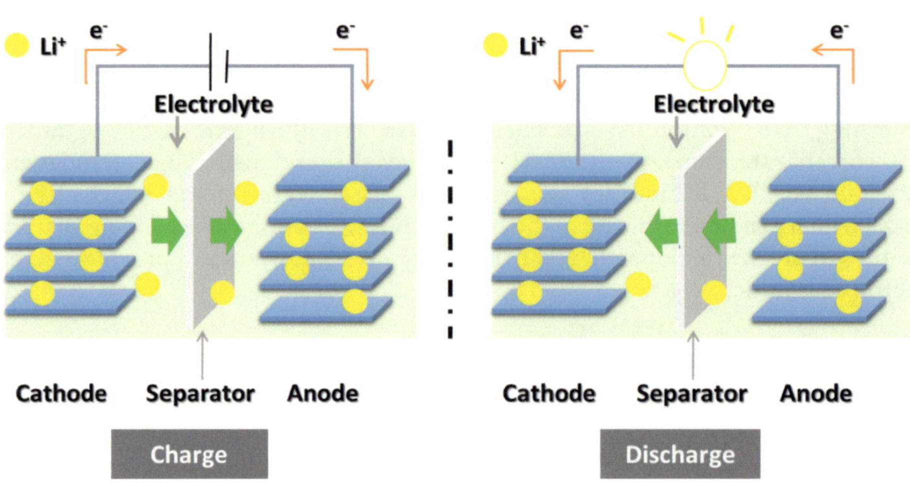

Figure 1.
Working principle of current rechargeable lithium-ion batteries.

- Low internal resistance which allows fast charging and discharging process,
- low lithium ions intercalation potential,
- low price and environmental friendliness.

Most commercial LIBs use transition metals oxides or phosphates such as $LiCoO_2$, $LiFePO_4$ and $LiMnPO_4$, as active materials for the cathode, while, the anode is typically made of graphite. Despite its wide commercial use, graphite suffers from a large volume variation during the charge/ discharge process, a low specific capacity, besides safety concerns. To overcome these concerns, TiO_2 is a promising alternative, as it possess excellent structural and cycling stability, high discharge voltage plateau (more than 1.7 V versus Li^+/Li), high safety, is environmentally friendly, and has a low cost [7, 8]. However, some of the limiting features of this material, including its low electrical conductivity, low capacity and poor rate capability need to be overcome. **Figure 2** shows the potential versus Li/Li^+ and the corresponding capacity density of some potential active anode materials and **Table 1** presents a brief comparison between TiO_2 and other anode active materials.

Several strategies have been developed to improve the capacity, the cycling stability, and the rate capability of TiO_2-based anodes, and are detailed in the next paragraphs.

1.1 Designing different nanostructured TiO_2

1.1.1 One-dimensional nanostructures (1D)

Nanostructured materials, such as nanotubes, nanowires, nanoneedles, nanofibers and nanorods have been designed for high performance anodes. The interesting performance of 1D TiO_2 was demonstrated by different groups; Tammawat and Meethong studied anatase TiO_2 nanofiber as an anode active material in LIBs,

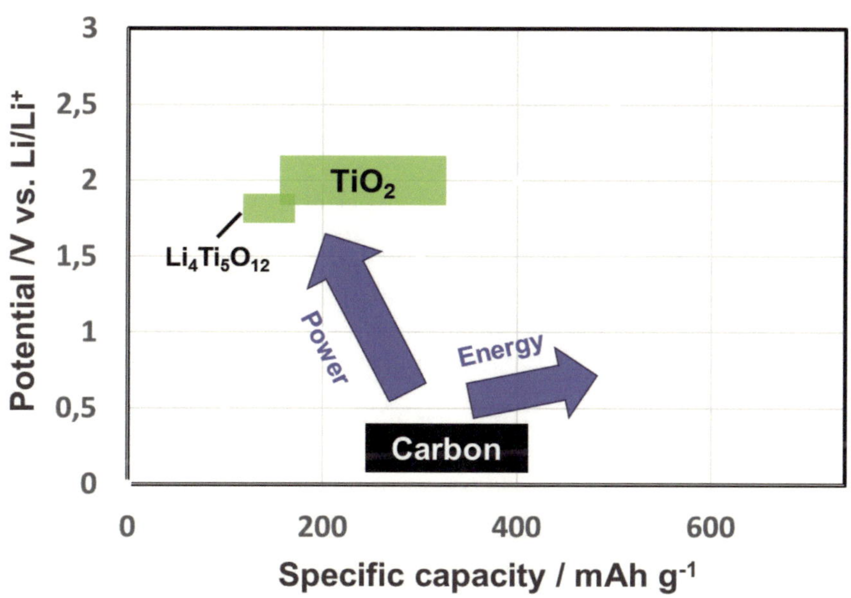

Figure 2.
Potential versus Li/Li^+ and the corresponding specific capacity of some potential active anode materials for lithium ion batteries.

Materials	Theoretical capacity (mAh/g)	Advantages	Drawbacks
Si	4200	High capacity	Poor cycling, large irreversible capacity
Metal oxides	500–1200	Low cost, High capacity	Low electrical conductivity, low capacity retention and coulombic efficiency
Sn	990	Low cost, good electrical conductivity, good safety	Poor cycling
Graphite	372	Low cost, good working potential	High irreversible capacity, low coulombic efficiency
TiO_2	330	Low cost, environmentally friendly, good safety	Low capacity, poor rate capability, low electrical conductivity

Table 1.
Brief comparison between TiO_2 and other anode active materials [5, 9].

showing a high lithium storage capacity with a stable cycle life and a good rate capability [10]. The excellent performances of these nanostructures could be explained by the increased electronic conductivity, the small nanocrystalline size, the large surface area of the nanofibers and the large Li nonstoichiometric parameters. Another study by Armstrong *et al.* demonstrated that TiO_2 nanowires exhibit a high capacity of 305 mAh g^{-1}, which is much higher than the capacity value achieved by the bulk TiO_2 (240 mAh g^{-1}) [11]. These improved results are attributed to the large surface area of the prepared nanowires and the good electronic conductivity.

1.1.2 Two-Dimensional Structure (2D)

Compared with Zero-Dimensional (0D) nanoparticles and One-Dimensional (1D) nanostructures, the 2D nanomaterials can store Li^+ ion in both sides of the structure which offer more exposed surfaces, open charge transport channel for electrolyte penetration and short ion diffusion length [12, 13]. Moreover, the 2D structure is an excellent choice for fast and high lithium storage.

To fabricate 2D TiO_2 materials, significant efforts have been made by several researchers. Li *et al.* have used hydrothermal methods to synthetize mesoporous TiO_2 nanoflakes (10–20 nm) and evaluate their performance as anode. The electrochemical tests showed that the prepared nanoflakes had a good cycling life and a high discharge specific capacity of 261 mAh g^{-1} [14]. Another team demonstrated a simple and green synthesis route of anatase petal-like TiO_2 nanosheets. The obtained TiO_2 materials presented a suitable surface area of 28.4 m^2 g^{-1}, which was proposed to be the reason behind the high capacity and the good cycling stability [15].

1.1.3 Three-Dimensional Porous Structure (3D)

In recent years, 3D porous structure materials have attracted much attention, due to their high porosity, high specific surface area, and low bulk density [16, 17]. Gerbaldi et al. synthesized highly crystalline, nonordered mesoporous anatase TiO_2 with excellent rate capability and cycling stability after prolonged cycling [18, 19]. Lou et al. demonstrated a significantly improved lithium storage capability of TiO_2 hollow spheres and sub-microboxes, together with a high specific capacity, an excellent rate capability, and a long-term cycling stability [20, 21].

1.2 Coating or combining TiO_2 with carbonaceous materials

To improve the electrochemical performance of TiO_2 materials, carbon coating was used in order to reduce the charge transfer resistance, improve the Li^+, buffer the large volume changes during lithium insertion/extraction, enhance electron transport and prevent the aggregation of active materials [9–22, 23]. Xia *et al.* have reported a carbon-coated TiO_2 anode material with an enhanced electronic conductivity and a low volume expansion during prolonged cycling [24]. In a different study, chemical vapor deposition was used to synthesize TiO_2/CNTs composites, exhibiting a high specific capacity and a long-term cycling stability [25]. This study demonstrated that the enhanced electrochemical performance of this material is due to the structural stability and the efficient conductive network of the TiO_2 particles offered by CNTs. Etacheri *et al.* mixed TiO_2 with graphene and the synthetized hybrid materials exhibited a high specific capacity, an improved capacity retention and a good rate capability, in comparison with the physical mixture of TiO_2 and reduced graphene oxide [26].

1.3 Selective doping with mono and heteroatoms

To improve the intrinsic conductivity and form more open channels and active sites for Li^+ transport, doping with cationic or anionic dopants has been shown to be advantageous [27, 28]. Liu *et al.* evaluated the performance of Ti^{3+} doped TiO_2 nanotube arrays as anode material for LIBs showing an enhanced lithium ion storage performance with an initial discharge capacity of 101 mAh g^{-1} at a high current density of 10 A g^{-1} [29]. Furthermore, Sn-doped TiO_2 nanotubes were synthetized by Kyeremateng and coworkers delivering higher capacity values compared to non-doped TiO_2 nanotubes [30]. Otherwise, TiO_2 materials with improved specific capacities were synthesized, by other researchers, using B and N and doping elements [31, 32].

In the following chapter, simple and cost-efficient strategies for the preparation of TiO_2 nanoparticles as anode material for LIBs are discussed. These strategies consist of using the Sol–Gel method, with a sodium alginate biopolymer as a templating agent, and studying the influence of the calcination temperature and the phosphorus doping on the structural, the morphological, the textural and the electrochemical properties of TiO_2 material.

2. Impact of calcination temperature on TiO_2 as anode for rechargeable Lithium-ion batteries

Our group has reported the synthesis of anatase TiO_2 as an anode material of LIBs by a facile synthesis method using a biopolymer as a templating agent. In order to stress the effect of the calcination temperature on the structural, morphological, textural and the electrochemical performances, two heating temperatures were selected: 300°C and 450°C [33]. Titanium dioxide was prepared by a sol–gel method. Sodium alginate powder (1 g) was dissolved by magnetic stirring in 100 mL of distilled water until a gel was formed. This gel was added dropwise to a 100 mL solution of titanium (IV) isopropoxide (0.32 M) and left under stirring for 3 h at room temperature. The obtained solid was collected by centrifugation, washed with distilled water, dried at 70°C overnight and calcined at 300°C and 450°C.

Concerning the structural, textural and morphological observations, all analysis technics resulted in the formation of a pure anatase TiO_2 with aggregated spherical particles. In fact, **Figure 3a** shows that the unannealed sample present an amorphous like structure, while the diffraction spectra recorded for TiO_2–300 and TiO_2–450 materials are clearly crystalline. The Raman spectroscopy, **Figure 3b**, confirmed these findings by the presence of three vibration peaks at 632, 508, and 390 cm^{-1}, attributed to Eg, A1g, and B1g modes, respectively, characteristic of TiO_2 anatase phase [34–36].

For the morphological characterization of TiO_2 particles, Scanning Electron Microscopy (SEM) was used. This is shown in **Figure 4**, where the shapes of the TiO_2–300 and the TiO_2–450 particles are spherical, with an inhomogeneous particles' size distribution (nano and submicrometric spherical particles). EDX spectroscopy demonstrated, on the other hand, the uniform distribution of Titanium and oxygen.

BET was used to confirm the effect of the calcination temperature on the average pore sizes, resulting in the respective value of 4.4 and 6.0 nm for TiO_2–300, and

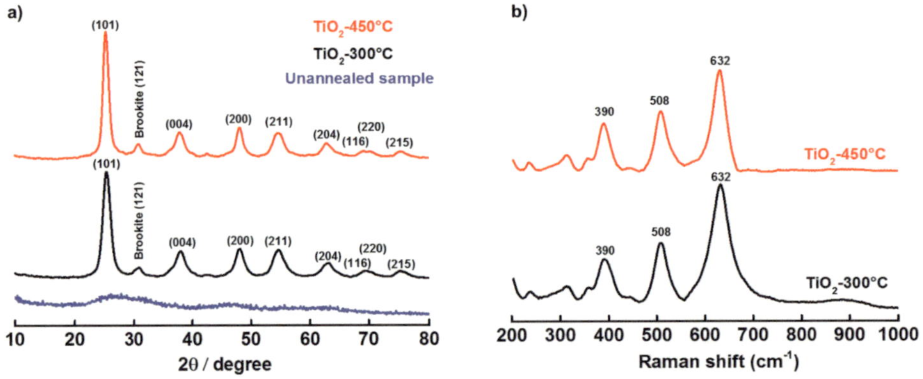

Figure 3.
(a) XRD patterns and (b) Raman spectra of TiO_2material obtained at 300°C (black) and 450°C (red).

Figure 4.
SEM images of TiO_2 materials calcined at (a, b) 300°C and (c, d) 450°C.

TiO_2–450 (**Figure 5**). Both samples were highly porous, which enhances the surface activity for Li^+ storage and facilitates the liquid electrolyte penetration [33].

In order to evaluate the electrochemical performances of TiO_2–300 and TiO_2–450 electrodes, the charge/discharge tests, cyclic voltammetry, Operando XRD of the TiO_2 electrodes were carried out. The charge/discharge profiles of the two electrodes at a current rate of 0.1C are illustrated in the **Figure 6**. The existence of the cathodic/anodic plateaus located at ~ 1.7 V (lithiation process) and 1.9 V (delithiation process) are characteristic of the TiO_2 anatase polymorph; tetragonal anatase TiO_2 for the non lithiated TiO_2 and orthorhombic $Li_{0.5}TiO_2$ for the Li-rich phase [37, 38]. Besides, the initial reversible capacity of the two electrodes was 266 and 275 mAh·g^{-1} for TiO_2–300 and TiO_2–450, respectively. TiO_2–300 and TiO_2–450 electrodes demonstrated a coulombic efficiency (CE) of 70% and 75% in the first cycle and a CE higher than 95% in the other cycles, respectively. From the potential

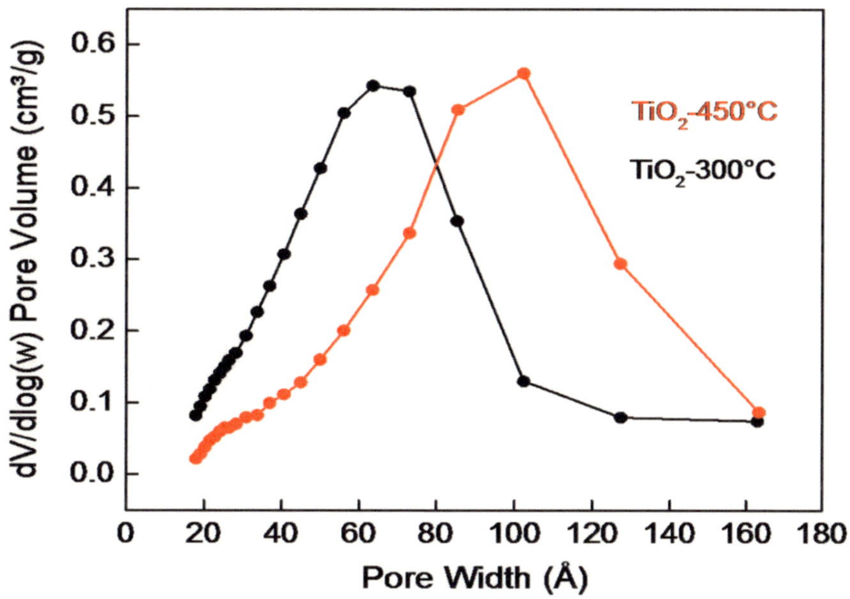

Figure 5.
The pore size distribution curves: (black) TiO_2–300°C, (red) TiO_2–450°C.

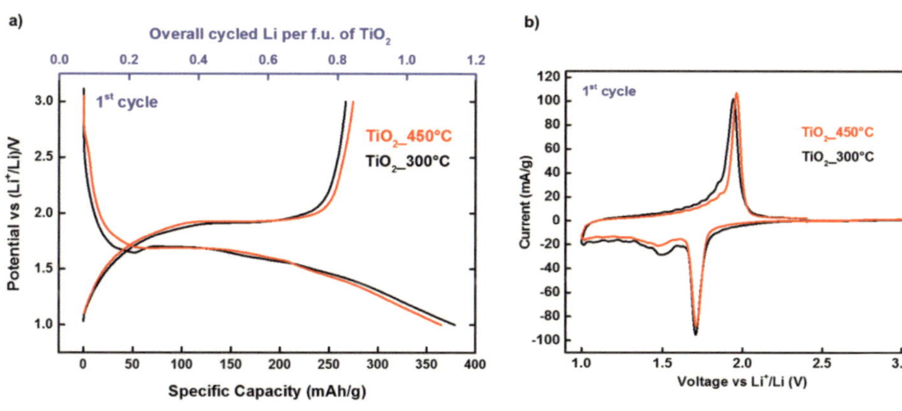

Figure 6.
(a) First charge/discharge profiles of TiO_2electrodes calcined at 450°C (red curve) and at 300°C (black curve) cycled between 3.0 and 1.0 V versus Li/Li$^+$ at C/10 current rate, (b) cyclic voltammograms of the firstcycle scanned at 0.02 mV s^{-1}.

vs. capacity profile, it is clearly observed that increasing the synthesis temperature from 300 to 450°C has no obvious impact on the cycling process since this profile is very similar.

Concerning the Cyclic voltammetry tests of the as-prepared electrodes (**Figure 6**), there are a pair of reduction/oxidation peaks at ~ 1.7 and 1.9 V for both materials, which could be attributed to the Li-ion intercalation/

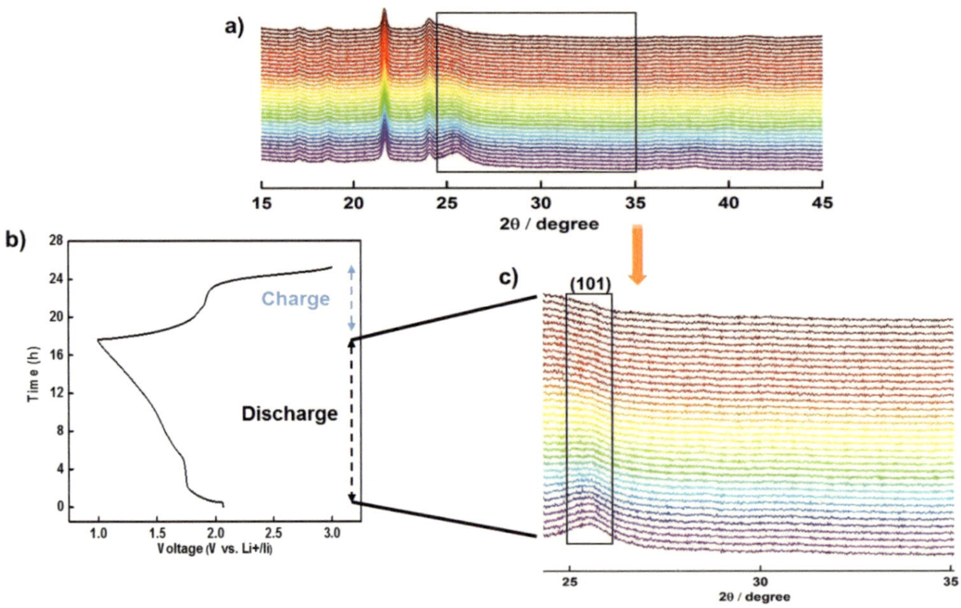

Figure 7.
(a) Operando XRD patterns of TiO_2during the 1st discharge from 3.0 to 1.0 V, (b) 1st discharge/charge galvanostatic data at 0.025C currentrate, (c) the 2Θ region from 24° to 35° showing the disappearance of the (101) reflection peak.

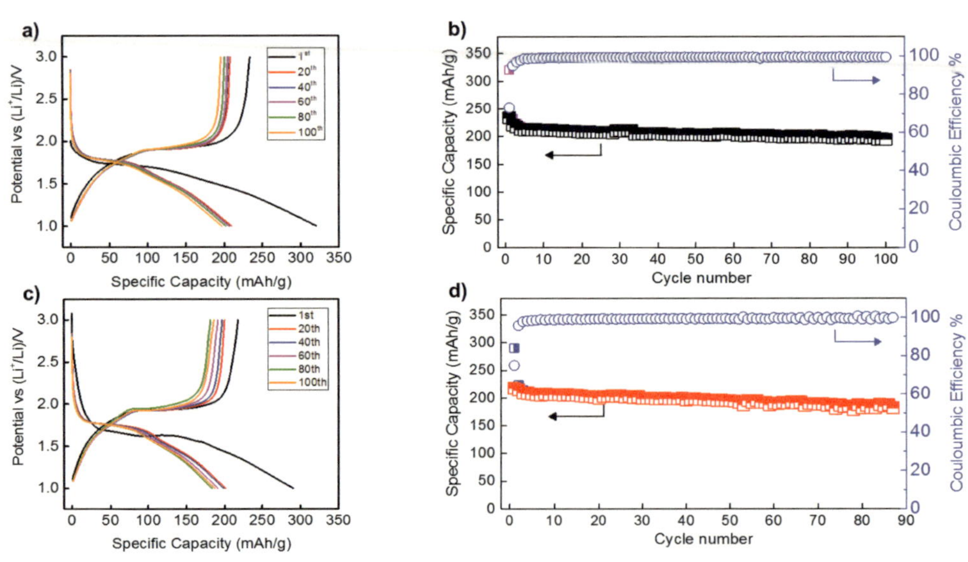

Figure 8.
Galvanostatic discharge/charge curves vs. $Li/Li^+$of (a) TiO_2–300 and (c) TiO_2–450 cycled at a rate of 0.1 C.Cycling performance and coulombic efficiency of (b) TiO_2–300 and (d) TiO_2–450 electrodes cycled between 3.0 and 1.0 V versus Li/Li^+ at 0.1 C current rate.

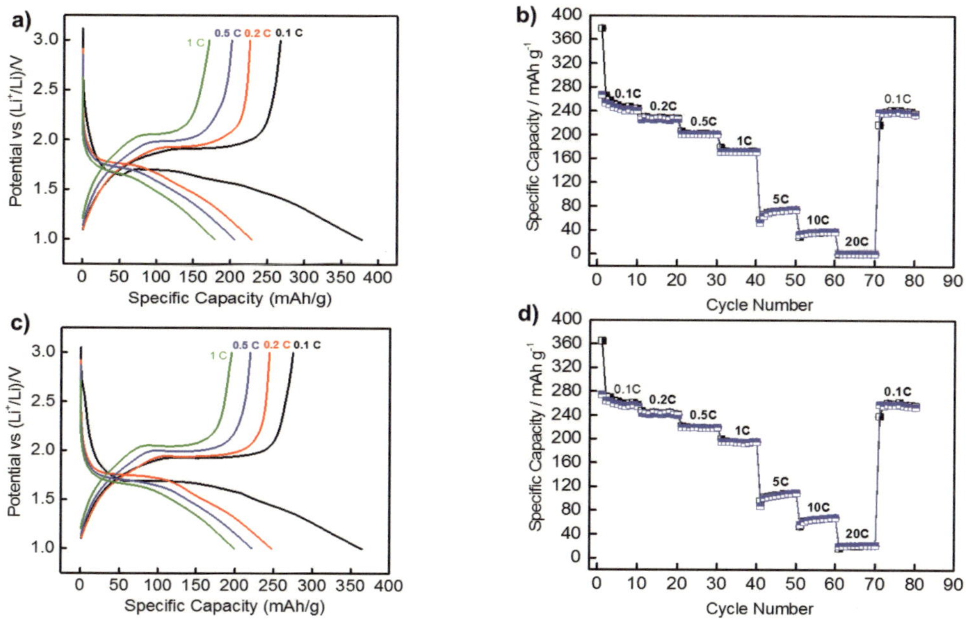

Figure 9.
Galvanostatic charge/discharge profiles at different rates of (a) TiO_2–300 and (c) TiO_2–450, rate capability of (b) TiO_2–300 and (d) TiO_2–450 electrodes at variant current rates from 0.1 C to 20 C (1C = 336 mA g^{-1}).

deintercalation in an anatase TiO_2 lattice (Ti^{4+} reduction/oxidation). The cathodic/anodic peaks were in accordance with the galvanostatic discharge/ charge profiles.

In order to follow the structural evolution of the anatase TiO_2 during the lithiation process, an operando XRD measurement during the discharge/charge of the TiO_2–300°C electrode was carried out. As shown in **Figure 7**, the (101) reflection peak characteristic of the anatase phase disappeared during the insertion process, which means that the starting material has been successfully lithiated.

The capacity retention of the two materials is presented in **Figure 8**. After 100 cycles, TiO_2–300 and TiO_2–450 electrodes showed an excellent capacity retention of 88% and 85%, respectively. **Figure 9** presents the rate capabilities of TiO_2 materials evaluated at different current rates at 1.0–3.0 V voltage range. The electrodes were discharged down to 1.0 V and recharged up to 3.0 V at different constant current density from 0.1 to 20 C (1 C = 336 mA g^{-1}). It is clearly observed that the reversible capacity declined gradually with the increase of the current, but still exceeds 73 mAh g^{-1} even at a rate of 5. These excellent electrochemical properties can be explained by the nanoparticle's aspect of TiO_2 prepared by biopolymers gelation method.

3. Impact of phosphorus doping on TiO_2 as anode for Lithium-ion batteries

Another study by our group have evaluated also the impact of phosphorus doping on the electrochemical performances of TiO_2 as anode material for lithium ion batteries. The phosphorus doped TiO_2 was synthesized using a simple and eco-friendly synthesis method, in which titanium tetra-isopropoxide was used as a titanium precursor and sodium alginate as a complexing agent. The effects of P-doping on the crystal structure, morphology and lithium insertion mechanism were investigated and compared with the undoped TiO_2. Moreover, the P-TiO_2 was tested electrochemically as anode material.

Concerning the synthesis process, TiO_2 and P-TiO_2 materials were synthetized via a gelation of biopolymers, following the synthesis technic proposed by El Ouardi *et al.* and using phosphoric acid as the phosphorus precursor. To prepare the working electrode, black carbon, PVDF and an active material were mixed in a 7:2:1 wt. ratio. The active material of the first electrode contained pure TiO_2 while the second contained TiO_2 doped at 2% phosphorus.

To identify the crystal structures of TiO_2 and P-TiO_2, XRD was carried out and the results are shown in **Figure 10a**. As it can be seen in the diffractograms, the diffraction peaks are centered at 25.3°, 37.9°, 48.1°, 54.7°, 55.0°, 62.7°, 68.9°, 75.04° and 83.0°. These are attributed to the (101), (004), (200), (105), (211), (204), (116), (215) and (312) diffraction planes of anatase TiO_2, respectively, indicating that the crystal phase of TiO_2 remained after phosphorylation treatments [39–41]. At 2θ = 30.7° there is a small peak (*) for the undoped sample, which can be attributed to the existence of the brookite phase, (121) formed during the synthesis [42]. No diffraction peaks that could be attributed to impurities are found in the XRD patterns of TiO_2 and P-TiO_2, suggesting that the sol–gel method can give highly purified anatase TiO_2 products. For the Raman spectra (**Figure 10b**), the obtained bands at 198, 400, 518, and 641 cm^{-1} represent the Raman active modes of anatase TiO_2. These results prove that the prepared nanoparticles have an anatase structure; the non-doped TiO_2 sample contained particles with uniform sizes and homogeneous granular surface, while the P-TiO_2 samples remained unchanged. The energy

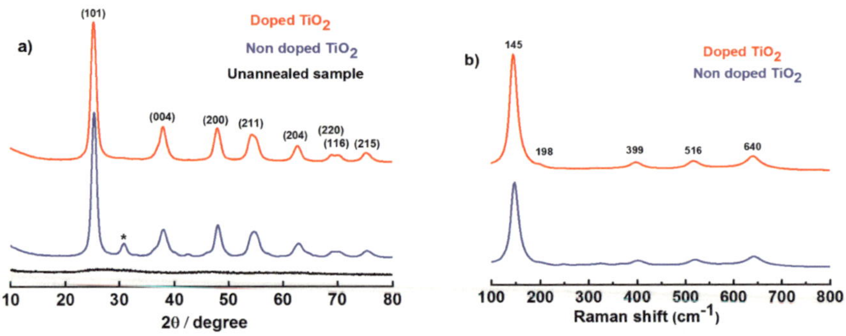

Figure 10.
(a) XRD patterns and (b) Raman spectra of non-doped TiO_2(bleu) and P doped TiO_2 (red) materials obtained at 450°C.

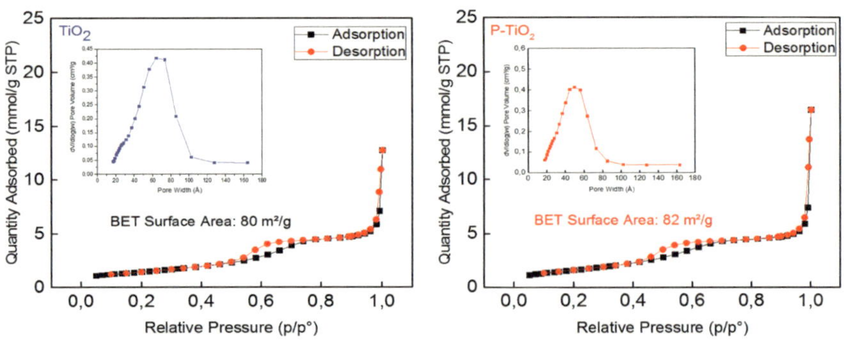

Figure 11.
Nitrogen adsorption–desorption isotherm curves and the pore size distribution curves of non-doped and P doped TiO_2materials obtained at 450°C.

dispersive X-ray (EDX) spectroscopic data of the P-doped TiO_2 demonstrate the uniform distribution of Ti, O and P with no other impurity elements.

The Brunauer–Emmett Teller (BET) method from N_2 adsorption and desorption isotherms carried out at 77 K (**Figure 11**) showed that both materials presented typical IV adsorption/desorption isotherms with mesoporous structures. Besides, both materials exhibited very similar BET surface, pore size distribution and mesopore diameter. For the absorbance measurement, UV-V spectroscopy showed that the phosphorus doping extended the wavelength response range of TiO_2 into the visible-light region (**Figure 12**). Moreover, the band gap of TiO_2 and P-TiO_2 was 2.90 and 2.87 eV, respectively. This result shows the effect of phosphorus doping to reduce the band gap and improve the electrotonic conductivity of TiO_2 [43–45].

Concerning the electrochemical tests, the charge/discharge curves (**Figure 13**) show the presence of two plateaus at 1.9 V and 1.7 V for both materials representing

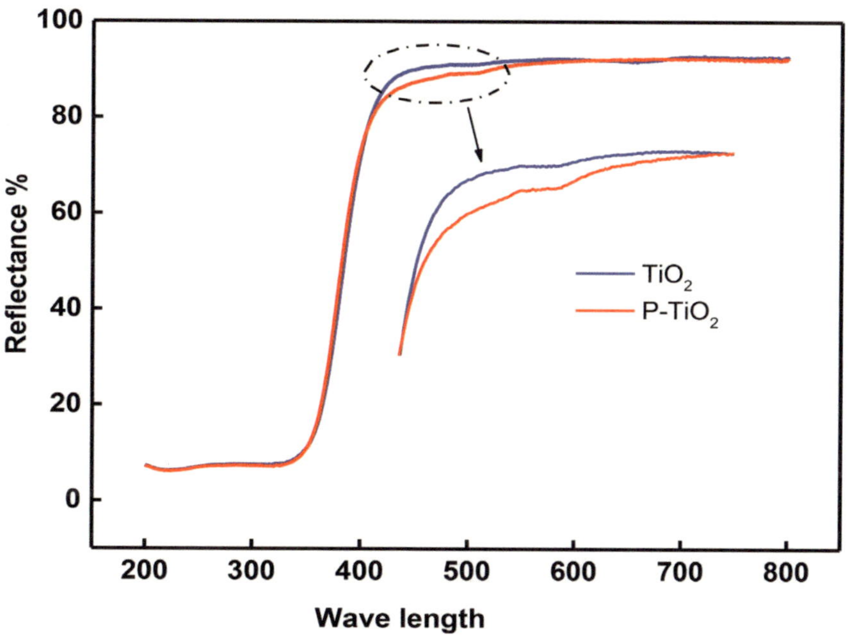

Figure 12.
UV-V spectra of non-doped TiO_2 (bleu) and P doped TiO_2 (red) materials obtained at 450°C.

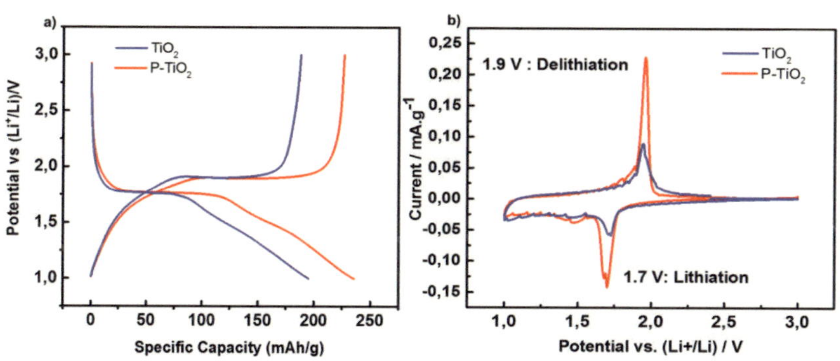

Figure 13.
(a) First charge/discharge profiles of TiO_2 (bleu curve) and P-TiO_2 (red curve) electrodes calcined at 450°C cycled between 3.0 and 1.0 V versus Li/Li^+ at C/10 current rate, (b) cyclic voltammograms of the first cycle scanned at 0.02 mV s^{-1}.

cathodic and anodic peaks of anatase TiO_2 nanoparticles, respectively. This charge/discharge process has shown also a good irreversible capacity which does not exceed 10 mAh/g for both materials. For the polarization, P-TiO_2 has shown improved characteristics compared to non-doped TiO_2. This result could be attributed to the improved electronic conductivity.

Cyclic voltammetry technique was used to study the insertion/extraction properties of lithium ions from the prepared electrodes in the 1.0 and 3.0 V potential window with a scanning speed of 0.02 mV s^{-1}. As it can be seen in **Figure 13**, there are two cathodic (reduction peak $I < 0$) / anodic (oxidation peak $I > 0$) peaks at 1.7 and 1.9 V, respectively, attributed to the insertion / extraction of lithium ions in the TiO_2 nanoparticles. This result is with agreement with the galvanostatic discharge/charge profiles.

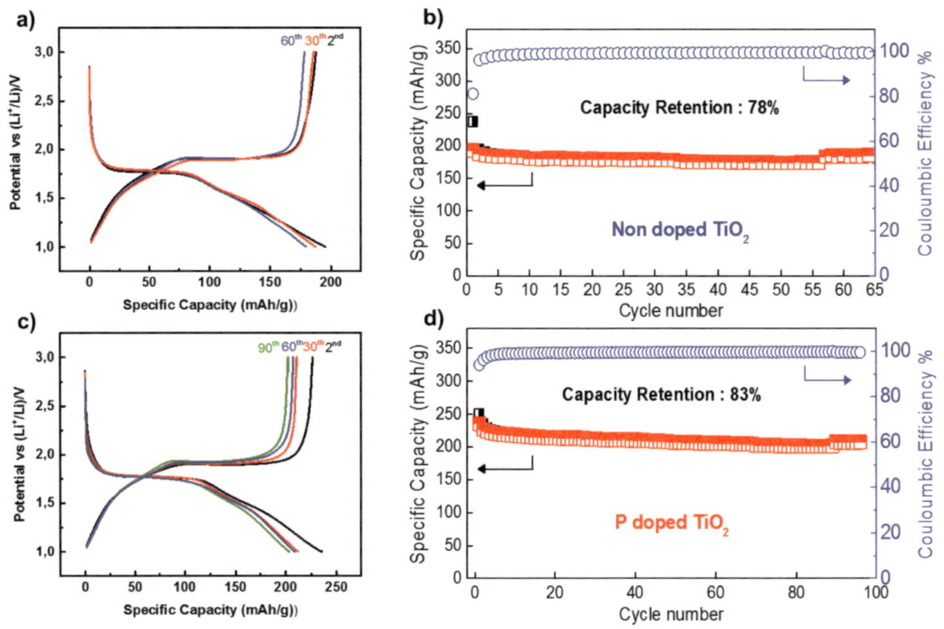

Figure 14.
Galvanostatic discharge/charge curves vs. Li/Li⁺ of (a) P doped TiO_2 and (c) non-doped TiO_2 cycled at a rate of 0.1 C. cycling performance and coulombic efficiency of (b) P doped TiO_2 and (d)non-doped TiO_2 electrodes cycled between 3.0 and 1.0 V versus Li/Li⁺ at 0.1 C current rate.

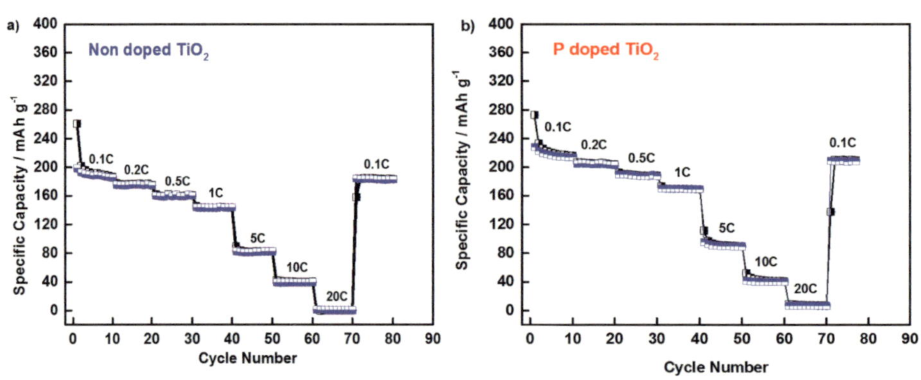

Figure 15.
Rate capability of (a) non doped TiO_2 and (b) P doped TiO_2 electrodes at variant current rates from 0.1 C to 20 C (1C = 336 mA g^{-1}).

For the cycling stability, **Figure 14** prove the excellent coulombic efficiency (about 100%) for both TiO_2 and P-TiO_2 electrodes. Besides, these electrodes showed a capacity retention of 78% after 70 cycles and 83% after 90 cycles, respectively. The reason behind these improved electrochemical properties for P-TiO_2 can be the smaller TiO_2 particle size which permits fast lithium insertion / disinsertion process.

Figure 15 present the rate capabilities of TiO_2 and P-TiO_2 evaluated at different current rates at 1.0–3.0 V voltage range. The electrodes were discharged down to 1.0 V and recharged up to 3.0 V at different constant current density from 0.1 to 20 C (1 C = 336 mA g^{-1}). It is clearly observed that the reversible capacity declined gradually with the increase of the current for both materials, but it still exceeds 80 mAh g^{-1} for non-doped TiO_2 and 98 mAh g^{-1} for P doped TiO_2 even at a rate of 5 C. The capacity at 0.1C rate after 70 cycles was recovered to about 185 mAh g^{-1} for non-doped TiO_2 and 213 mAh/g for P doped TiO_2 after 90 cycles. Thus, indicating the high stability of the anatase TiO_2 nanoparticles and confirming the better performance of P-TiO_2 compared to TiO_2.

4. Conclusion

In summary, this chapter show the huge interest in the development and improvement of TiO_2 as anode for high performance rechargeable lithium ion batteries. Several strategies have been developed to improve the conductivity, the capacity, cycling stability, and rate capability of this material, such as designing different nanostructured (1D, 2D and 3D), Coating or combining TiO_2 with carbonaceous materials, and Selective doping with mono and heteroatoms.

Biopolymer gelation is a simple and economically favorable approach for the preparation of TiO_2 nanoparticles. This method that consists of using the Sol–Gel method with a sodium alginate biopolymer as a templating agent showed enhanced performances in comparison with other synthesis techniques. In fact, the prepared TiO_2 electrodes displayed a high specific capacity above 275 mAh g^{-1} and excellent cycling stability with over 85% capacity retention after 100 cycles. Besides, combining phosphorus doping with this synthesis strategy demonstrated an important discharge capacity of 200 mAh g^{-1} after 90 cycles under C/10 current rate and has an excellent rate performance. The improved electrochemical performance can be explained based on the P-TiO_2 particles size and band gap modifications upon doping proved by UV-V measurement.

Finally, from this chapter, we can conclude that the use of TiO_2-based materials as anode for commercial lithium ion batteries requires more efforts to overcome the problems encountered, especially the low electrical conductivity, the low energy density, the poor cycling life and the low efficiency.

Acknowledgements

The authors wish to acknowledge Office Chérifien des Phosphates (OCP S.A.) for financial support.

Author details

Nabil El Halya[1], Karim Elouardi[1,2], Abdelwahed Chari[1], Abdeslam El Bouari[2], Jones Alami[1] and Mouad Dahbi[1*]

1 Materials Science, Energy, and Nano-Engineering Department, Mohammed VI Polytechnic University, Ben Guerir, Morocco

2 Laboratory of Physic-Chemical of Applied Materials, Sciences Faculty of Ben M'sik, Hassan II University Casablanca, Morocco

*Address all correspondence to: mouad.dahbi@um6p.ma

References

[1] Z. Yang *et al.*, "Nanostructures and lithium electrochemical reactivity of lithium titanites and titanium oxides: A review," *J. Power Sources*, vol. 192, no. 2, pp. 588-598, 2009, doi: 10.1016/j.jpowsour.2009.02.038.

[2] L. Jabbour, R. Bongiovanni, D. Chaussy, C. Gerbaldi, and D. Beneventi, "Cellulose-based Li-ion batteries: A review," *Cellulose*, vol. 20, no. 4, pp. 1523-1545, 2013, doi: 10.1007/s10570-013-9973-8.

[3] P. Xiong, L. Peng, D. Chen, Y. Zhao, X. Wang, and G. Yu, "Two-dimensional nanosheets based Li-ion full batteries with high rate capability and flexibility," *Nano Energy*, vol. 12, pp. 816-823, 2015, doi: 10.1016/j.nanoen.2015.01.044.

[4] D. Ma, Z. Cao, and A. Hu, "Si-based anode materials for li-ion batteries: A mini review," *Nano-Micro Lett.*, vol. 6, no. 4, pp. 347-358, 2014, doi: 10.1007/s40820-014-0008-2.

[5] S. Goriparti, E. Miele, F. De Angelis, E. Di Fabrizio, R. Proietti Zaccaria, and C. Capiglia, "Review on recent progress of nanostructured anode materials for Li-ion batteries," *J. Power Sources*, vol. 257, pp. 421-443, 2014, doi: 10.1016/j.jpowsour.2013.11.103.

[6] L. Bai *et al.*, "A sandwich structure of mesoporous anatase TiO2 sheets and reduced graphene oxide and its application as lithium-ion battery electrodes," *RSC Adv.*, vol. 4, no. 81, pp. 43039-43046, 2014, doi: 10.1039/c4ra04979a.

[7] X. Li, Y. Zhang, T. Li, Q. Zhong, H. Li, and J. Huang, "Graphene nanoscrolls encapsulated TiO2 (B) nanowires for lithium storage," *J. Power Sources*, vol. 268, pp. 372-378, 2014, doi: 10.1016/j.jpowsour.2014.06.056.

[8] J. Jin *et al.*, "Design of new anode materials based on hierarchical, three dimensional ordered macro-mesoporous TiO2 for high performance lithium ion batteries," *J. Mater. Chem. A*, vol. 2, no. 25, pp. 9699-9708, 2014, doi: 10.1039/c4ta01775g.

[9] F. Paquin, J. Rivnay, A. Salleo, N. Stingelin, and *C. Silva*, "Multi-phase semicrystalline microstructures drive exciton dissociation in neat plastic semiconductors," *J. Mater. Chem. C*, vol. 3, pp. 10715-10722, 2015, doi: 10.1039/b000000x.

[10] P. Tammawat and N. Meethong, "Synthesis and characterization of stable and binder-free electrodes of TiO2 nanofibers for li-ion batteries," *J. Nanomater.*, vol. 2013, 2013, doi: 10.1155/2013/413692.

[11] A. R. Armstrong, G. Armstrong, J. Canales, R. García, and P. G. Bruce, "Lithium-ion intercalation into TiO2-B nanowires," *Adv. Mater.*, vol. 17, no. 7, pp. 862-865, 2005, doi: 10.1002/adma.200400795.

[12] T. Hu *et al.*, "Flexible free-standing graphene-TiO 2 hybrid paper for use as lithium ion battery anode materials," *Carbon N. Y.*, vol. 51, no. 1, pp. 322-326, 2013, doi: 10.1016/j.carbon.2012.08.059.

[13] Y. Liu, W. Wang, H. Huang, L. Gu, Y. Wang, and X. Peng, "The highly enhanced performance of lamellar WS2 nanosheet electrodes upon intercalation of single-walled carbon nanotubes for supercapacitors and lithium ions batteries," *Chem. Commun.*, vol. 50, no. 34, pp. 4485-4488, 2014, doi: 10.1039/c4cc01622j.

[14] T. D. Nguyen-Phan *et al.*, "Uniform distribution of TiO 2 nanocrystals on reduced graphene oxide sheets by the chelating ligands," *J. Colloid Interface Sci.*, vol. 367, no. 1, pp. 139-147, 2012, doi: 10.1016/j.jcis.2011.10.021.

[15] F. Wu, Z. Wang, X. Li, and H. Guo, "Simple preparation of petal-like TiO2 nanosheets as anode materials for lithium-ion batteries," *Ceram. Int.*, vol. 40, no. PB, pp. 16805-16810, 2014, doi: 10.1016/j.ceramint.2014.07.060.

[16] X. Lü, F. Huang, J. Wu, S. Ding, and F. Xu, "Intelligent hydrated-sulfate template assisted preparation of nanoporous TiO2 spheres and their visible-light application," *ACS Appl. Mater. Interfaces*, vol. 3, no. 2, pp. 566-572, 2011, doi: 10.1021/am101137w.

[17] X. Lü, S. Ding, Y. Xie, and F. Huang, "Non-aqueous preparation of high-crystallinity hierarchical TiO2 hollow spheres with excellent photocatalytic efficiency," *Eur. J. Inorg. Chem.*, no. 18, pp. 2879-2883, 2011, doi: 10.1002/ejic.201100151.

[18] F. Di Lupo *et al.*, "Mesoporous TiO2 nanocrystals produced by a fast hydrolytic process as high-rate long-lasting Li-ion battery anodes," *Acta Mater.*, vol. 69, pp. 60-67, 2014, doi: 10.1016/j.actamat.2014.01.057.

[19] S. Casino, F. Di Lupo, C. Francia, A. Tuel, S. Bodoardo, and C. Gerbaldi, "Surfactant-assisted sol gel preparation of high-surface area mesoporous TiO2 nanocrystalline Li-ion battery anodes," *J. Alloys Compd.*, vol. 594, pp. 114-121, 2014, doi: 10.1016/j.jallcom.2014.01.111.

[20] S. Fang, L. Shen, Z. Tong, H. Zheng, F. Zhang, and X. Zhang, "Si nanoparticles encapsulated in elastic hollow carbon fibres for Li-ion battery anodes with high structural stability," *Nanoscale*, vol. 7, no. 16, pp. 7409-7414, 2015, doi: 10.1039/c5nr00132c.

[21] S. Wang, Y. Xing, C. Xiao, X. Wei, H. Xu, and S. Zhang, "Hollow carbon-shell/carbon-nanorod arrays for high performance Li-ion batteries and supercapacitors," *RSC Adv.*, vol. 5, no. 11, pp. 7959-7963, 2015, doi: 10.1039/c4ra14683b.

[22] T. Xia *et al.*, "Amorphous carbon-coated TiO2 nanocrystals for improved lithium-ion battery and photocatalytic performance," *Nano Energy*, vol. 6, pp. 109-118, 2014, doi: 10.1016/j.nanoen.2014.03.012.

[23] V. Etacheri, J. E. Yourey, and B. M. Bartlett, "Chemically bonded TiO2-Bronze nanosheet/reduced graphene oxide hybrid for high-power lithium ion batteries," *ACS Nano*, vol. 8, no. 2, pp. 1491-1499, 2014, doi: 10.1021/nn405534r.

[24] X. Lü *et al.*, "Enhanced electron transport in Nb-doped TiO2 nanoparticles via pressure-induced phase transitions," *J. Am. Chem. Soc.*, vol. 136, no. 1, pp. 419-426, 2014, doi: 10.1021/ja410810w.

[25] Y. X. Wang, J. Xie, G. S. Cao, T. J. Zhu, and X. B. Zhao, "Electrochemical performance of TiO 2/carbon nanotubes nanocomposite prepared by an in situ route for Li-ion batteries," *J. Mater. Res.*, vol. 27, no. 2, pp. 417-423, 2012, doi: 10.1557/jmr.2011.406.

[26] X. Lü *et al.*, "Improved-Performance Dye-Sensitized solar cells using Nb-Doped TiO 2 electrodes: Efficient electron Injection and transfer," *Adv. Funct. Mater.*, vol. 20, no. 3, pp. 509-515, 2010, doi: 10.1002/adfm.200901292.

[27] D. Liu *et al.*, "TiO2 nanotube arrays annealed in CO exhibiting high performance for lithium ion intercalation," *Electrochim. Acta*, vol. 54, no. 27, pp. 6816-6820, 2009, doi: 10.1016/j.electacta.2009.06.090.

[28] N. A. Kyeremateng *et al.*, "Effect of Sn-doping on the electrochemical behaviour of TiO2 nanotubes as potential negative electrode materials for 3D Li-ion micro batteries," *J. Power Sources*, vol. 224, pp. 269-277, 2013, doi: 10.1016/j.jpowsour.2012.09.104.

[29] J. H. Jeong, D. W. Jung, E. W. Shin, and E. S. Oh, "Boron-doped TiO2 anode

materials for high-rate lithium ion batteries," *J. Alloys Compd.*, vol. 604, pp. 226-232, 2014, doi: 10.1016/j.jallcom.2014.03.069.

[30] G. Hasegawa, T. Sato, K. Kanamori, K. Nakanishi, and T. Abe, "Synthesis and electrochemical performance of hierarchically porous N-doped TiO2 for Li-ion batteries," *New J. Chem.*, vol. 38, no. 4, pp. 1380-1384, 2014, doi: 10.1039/c3nj01332d.

[31] Y. Zhou, Q. Bao, L. A. L. Tang, Y. Zhong, and K. P. Loh, "Hydrothermal dehydration for the 'green' reduction of exfoliated graphene oxide to graphene and demonstration of tunable optical limiting properties," *Chem. Mater.*, vol. 21, no. 13, pp. 2950-2956, 2009, doi: 10.1021/cm9006603.

[32] S. Chen, W. Chu, Y. Y. Huang, X. Liu, and D. G. Tong, "Preparation of porous nitrogen-doped titanium dioxide microspheres and a study of their photocatalytic, antibacterial and electrochemical activities," *Mater. Res. Bull.*, vol. 47, no. 12, pp. 4514-4521, 2012, doi: 10.1016/j.materresbull.2012.09.031.

[33] K. El Ouardi *et al.*, "Facile synthesis of nanoparticles titanium oxide as high-capacity and high-capability electrode for lithium-ion batteries," *J. Appl. Electrochem.*, vol. 50, pp. 583-595, 2020, doi: 10.1007/s10800-020-01419-y.

[34] M. Buaki-Sogo, M. Serra, A. Primo, M. Alvaro, and H. Garcia, "Alginate as Template in the Preparation of Active Titania Photocatalysts," *ChemCatChem*, vol. 5, no. 2, pp. 513-518, 2013, doi: 10.1002/cctc.201200386.

[35] J. Wang, Y. Zhou, B. Xiong, Y. Zhao, X. Huang, and Z. Shao, "Fast lithium-ion insertion of TiO2 nanotube and graphene composites," *Electrochim. Acta*, vol. 88, pp. 847-857, 2013, doi: 10.1016/j.electacta.2012.10.010.

[36] H. Cao, B. Li, J. Zhang, F. Lian, X. Kong, and M. Qu, "Synthesis and superior anode performance of TiO 2@ reduced graphene oxide nanocomposites for lithium ion batteries," *J. Mater. Chem.*, vol. 22, no. 19, pp. 9759-9766, 2012, doi: 10.1039/c2jm00007e.

[37] J. S. Chen, Y. L. Tan, C. M. Li, Y. L. Cheah, and D. Luan, "Ja100102Y.Pdf," no. 001, pp. 6124-6130, 2010.

[38] J. Wang, Y. Bai, M. Wu, J. Yin, and W. F. Zhang, "Preparation and electrochemical properties of TiO2 hollow spheres as an anode material for lithium-ion batteries," *J. Power Sources*, vol. 191, no. 2, pp. 614-618, 2009, doi: 10.1016/j.jpowsour.2009.02.056.

[39] A. K. Rai *et al.*, "Simple synthesis and particle size effects of TiO2 nanoparticle anodes for rechargeable lithium ion batteries," *Electrochim. Acta*, vol. 90, pp. 112-118, 2013, doi: 10.1016/j.electacta.2012.11.104.

[40] M. Das *et al.*, "One step hydrothermal synthesis of a rGO-TiO2 nanocomposite and its application on a Schottky diode: Improvement in device performance and transport properties," *RSC Adv.*, vol. 5, no. 123, pp. 101582-101592, 2015, doi: 10.1039/c5ra17795b.

[41] J. Ma *et al.*, "Lithium Intercalation in Anatase Titanium Vacancies and the Role of Local Anionic Environment," *Chem. Mater.*, vol. 30, no. 9, pp. 3078-3089, 2018, doi: 10.1021/acs.chemmater.8b00925.

[42] R. Fernández Acosta, E. Peláe Abellán, J. R. Correa, and U. Jáuregui Haza, "Nanostructured TiO2 Obtained by Electrolysis and its Application in the Remediation of Water Polluted with Paracetamol," *Int J Chem Mater Env. Res*, vol. 3, no. 2, pp. 20-28, 2016.

[43] A. Bhaumik and S. Inagaki, "Mesoporous titanium phosphate molecular-sieves with ion-exchange

capacity," *J. Am. Chem. Soc.*, vol. 123, no. 4, pp. 691-696, 2001, doi: 10.1021/ja002481s.

[44] R. Sasikala *et al.*, "Enhanced photocatalytic activity of indium and nitrogen co-doped TiO2-Pd nanocomposites for hydrogen generation," *Appl. Catal. A Gen.*, vol. 377, no. 1-2, pp. 47-54, 2010, doi: 10.1016/j.apcata.2010.01.039.

[45] X. Wang, Z. Hu, Y. Chen, G. Zhao, Y. Liu, and Z. Wen, "A novel approach towards high-performance composite photocatalyst of TiO 2 deposited on activated carbon," *Appl. Surf. Sci.*, vol. 255, no. 7, pp. 3953-3958, 2009, doi: 10.1016/j.apsusc.2008.10.083.

Chapter 9

Applications of Titanium Dioxide Materials

Xiaoping Wu

Abstract

Titanium dioxide (TiO_2) is a stable, non-toxic inorganic material. Because of very high refractive index, TiO_2 has been widely used as a white pigment. The optimal particle sizes of TiO_2 for pigment applications are around 250 nm. The pigmentary applications of TiO_2 can be found in many common products such as paints, plastics, paper and ink. Global titanium dioxide pigment sales have reached several million tons annually. Titanium dioxide is also a semiconducting material. When excited by photons which have energy equal to or higher than the band gap of TiO_2, electron/hole pairs can be generated. The dynamics of the photo-generated electron/hole pairs of TiO_2 is fundamentally important to its photocatalytic properties. More recently, nano-structured TiO_2 has raised a great deal of interests in research after the discoveries of the important potentials for applications. The enormous efforts have been put in the preparation, characterization, scientific understandings, and modifications of the photocatalytic properties of TiO_2. The applications of nano-structured TiO_2 can be now found in a wide range of areas including electronic materials, energy, environment, health & medicine, catalysts, etc. This chapter has discussed and highlighted the development of the applications of titanium dioxide materials in many of those areas.

Keywords: Titanium dioxide, Applications, Pigment, Nano-structured

1. Introduction

Titanium dioxide (TiO_2) is an inorganic substance that is used extensively as a white pigment. Compared with many other inorganic pigments, TiO_2 has the advantages of high stability, being non-toxic, and low cost. TiO_2 have three polymorphs: anatase, rutile and brookite, but only anatase and rutile crystal forms have been useful as pigment. Both anatase and rutile crystals have very high refractive indices, and their particles can scatter visible light almost completely [1]. The optimum particle size of TiO_2 for pigment applications is around 250 nm. In the early application of TiO_2 as pigment, it was found that paint faded more rapidly than others when painted films were exposed to the Sun and ultraviolet (UV) light. Coating with inorganic compounds such as alumina or silica suppressed the catalytic activity on the surface and improve the weather resistance, leading titanium dioxide in wide applications as white pigment. Global titanium dioxide pigment sales were about 6 million tons in 2017 and the growth trend of global titanium dioxide pigment sales is continuing over the recent years [2].

In the 60s of the last century, scientists studied the photo-induced phenomena on the solids of TiO_2 and ZnO under UV light irradiation [3–5]. In the early 1970s, research on photocatalysis by TiO_2 got wide attention due to the historic discovery of the electrochemical water splitting by use of TiO_2 [6]. In the 1990s, photocatalytic research of TiO_2 had made progress in the practical applications of TiO_2 in the decomposition of harmful organic materials [7, 8]. A function of super hydrophilicity of TiO_2 was also discovered [9]. Since the beginning of this century, nano-structured TiO_2 has attracted extensive interests. When the particle sizes of TiO_2 are reduced down to the nano-meter scale (generally in 1–100 nm), the surface characteristics and surface areas of TiO_2 have changed dramatically. The new or enhanced physical and chemical properties of nano-structured TiO_2 begin to emerge. The photocatalytic property of nano-structured TiO_2 has been greatly enhanced because of the changes in the surface characteristics and surface areas. Quantum effects of nano-structured TiO_2 can also have a role to play, affecting its photocatalytic, optical or electronic properties. As the results of academic and industrial research in recent years, enormous progresses have been made in the preparation, characterization, and scientific understandings of nano-structured TiO_2. Nano-structured TiO_2 have begun to find applications in a wide range of areas including electronic materials, energy, environment, health & medicine, sensors, catalysts, etc.

TiO_2 pigment is industrially produced from titanium containing ores by using a Chloride or Sulfate process [1, 10]. Nano-structured TiO_2 are made in the different ways, depending on the material characteristics required for the specific applications. A number of innovative fabrication methods of nano-structured TiO_2 materials have been developed and are used for different applications. These methods can be broadly classified as the liquid phase or the gas phase methods. Nano-particles, nano-wires, nano-tubes, two or three-dimensional nano-structured TiO_2 materials can also be fabricated for different applications [11–18].

This chapter will discuss and highlight the recent development of applications of titanium dioxide as pigment and as functional materials in the areas of energy, environment, catalyst, and biomedicine.

2. Application of titanium dioxide as pigment

2.1 Light scattering of pigmentary TiO_2

The main use of titanium dioxide is white pigment, because it absorbs almost no incident light in the visible region of the spectrum (380–700 nm). Titanium dioxide has a strong light scattering power, and scatters incident light in three ways: surface reflection, refraction and diffraction in the crystal [1]. When the refractive index difference between titanium dioxide and medium increases, the reflected light increases and complies with Eq. (1):

$$R = \frac{(n_p - n_m)^2}{(n_p + n_m)^2} \tag{1}$$

n_p and n_m are the refractive index of pigment and medium, respectively [1]. Titanium dioxide has a high refractive index (refractive index of rutile and anatase titanium dioxide is 2.70 and 2.55 respectively) [19]. These high refractive index values enable the rutile and anatase TiO_2 pigments to have much greater hiding power in coatings or in plastics, making TiO_2 to be a much better pigment than the other chemical substances. Therefore, under the same conditions, only less titanium dioxide is needed to form a coating, which is white and opaque. Studies have shown that the optical properties of titanium dioxide pigments are related to their particle size, and the optimum of particle size of pigmentary titanium dioxide is around 250 nm [1].

2.2 Application areas

Pigmentary TiO_2 is inert, non-toxic, stable and less costly. Over 50 percent of all TiO_2 pigment produced is consumed by the coatings industry, and approximately a quarter by the paper industry. Eleven per cent goes into plastics; remaining a few percent into inks and other end-uses [20]. Titanium dioxide particles optimized with particle size and surface treatments have excellent hiding power, brightness, and other important features such as resistance to chemical degradation. Rutile pigment is more resistant to UV light than anatase, and is preferred for paints, plastics, especially for the applications in outdoor conditions. Anatase pigment is less abrasive and is used mainly in indoor paints and in paper manufacture. TiO_2 is surface treated with one or more inorganic oxides such as alumina, silica, zirconia or a combination of these inorganic oxides, and organic compounds such as polyhydric alcohol to have the required properties of dispersion, photoactivity, and opacity required for a specific application [21].

In coating applications, a relatively high quantity of TiO_2 pigment must be used to achieve desirable hiding effect on the coating subtracts, because coatings of titanium dioxide are usually in the form of very thin layers. The pigment volume concentration (PVC) is practically used to specify the amount of TiO_2 in a coating. Different types of paints containing TiO_2 pigment will have different levels of PVC, depending on different coating applications. TiO_2 coatings are used to cover a wide range of surfaces, including indoor and outdoor building, wood products, metal objects, domestic and industrial equipment [21].

In plastics applications, titanium dioxide pigment is used to opacify plastic materials. In some applications, TiO_2 is used to improve photodurability. The requirements for TiO_2 in plastics are good dispersibility in a polymer system and good heat stability. Hydrophobic organic surface treatments on the pigments are utilized to facilitate their dispersion in the viscous molten plastic resin. These are often silicone oils and other organic compounds for specialized uses. In many plastics applications, a blue undertone is also desirable to mask an intrinsic yellowness in the color of the resin or a slight degradation that occurs during the high-temperature processing. For this reason, plastics pigments often have a smaller crystal size than those for coatings applications [19].

The amount of titanium dioxide used in paper industry is the third largest after coating and plastic industries [20]. Although other white pigments can be used in paper industry, the production of high quality papers must use titanium dioxide as pigment. Titanium dioxide imparts desirable brightness and opacity to high-quality papers. Papers containing titanium dioxide pigment have high strength and have appearance to be white, shiny, thin and smooth. Because photochemical stability is not as critical in paper as in paint, both anatase and rutile pigments are widely be used in paper industry.

In inks applications, performance requirements for TiO_2 pigment are different from coatings, plastics and paper. Inks are usually applied to produce a much thinner film on a surface than a general coating. It is very important to choose titanium dioxide particles with good shape, suitable size and size-distribution, smooth surface and non-angular. The type of TiO_2 can also affect the rheology, abrasiveness, gloss and redispersibility for ink products and applications.

TiO_2 is also widely used as a pigment for coloring of different products in pharmaceuticals and cosmetics industries. The characteristics of titanium dioxide provide interesting colors and allow new properties to pharmaceuticals with very small amounts of pigments. There are many products in this field that contain titanium dioxide, including: shampoos, creams, sunscreens, toothpaste, etc. [10].

3. Applications TiO_2 in energy generation and storage

With the special physical and chemical properties, nano-structured titanium dioxide has shown a number of promising application prospects in energy generation and storage. These include: solar cells, hydrogen production, and lithium battery [22–24].

3.1 Application of TiO_2 in dye-sensitized solar cell (DSSC)

The solar energy is a clean, abundant and renewable energy [25]. The current technology for the conversion of sunlight to electrical power is predominately silicon-based solid state solar cells. In recent years, the new semiconducting material-based solar cells have emerged to offer the possible alternative photovoltaic technology with prospect of cheap fabrication and flexibility [26–28]. Nano-structured TiO_2 has been the main semiconducting material for this new generation of solar cells. In this technology, an electron sensitizer absorbing in the visible is used to inject charge carriers across the semiconductor-electrolyte junction into TiO_2 to enhance the conversion efficiency from solar energy, because TiO_2 with its band gap of 3.2 electronvolt (eV) absorbs only the ultraviolet part of the solar energy. This type of solar cells is therefore called dye-sensitized solar cells (DSSCs). The dye-sensitized solar cells (DSSCs) have exhibited high performance and have the potential to be low-cost [29–33].

Figure 1 illustrates the working principle of a dye-sensitized solar cell. The dye-sensitized solar cell consists of two electrodes, a dye-sensitized nano-structured TiO_2 mesoporous layer, and a liquid electrolyte containing redox system (I^-/I_3^-). The nano-structured TiO_2 mesoporous layer with a monolayer of the charge transfer dye at its surface is placed in contact with a redox electrolyte. Under solar

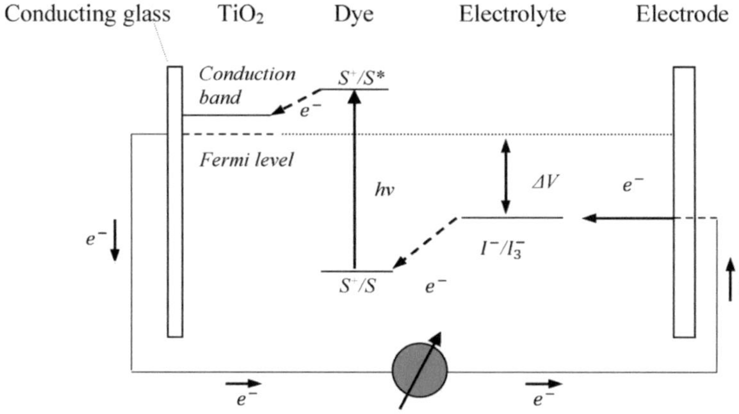

Figure 1.
Working principle of a dye-sensitized solar cell. S, S^+, and S^* represent dye sensitizer, oxidized dye sensitizer, and excited dye sensitizer, respectively. ΔV is the difference between Fermi level and electrochemical potential of the electrolyte.

irradiation, the charge transfer dye injects electrons into the conduction band of TiO_2, and the electrons are conducted to the external circuit to produce electric power. The original state of the dye is subsequently restored by an electron donation from the electrolyte (for example, an organic solvent containing a redox system of iodide/triiodide couple).

The nano-structured TiO_2 mesoporous layer in a dye-sensitized solar cell has a much larger surface area available for the dye-chemisorptions. The kinetic processes occurring in a dye-sensitized solar cell have been profoundly changed as a result of using nano-structured TiO_2. Solar energy-to-electricity conversion efficiencies of DSSCs have been increased. The record for the highest certified single cell and DSSCs module efficiencies are 11.9% and 8.8%, respectively [34].

More recently, TiO_2 is used in a new type of solar device so-called pervoskite solar cells. As in DSSCs, TiO_2 is used as a mesoporous layer. However, instead of using organic dye in DSSCs, organic lead complex (for example, $CH_3NH_2PbI_3$) is used to inject electrons into the conduction band of TiO_2. In a short period of the recent few years, the reported efficiency of pervoskite solar cells was 9.7% initially, and then 12.0% [35, 36]. Further progress was made with efficiencies above 15.0% [37]. The record for the highest certified single cell and minimodule efficiencies are 20.9% and 16.0%, respectively [34].

3.2 Application of TiO_2 in hydrogen production

In 1972, Fujishima and Honda discovered the phenomenon of photocatalytic splitting of water on a TiO_2 electrode under UV light [6, 38, 39]. Compared with other photocatalysts, TiO_2 is much more promising as it is stable, non-corrosive, environmentally friendly, abundant and cost effective. **Figure 2** illustrates the mechanism of the photocatalytic hydrogen production by TiO_2 semiconducting materials. When excited by photons which have energy equal to or higher than their band gap (*Eg*), electrons (e^-) in valence band (VB) of TiO_2 are promoted from to the conduction band (CB). Simultaneously, Holes (h^+) create in VB of TiO_2. The process of generating (e^-) and (h^+) in TiO_2 with excitation by photons is described with Eq. (2):

$$TiO_2 + hv \rightarrow e^- + h^+ \tag{2}$$

The photo-generated (e^-) and (h^+) in TiO_2 can recombine, releasing energy in the form of heat or photons. The photo-generated (e^-) and (h^+) that migrate to the surface of TiO_2 without recombination can reduce and oxidize H_2O molecules adsorbed on the surface of TiO_2 to generate H_2 and O_2.

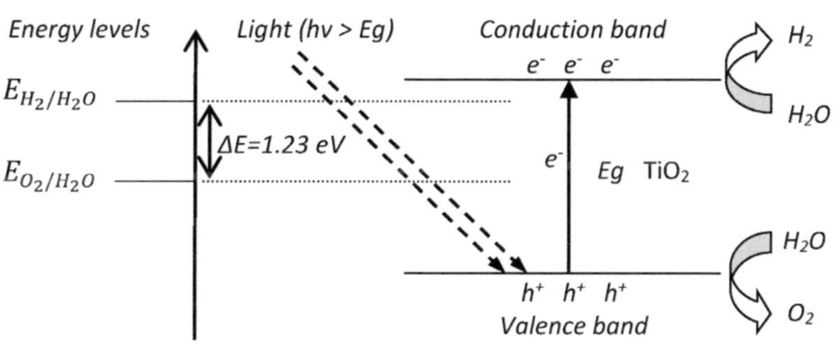

Figure 2.
Illustration of mechanism of photocatalytic hydrogen production by TiO_2.

As can be seen in **Figure 2**, the conduction band level of TiO_2 is more negative than the hydrogen production level (E_{H_2}/H_2O), and the valence band is more positive than water oxidation level (E_{O_2}/H_2O). ΔE, representing the energy difference between hydrogen production level and water oxidation level, is 1.23 eV. The conduction and valence band levels of TiO_2 meet the requirement for hydrogen production. As in dye-sensitized solar cell, nano-structured TiO_2 can enhance the photocatalytic reactions for the generation of hydrogen by a number of ways. For example, much more surface area per mass is available for adsorption of water due to the decreased particle sizes of TiO_2 nano-particles. The surface of TiO_2 nano-particles can also become more reactive because much higher portion of atoms exists at the surface. Furthermore, the quantum effect of TiO_2 nano-particles becomes more significant as particle size of TiO_2 is getting very small. Nano-structured TiO_2 has been known to be stable, chemically inert, and low cost.

Despite many advantages of using TiO_2 for photocatalytic hydrogen production, the efficiency using solar energy for water-splitting by TiO_2 is still low, and is currently not been used for industrial scale of hydrogen production. The low energy conversion efficiency of TiO_2 in water-splitting is believed to be caused by the wasteful recombination of electron/hole pairs, backward reaction of combining hydrogen and oxygen into water, and limitations for TiO_2 to utilize visible light due to its large band gap. Research has been carried out to produce nano-structured TiO_2 with a narrower band gap in order to utilize visible-light energy more efficiently. Progresses have been made in modifying the band gap of nano-structured TiO_2 by means of metal loading, ion doping, metal ion-implantation, dye sensitization and composite TiO_2. Noble metals, such as Pt, Au, Pd, and Ag, have been reported to be very effective in enhancing TiO_2 photocatalysis [40–43]. Carbon-doped nano-structured TiO_2 have showed much more efficient water splitting under visible-light illumination [44]. A study using a dye sensitizer for photocatalytic hydrogen production was investigated [45]. A visible light absorber, C_3N_4, has been coupled to many wide-band gap semiconductors to improve solar harvesting. A 50 wt % C_3N_4/TiO_2 junction was found to double H_2 evolution compared to pure C_3N_4 under visible irradiation [46].

3.3 Application of TiO_2 in energy storage

Lithium-ion batteries are a type of rechargeable batteries commonly used in consumer electronics. Lithium ion battery system and technology has been a revolutionary change in the field of power supply battery. Anode materials based on titanium oxides are the promising candidates as alternative materials to carbonaceous anodes due to advantages in terms of cost, safety and toxicity [47, 48]. TiO_2 also exhibits excellent structural stability, high discharge voltage plateau (more than 1.7 V versus Li^+/Li), and excellent cycling stability [49, 50].

Typically the Li^+ insertion–extraction reaction for TiO_2 polymorphs occurs according to reaction (3):

$$xLi^+ + TiO_2 + xe^- = LixTiO_2 \tag{3}$$

x can range between 0 and 1, depending strongly on the polymorph, particle size, and morphology of TiO_2. The maximum theoretical capacity is 335 mAh g^{-1} which corresponds to x = 1. This makes TiO_2 a highly competitive alternative to graphite anodes having a theoretical capacity of 372 mAh g^{-1} [51–53]. However, TiO_2 has limitations, such as low capacity, low electrical conductivity, and poor rate capability. Strategies have been developed to address the issues of TiO_2-based anodes. These include the use of multi-dimensional nanostructured TiO_2, composite and coating materials, and element doping.

One dimensional anatase TiO_2 nanofiber anodes were used as an anode active material in Li ion batteries and exhibited a high lithium storage capacity, a stable cycle life, and good rate capability [54]. Two dimensional TiO_2 nanosheets have been shown to exhibit the superior capacities, improved cycling stability and rate capabilities, owing to unique exposed facets, shortened path, and reserved porous structures [55–57]. Nanostructured TiO_2 is a low voltage insertion host for Li and a fast Li insertion/extraction host [58, 59]. These characteristics provide nanostructured TiO_2 a potential anode material for high-power Li-ion batteries. Studies on the use of nanostructured TiO_2 as anode with $LiCoO_2$ cathode demonstrated specific capacity of 169 mAh g^{-1} [60]. Xu, et al. investigated electrochemical performance of TiO_2-coated $LiCoO_2$ and $LiMn_2O_4$ in different potential regions [61]. Mechanically blended composite of nanosized TiO_2 and carbon nanotubes (CNTs) has been used as potential anode materials for Li-ion batteries. It was found that the TiO_2/CNTs nanocomposites exhibited an improved cycling stability and higher reversible capacity than CNTs [62, 63]. Metal oxide coatings containing TiO_2 can efficiently improve the capacitive performance of the materials through synergistic effects in an electrode system [64–67].

4. Application of TiO_2 in environment protection

4.1 Fundamental

Excitation of TiO_2 with UV light with energy greater than the band gap (>3.2 eV) promotes electrons from valence band into the conduction band and generates electron/hole pairs [68, 69]. **Figure 3** illustrates the mechanism of generating reactive radicals from TiO_2 under irradiation of UV light. The conduction band electrons e^- can reduce molecular oxygen to generate ($O_2^{\bullet -}$) superoxide radicals, and valance band holes h^+ is positive enough to generate ($OH^\bullet$) radicals from H_2O or OH^- on TiO_2 surface. $OH^\bullet$ radicals have the strongest oxidation potential. Superoxide radicals ($O_2^{\bullet -}$) have moderate oxidation potentials, but their diffusion distances can reach up to hundreds of micrometers [70]. Both radicals are very reactive, and they attack the organic matter present or near the surface of TiO_2 to degrade toxic and bio-resistant compounds or species into CO_2, H_2O, etc. [69, 71].

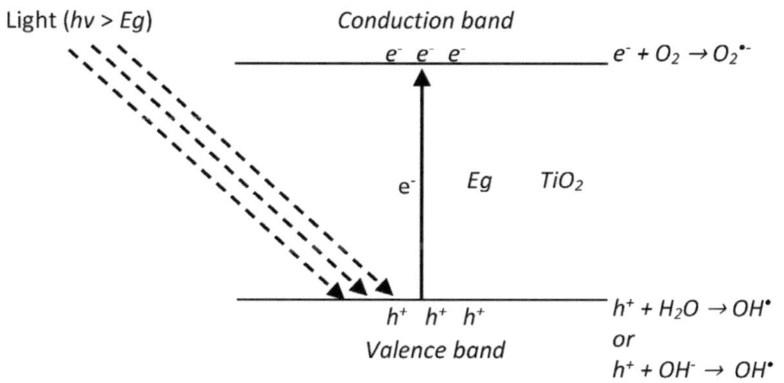

Figure 3.
Mechanism of generating reactive radicals ($OH^\bullet$) and superoxide ($O_2^{\bullet -}$) from TiO_2 under irradiation of UV light.

The generation of reactive radicals ($OH^{\bullet}$) and ($O_2^{\bullet -}$) is affected by the crystalline state, and properties such as surface area and particle size. Although anatase and rutile have the similar band gaps, anatase has shown to have more rapid rate in photo-degradation of organic or bio-resistant compound than rutile [72, 73]. Therefore, nano-structured anatase TiO_2 is often used as a catalyst in photo-degradation applications.

4.2 Self-clean and antibacterial uses of TiO_2

One application, which is commercially successful, is the nano-structured TiO_2 material for self-clean and antibacterial uses [68, 74, 75]. Many nano-structured TiO_2 material based products have been used as construction materials [76–82]. Self-clean application is based on the actions of sunlight, rainwater, and photocatalytic properties of TiO_2. Under the irradiation of sunlight, adsorbed organic materials like oil can be decomposed by hydroxyl radicals on the surface of TiO_2. Because of the hydrophilic property of TiO_2 surface, contaminates and dust can be washed away off by rainwater. Tiles containing nano-structured TiO_2 have been used to construct photocatalytic surface to decompose bacteria and viruses on the surface or bacteria floating in the air as they come in contact with surface.

Studies have shown that the photocatalytic properties of TiO_2 can sometimes be enhanced by doping TiO_2 with different elements. For example, TiO_2 nano-particles containing Ag^+ have been widely used in antibacterial plastics and coatings [83–85]. Fe or Sb-doped TiO_2 have been used to make coatings with high antibacterial property [79, 80].

Nano-structured self-clean glass is now an important commercial product. Pilkington Glass has developed the first self-cleaning windows. The window glass is coated with a very thin and transparent TiO_2 layer to have the properties of photocatalysis and hydrophilicity on the glass surface. Photocatalysis of TiO_2 break down the organic dirt adsorbed onto the window in sunlight, and the decomposed organic species is washed away efficiently by rain or other water in the form of thin layer instead of droplets [86].

4.3 Application of TiO_2 in water-treatment

Another important application of nano-structured TiO_2 is in the water-treatment, utilizing its photocatalytic properties [87–92]. Research of using nano-structured TiO_2 for water-treatment has been very active in recent years. TiO_2 has been used in the photocatalytic decomposition of organic dyes in waste water, and organic pollutants such as pesticides, dyes and pharmaceuticals in other contaminated water [93, 94]. The photocatalytic decomposition of organic matters in water are all based on the mechanism of the generation of highly reactive radicals ($OH^{\bullet}$) and superoxide ions ($O_2^{\bullet -}$) in TiO_2 under UV irradiation, as illustrated in **Figure 3**. TiO_2 has been considered to be the best choice to be used as photo-catalysts, as TiO_2 is chemically inert, and cheap to manufacture and to apply.

The complete separation and recycling of TiO_2 fine particles is important for the practical applications. A number of innovative methods have been developed for this purpose. For example, fixing TiO_2 nano-particles on supports such as glass plates, aluminum sheets, and activated carbon are investigated to recycle the catalyst [95], or developing TiO_2 catalyst system which can be separated from reaction liquid by applying external magnetic field [96, 97].

Because TiO_2 and many other semiconductors have the large band gaps, the application of photocatalytic water treatment using TiO_2 is limited by its relatively low efficiency. To improve photocatalytic efficiency of TiO_2 for water treatment, as well as other photocatalytic applications, Enormous research has been carried out to extend the photocatalytic response of TiO_2 into the visible range [98]. One of the strategies for improving photocatalytic efficiency for water treatment is to modify the band gap of TiO_2 by incorporation of other ions into TiO_2 structure, through metal and non-metal doping, metal implantation, noble metal loading, and others [99–104].

5. Application of TiO_2 in catalyst

TiO_2-based composite materials have been widely used as catalysts [105–107]. TiO_2 is used as support in commercial V_2O_5-WO_3/TiO_2 catalysts for the selective catalytic reduction (SCR) of NOx. In SCR technology, highly undesirable NOx acid gas emissions from various industrial sources are reduced to harmless N_2 and H_2O. The V_2O_5-WO_3/TiO_2 catalysts are widely used in commercial applications because of their excellent thermal stability and lower oxidation activity for the conversion of SO_2 to SO_3 [108, 109]. The V_2O_5-WO_3/TiO_2 catalysts have become the most widely used industrial catalysts for these SCR applications since the introduction of this technology in the early of 1970s [110].

TiO_2 has the potential to induce the reductive chemical transformation. The reductive photocatalysis of ethyne and ethene have been reported [111–113]. TiO_2 have been used as a useful catalyst for the reduction of carbonyl compounds such as aldehydes or ketones, nitro compounds, imines and for the some of the chemical transformations involving redox processes. Photocatalysis on TiO_2 is a light-driven redox reaction. Redox reactions can be induced by electrons (e^-) generated in conduction band (CB) and holes (h^+) simultaneously generated in valance band (VB) under the irradiation of light. The electrons in the conduction band are readily available for transferring while the holes in the valence band are open for donations [114]. The photocatalytic reduction of an electron acceptor can be carried out in the presence of a large excess amount of electron donors such as alcohols or amines, which are used to scavenge (h^+). Oxygen (O_2) is a competitive electron acceptor, and can influence the reduction reaction. Therefore, the reductive chemical transformation should be generally performed in an O_2 free environment. Under these conditions, a photocatalytic reduction proceeds through transferring electrons (e^-) in CB or trapped at surface defects of TiO_2 into the organic molecules adsorbed on TiO_2 surface. The photocatalytic reduction of aldehydes, nitro compounds, and imines have been reported. Aromatic aldehydes and ketones were reduced to the corresponding alcohols using TiO_2 as a photocatalyst [115, 116]. Aromatic and aliphatic nitro compounds were reduced to corresponding amines using TiO_2 as catalyst [117]. The direct reduction of imines to corresponding secondary amines was studied [118].

6. TiO_2 in health and biomedicine

6.1 TiO_2 in sunscreen

The sunlight reaching the earth's surface contains UV, visible and infrared wavelength. The Sun releases ultraviolet (UV) radiation in three different wavelengths, and all are harmful in different ways. These wavelengths in sunlight are called UVA (315–400 nm), UVB (280–315 nm) and UVC (100–280 nm) [119]. Because the earth's atmosphere blocks most UVC rays, UVC does not generally reach the earth's surface to a significant degree. Therefore, they are not thought to be important contributors to the biological effects on human skin [120]. UVA wavelength penetrates more deeply into the skin causing photo-aging and the formation of skin cancer. UVB is shorter, and damages the surface of the skin. The damage from UVB can cause sunburn and cancer [121–124].

TiO_2 is a semiconducting material with very high refractive index. The high refractive index is what allows the substance to scatter visible light. The current method of preventive treatment again harmful UV radiation involves suspending a substance that either absorbs or scatters UV radiation in a thick emulsion, called sunscreen. Titanium dioxide (TiO_2) is an ingredient in sunscreens where its loading is frequently 2–15%. Sunscreen typically contain chemical filters that are organic compounds that absorb strongly the UV (most often UVB) and physical filters such as TiO_2 and ZnO that block UVA and UVB sunlight through absorption, reflection and scattering.

6.2 TiO_2 for cancer treatment

In biomedicine, TiO_2 nanoparticles with their extraordinary stability, exceptional photo-reactivity, and biocompatibility have a special place in biomedical solutions. The therapeutic potential of TiO_2 lies in the ability of these particles in response to light to produce reactive oxygen species (ROS). Production of ROS is the main factor in causing detrimental effects on cells. This effect was first applied by Cai *et al* in the immortal HeLa cell lines [125]. Distinct cell death was detected after HeLa cells were illuminated with UV light in the presence of TiO_2 (100 μg/mL). TiO_2 particles in the absence of light showed little cytotoxicity for concentration as high as 360 μg/mL. This demonstrated that the cells were killed by radicals produced from water upon illumination of TiO_2 particles and also oxidized by the photogenerated holes in TiO_2. Because the size and shape of TiO_2 nanoparticles have strongly influence in crystallinity, surface characteristics, electron/hole transportation and charge separation, it is important to be able to control the shape and size of TiO_2 nanoparticles to optimize their electronic and chemical properties, resulting in more efficient site-selective reactions. Various functionalized TiO_2 nanoparticles have been designed to be used in nanomedicine, as agents for photosensitization or sonosensitization and as drug carriers [126].

In both photodynamic therapy (PDT) and sonodynamic therapy (SDT), nanostructured titanium dioxide is used as an agent to produce reactive oxygen species (ROS). Photodynamic therapy (PDT) is an anti-tumor method in which photosensitive agent is applied and target area is illuminated for the activation of the agent. TiO_2 is normally a photocatalyst that produces oxidizing radicals by reacting with water during UV exposure and can damage nearby cells [127, 128]. Titanium dioxide and zinc oxide are two of the most effective photosensitizers for PDT applications. In sonodynamic therapy, TiO_2 acts as a sonocatalyst. Studies have shown that TiO_2 particles can promote the production of hydroxyl ($OH^\bullet$) radicals by ultrasound irradiation even in dark conditions [129, 130]. Ultrasound technology has been already used for some cancer therapies, either by generating localized heating using high intensity ultrasound or by activating a drug release using low intensity ultrasound. Ultrasound can penetrate inches below the skin. Therefore, it can be used to activate TiO_2 nanoparticles deep below the skin surface.

TiO_2 has been considered to be a good material for the design of drug carriers, for the reasons that the shape and size of TiO_2 nanoparticles can be engineered to control their electronic and chemical properties, and the surface of TiO_2 nanoparticles can be functionalized with various drug molecules [131, 132]. These capabilities bring new opportunities for more efficient site-selective chemistry of TiO_2, and form the vehicles for drug delivery applications.

7. Conclusions

Titanium dioxide is a stable, non-toxic inorganic material with very high refractive index, and can scatter visible light almost completely. The particle sizes for pigmentary TiO_2 are generally engineered to be around 250 nm to have optimized light scattering property. After coating with inorganic compounds such as alumina or silica, the catalytic activity on the surface of TiO_2 particles is suppressed and the weather resistance is improved. Because of the superior optical properties and chemical stability, TiO_2 has been developed and used as white pigment over several decades. Pigmentary titanium dioxide has excellent ability to impart brightness and opacity. Titanium dioxide has now been a well established inorganic white pigment and is widely applied in the coatings, plastics, paper manufacturing, and in many common products. Global sales of titanium dioxide pigment were about 6 million tons in 2017 and the growth trend of global titanium dioxide pigment sales is continuing over the recent years.

Titanium dioxide is also a semiconducting material which is characterized by a filled valence band and an empty conduction band. When excited by photons which have energy equal to or higher than their band gap, electrons (e^-) in valence band of TiO_2 are promoted to the conduction band and holes (h^+) are created in the valence band of TiO_2. Because of the discovery of photocatalytic properties of titanium dioxide, and the ability to engineer TiO_2 nanomaterials for controlling their electronic and chemical properties, the applications of titanium dioxide as functional materials have become the focus of enormous research and development in the recent years. The applications of nano-structured TiO_2 can now be found in a wide range of areas including electronic materials, energy, environment, health & medicine, and catalysts. A number of materials containing nano-structured TiO_2 have become the important commercial products. Further research is continuing to modify the electronic and chemical properties, as well as surface characteristics of TiO_2 for the creation of more efficient TiO_2 functional materials in more specific application areas.

Acknowledgements

This work was financially supported by the Pangang Group under a Basic Research Grant.

Conflict of interest

The authors declare no conflict of interest.

Author details

Xiaoping Wu
Pangang Group Research Institute Co. Ltd., State Key Laboratory of Vanadium and Titanium Resources Comprehensive Utilization, Panzhihua, Sichuan, China

*Address all correspondence to: 13308173290@163.com

References

[1] Winkler J. Titanium Dioxide, Production, Properties and Effectives Usage. 2nd ed. Hanover: Vincentz Network; 2013. 42p

[2] Chemours. Titanium Technologies Presentation [Internet]. 2018. Available from: https://s21.q4cdn.com/411213655/files/doc_presentations/2018/09/FINAL-September-2018-TT-Investor-Deck.pdf [Accessed: 2021-04-12]

[3] Kennedy D, Ritchie M, Mackenzie J. The photosorption of oxygen and nitric oxide on titanium dioxide. Transactions of the Faraday Society. 1958;54:119-129. DOI: 10.1039/TF9585400119

[4] Barry T, Stone F. The reactions of oxygen at dark and irradiated zinc oxide surfaces. Proceedings of the Royal Society A. 1960;255:124-144. DOI: 10.1098/rspa.1960.0058

[5] Doerffler W, Hauffe K. Heterogeneous photocatalysis I. The influence of oxidizing and reducing gases on the electrical conductivity of dark and illuminated zinc oxide surfaces. Journal of Catalysis. 1964;3:156-170. DOI: 10.1016/0021-9517(64)90123-X

[6] Fujishima A, Honda K. Electrochemical photolysis of water at a semiconductor electrode. Nature. 1972; 238:37-38. DOI: 10.1038/238037a0

[7] Linsebigler AL, Lu G, and Yates JT. Photocatalysis on TiO_2 surfaces: principles, mechanisms, and selected results. Chemical Reviews. 1995;95: 735-758. DOI: 10.1021/cr00035a013

[8] Hoffmann MR, Martin ST, Choi W, Bahnemann DW. Environmental applications of semiconductor photocatalysis. Chemical Reviews. 1995; 95:69-96. DOI: 10.1021/cr00033a004

[9] Wang R, Hashimoto K, Fujishima A, Chikuni M, Kojima E, Kitamura A, Shimohigoshi M, Watanabe T. Light-induced amphiphilic surfaces. Nature. 1997;388:431-432. DOI: 10.1038/41233

[10] Lakshmanan VI, Bhowmick A, Ha lim MA. In: Brown J, editor. Titanium Dioxide: Chemical Properties, Applications and Environmental Effects. New York: Science Publishers; 2014. p. 75-130

[11] Pierre AC, Pajonk GM. Chemistry of aerogels and their applications. Chemical Reviews. 2002;102:4243-4266. DOI: 10.1021/cr0101306

[12] Hench LL, West JK. The sol-gel process. Chemical Reviews. 1990;90: 33-72. DOI: 10.1021/cr00099a003

[13] Lee M, Lee G, Ju C, Hong S. Preparations of nanosized TiO_2 in reverse microemulsion and their photocatalytic activity. Solar Energy Materials and Solar Cells. 2005;88: 389-401. DOI: 10.1016/j.solmat.2004.11.010

[14] Schaf O, Ghobarkar H, Knauth P. Hydrothermal synthesis of nanomaterials. In: Knauth P, Shoonman J, editors. Nanostructured materials: selected synthesis methods, properties and applications. New York: Kluwer Academic Publishers; 2004. p. 23-41

[15] Wahi RK, Liu Y, Falkner JC, Colvin VL. Solvothermal synthesis and characterization of anatase TiO_2 nanocrystals with ultrahigh surface area. Journal of Colloid and Interface Science. 2006;302:530-536. DOI: 10.1016/j.jcis.2006.07.003

[16] Sivalingam G, Priya MH, Madra G. Kinetics of the photodegradation of substituted phenols by solution combustion synthesized TiO_2. Applied Catalysts B: Environmental.

2004;51:67-76. DOI: 10.1016/j.apcatb.2004.02.006

[17] Byun D, Jin B, Kim J, Lee JK, Park D. Photocatalytic TiO_2 deposition by chemical vapor deposition. Journal of Hazardous Materials. 2000;73:199-206. DOI: 10.1016/S0304-3894(99)00179-X

[18] Miyata T, Tsukada S, Minami T. Preparation of anatase TiO_2 thin films by vacuum arc plasma evaporation Thin Solid Films. 2006;496:136-140. DOI: 10.1016/j.tsf.2005.08.294

[19] Gazquez M, Bolivar J, Tenorio R, Vaca F. A Review of the production cycle of titanium dioxide pigment. Materials Science and Applications. 2014;5:441-458. DOI: 10.4236/msa.2014.57048

[20] Linak E, Inoguchi Y. Chemical economics handbook: titanium dioxide. SRI Consulting, Menlo Park. 2005

[21] Braun JH, Baidins A, Marganski RE. TiO_2 pigment technology: a review. Progress in Organic Coatings. 1992;20:105-138. DOI: 10.1016/0033-0655(92)80001-D

[22] Hadjiivanov KI, Klissurski DG. Surface chemistry of titania (anatase) and titania-supported catalysts. Chemical Society Reviews. 1996;25:61-69. DOI: 10.1039/CS9962500061

[23] M. Grätzel. Conversion of sunlight to electric power by nanocrystalline dye sensitized solar cells. Journal of Photochemistry and Photobiology A: Chemistry. 2004;164:3-14. DOI: 10.1016/j.jphotochem.2004.02.023

[24] Ni M, Leung M, Sumathy K. A review and recent developments in photocatalytic water-splitting using TiO_2 for hydrogen production. Renewable and Sustainable Energy Reviews. 2007;11:401-425. DOI: 10.1016/j.rser.2005.01.009

[25] Grätzel M. Photoelectrochemical cells. Nature. 2001;414:338344. DOI: 10.1038/35104607

[26] O'Regan B, Grätzel M. A low-cost, high-efficiency solar cell based on dye-sensitized colloidal TiO_2 films. Nature. 1991;353:737-740. DOI: 10.1038/353737a0

[27] Chiba Y, Islam A, Watanabe Y, Komiya R, Koide N, Han L. Dye-sensitized solar cells with conversion efficiency of 11.1%. Japanese Journal of Applied Physics. 2006;45:L638-640. DOI: 10.1143/JJAP.45.L638

[28] Yu Q, Wang Y, Yi Z, Zu N, Zhang J, Zhang M, Wang P. high-efficiency dye-sensitized solar cells: the influence of lithium ions on exciton dissociation, charge recombination, and surface states. ACS Nano. 2010;4:6032-6038. DOI: 10.1021/nn101384e

[29] Smestad G, Bignozzi C, Argazzi R. Testing of dye sensitized TiO_2 solar cells I: Experimental photocurrent output and conversion efficiencies. Solar Energy Materials and Solar Cells. 1994;32:259-272. DOI: 10.1016/0927-0248(94)90263-1

[30] Kay A, Grätzel M. Low cost photovoltaic modules based on dye sensitized nanocrystalline titanium dioxide and carbon powder. Solar Energy Materials and Solar Cells. 1996;44:99-117. DOI: 10.1016/0927-0248(96)00063-3

[31] Hagfeldt A, Grätzel M. Molecular Photovoltaics. Accounts of Chemical Research. 2000;33:269-277. DOI: 10.1021/ar980112j

[32] Nazeeruddin MK, Angelis FD, Fantacci S, Selloni A, Viscardi G, Liska P, Ito S, Takeru B, Grätzel M. Combined experimental and DFT-TDDFT computational study of photoelectrochemical cell ruthenium sensitizers. Journal of American

Chemical Society. 2005;127: 16835-16847. DOI: 10.1021/ja052467l

[33] Cao Y, Bai Y, Yu Q, Cheng Y, Liu S, Shi D, Gao F, Wang P. Dye-sensitized solar cells with a high absorptivity ruthenium sensitizer featuring a 2-(hexylthio)thiophene conjugated bipyridine. The Journal of Physical Chemistry C. 2009;113:6290-6297. DOI: 10.1021/jp9006872

[34] Green MA, Hishikawa Y, Dunlop ED, Levi DH, Hohl-Ebinger J, Ho-Baillie A. Solar cell efficiency tables (version 51). Progress in Photovoltaics. 2018;26:3-12. DOI: 10.1002/pip.2978

[35] Kim HS, Kim HS, Lee CR, Im JH, Lee KB, Moehl T, Marchioro A, Moon SJ, Humphry BR, Yum JH, Moser JE, Grätzel M, Park N.G. lead iodide perovskite sensitized all-solid-state submicron thin film mesoscopic solar cell with efficiency exceeding 9%. Scientific Reports. 2012;2:591-597. DOI: 10.1038/srep00591

[36] Heo J, Im S, Noh J, et al. Efficient inorganic–organic hybrid heterojunction solar cells containing perovskite compound and polymeric hole conductors. The Nature Photonics. 2013;7:486-491. DOI: 10.1038/nphoton.2013.80

[37] Burschka J, Burschka J, Pellet N, Moon S, Humphry-Baker R, Gao P, Nazeeruddin MK, Grätzel M. Sequential deposition as a route to high-performance perovskite-sensitized solar cells. Nature. 2013;499:316-319. DOI: 10.1038/nature12340

[38] Fujishima A, Rao TN, Tryk DA. Titanium Dioxide Photocatalysis. Journal of Photochemistry and Photobiology. C. 2000;1:1-21. DOI: 10.1016/S1389-5567(00)00002-2

[39] Tryk DA, Fujishima A, Honda K. Recent topics in photoelectrochemistry: achievements and future prospects. Electrochimica Acta. 2000;45: 2363-2376. DOI: 10.1016/S0013-4686(00)00337-6

[40] Bamwenda GR, Tsubota S, Nakamura T, Haruta M. Photoassisted hydrogen production from a water ethanol solution: a comparison of activities of Au-TiO_2 and Pt-TiO_2. Journal of Photochemistry and Photobiology A: Chemistry. 1995;89:177-189. DOI: 10.1016/1010-6030(95)04039-I

[41] Sakthivel S, Shankar MV, Palanichamy M, Arabindoo B, Bahnemann DW, Murugesan V. Enhancement of photocatalytic activity by metal deposition: characterization and photonic efficiency of Pt, Au and Pd deposited on TiO_2 catalyst. Water Research. 2004;38:3001-3008. DOI: 10.1016/j.watres.2004.04.046

[42] Li FB, Li XZ. The enhancement of photodegradation efficiency using Pt-TiO_2 catalyst. Chemosphere, 2002;48: 1103-1111. DOI: 10.1016/S0045-6535(02)00201-1

[43] Kim S, Choi W. Dual photocatalytic pathways of trichloroacetate degradation on TiO_2: effects of Nanosized platinum deposition on kinetics and mechanism. The Journal of Physical Chemistry B. 2002;106: 13311-13317. DOI: 10.1021/jp0262261

[44] Park JH, Kim S, Bard AJ. Novel carbon-doped TiO_2 nanotube arrays with high aspect ratios for efficient solar water splitting. Nano Letter. 2006;6: 24-28. DOI: 10.1021/nl051807y

[45] Dhanalakshmi KB, Latha S, Anandan S, Maruthamuthu P. Dye sensitized hydrogen evolution from water. International Journal of Hydrogen Energy. 2001;26:669-674. DOI: 10.1016/S0360-3199(00)00134-8

[46] Yan H, Yang H. TiO_2–g-C_3N_4 composite materials for photocatalytic H_2 evolution under visible light

irradiation. Journal of Alloys and Compounds. 2011;509: L26-L29. DOI: 10.1016/j.jallcom.2010.09.201

[47] Kavan L. Electrochemistry of titanium dioxide: some aspects and highlights. Chem. Record. 2012;12:131-142. DOI: 10.1002/tcr.201100012

[48] Yang Z, Choi D, Kerisit S, Rosso KM, Wang D, Zhang J, Graff G, Liu J. Nanostructures and lithium electrochemical reactivity of lithium titanites and titanium oxides: A review. Journal of Power Sources. 2009;192: 588-598. DOI: 10.1016/j.jpowsour.2009.02.038

[49] Li X, Zhang Y, Li T, Zhong Q, Li H, Huang J. Graphene nanoscrolls encapsulated TiO_2 (B) nanowires for lithium storage. Journal of Power Sources. 2014;268:372-378. DOI: 10.1016/j.jpowsour.2014.06.056

[50] Jin J, Huang S, Liu J, Li Y, Chen D, Wang H, Yu Y, Chen L, Su B. Design of new anode materials based on hierarchical, three dimensional ordered macromesoporous TiO_2 for high performance lithium ion batteries. Journal of Materials Chemistry A. 2014; 2:9699-9708, 2014. DOI: 10.1039/c4ta01775g

[51] Murphy DW, Cava RJ, Zahurak SM, Santoro A. Ternary $LixTiO_2$ phases from insertion reactions. Solid State Ionics. 1983;9-10:413-417. DOI: 10.1016/0167-2738(83)90268-0

[52] Zachau-Christiansen B, West K, Jacobsen T, Atlung S. Lithium insertion in different TiO_2 modifications. Solid State Ionics. 1988;28:1176-1182. DOI: 10.1016/0167-2738(88)90352-9

[53] Dambournet D, Belharouak I, Amine K. Tailored preparation methods of tio_2 anatase, rutile, brookite: mechanism of formation and electrochemical properties. Chemistry of Materials. 2010;22:1173-1179. DOI: 10.1021/cm902613h

[54] Tammawat P, Meethong N. Synthesis and characterization of stable and binder-free electrodes of TiO_2 nanofibers for Li ion batteries. Journal of Nanomaterials. 2013. Article ID 413692. 8 pages. DOI: 10.1155/2013/413692

[55] Chen J, Liu H, Qiao S, and Lou X. Carbon-supported ultra-thin anatase TiO_2 nanosheets for fast reversible lithium storage. Journal of Materials Chemistry. 2011;21:5687-5692. DOI: 10.1039/C0JM04412A

[56] Chen J, Lou X. Anatase TiO_2 nanosheet: an ideal host structure for fast and efficient lithium insertion/extraction. Electrochemistry Communications. 2009;11:2332-2335. DOI: 10.1016/j.elecom.2009.10.024

[57] Nguyen-Phan TD, Pham VH, Kweon H, et al. Uniform distribution of TiO_2 nanocrystals on reduced graphene oxide sheets by the chelating ligands. Journal of Colloid and Interface Science. 2012;367:139-147. DOI: 10.1016/j.jcis.2011.10.021

[58] Winter M, Brodd RJ. What are batteries, fuel cells and supercapacitors? . Chemical Reviews. 2004;104: 4245-4244. DOI:10.1021/cr020730k

[59] Whittingham MS, Savinell RF, Zawodzinski T. Introduction: Batteries and Fuel Cells. Chemical Reviews. 2004; 104:4243-4244. DOI: 10.1021/cr020705e

[60] Subramamian V, Karki A, Gnansekra KI, Eddy FP, Rambabu B. Nanocrystalline TiO_2 (anatase) for Li-ion batteries. Journal of Power Sources. 2006;159:219-222. DOI: 10.1016/j.jpowsour.2006.04.027

[61] Zang Y, Zhang Z, Gong Z. Electrochemical performance and surface properties of bare and

TiO_2-coated cathode materials in lithium-ion batteries. Journal of Physical Chemistry B. 2004;108:17546-17552. DOI: 10.1021/jp046980h

[62] Huang H, Zhang WK, Gan XP, Wang C, Zhang L. Electrochemical investigation of TiO_2/carbon Nanotubes nanocomposite as anode materials for lithium-ion batteries. Material Letters. 2007;61:296-299. DOI: 10.1016/j.matlet.2006.04.053

[63] Fu L, Liu H, Zhang H, Li C, Zhang T, Wu Y, Wu H. Novel TiO_2/C nanocomposites for anode materials of lithium-ion batteries. Journal of Power Sources. 2006;159:219-222. DOI: 10.1016/j.jpowsour.2006.04.081

[64] Park S, Seo S, Lee S, Seo S. Sb: SnO_2@TiO_2 heteroepitaxial branched nanoarchitectures for Li ion battery electrodes. The Journal of Physical Chemistry C. 2012; 116:21717-21726. DOI: 10.1021/jp308077a

[65] Wu X, Zhang S, Wang L, Du Z, Fang H, Ling Y, Huang Z. Coaxial SnO_2@TiO_2 nanotube hybrids: from robust assembly strategies to potential application in Li^+ storage. Journal of Materials Chemistry. 2012; 22: 11151-11158. DOI: 10.1039/C2JM30885A

[66] Guan C, Wang X, Zhang Q, Fan Z, Zhang H, Fan H. Highly stable and reversible lithium storage in SnO_2 nanowires surface coated with a uniform hollow shell by atomic layer deposition. Nano Letters. 2014;14: 4852-4858. DOI: 10.1021/nl502192p

[67] Jeun J, Park K, Kim D, Kim W, Kim H, Lee B, Kim H, Yu W, Kanga K, Hong S. SnO_2@TiO_2 double shell nanotubes for a lithium ion battery anode with excellent high rate cyclability. Nanoscale. 2013;5: 8480-8483. DOI: 10.1039/C3NR01964K

[68] Daneshvar N, Salari D, Khataee AR. Photocatalytic degradation of azo dye acid red 14 in water: investigation of the effect of operational parameters. Journal of Photochemistry and Photobiology A: Chemistry. 2003;157:111-116. DOI: 10.1016/S1010-6030(03)00015-7

[69] Daneshvar N, Salari D, Niaie A, Rasoulifard MH. A. R. Khataee. Immobilization of TiO_2 nanopowder on glass beads for the photocatalytic decolorization of an azo dye c.i. direct red 23. Journal of Environmental Science and Health, Part A. 2005;40: 1605-1617. DOI: 10.1081/ESE-200060664

[70] Tachikawa T, Majima T. Single-molecule fluorescence imaging of tio_2 photocatalytic reactions. Langmuir. 2009;25:7791-7802. DOI: 10.1021/la900790f

[71] Daneshvar N, Aleboyeh A, Khataee AR. The evaluation of electrical energy per order (E_{Eo}) for photooxidative decolorization of four textile dye solutions by the kinetic model. Chemosphere. 2005;59:761-767. DOI: 10.1016/j.chemosphere.2004.11.012

[72] Loddo V, Marcì G, Martín C, Palmisano L, Rives V, Sclafania A. Preparation and characterisation of TiO_2 (anatase) supported on TiO_2 (rutile) catalysts employed for 4-nitrophenol photodegradation in aqueous medium and comparison with TiO_2 (anatase) supported on Al_2O_3. Applied Catalysis B: Environmental. 1999;20:29-45. DOI: 10.1016/S0926-3373(98)00089-7

[73] Bakardjieva S, Subrt J, Stengl V, Dianez M, Sayagues M. Photoactivity of anatase-rutile TiO_2 nanocrystalline mixtures obtained by heat treatment of homogeneously precipitated anatase. Applied Catalysis B: Environmental. 2005;58:193-202. DOI: 10.1016/j.apcatb.2004.06.019

[74] Beydoun D, Amal DR, Low G, McEvoy S. Role of nanoparticles in

photocatalysis. Journal of Nanoparticles Research. 1999;1:439-458. DOI: 10.1023/A:1010044830871

[75] Henglein A. Small-particle research: physicochemical properties of extremely small colloidal metal and semiconductor particles. Chemical Reviews. 1989;89:1861-1873. DOI: 10.1021/cr00098a010

[76] Guan K. Relationship between photocatalytic activity, hydrophilicity and self-cleaning effect of TiO_2/SiO_2 films. Surface and Coating Technology. 2005;191:155-160. DOI: 10.1016/j.surfcoat.2004.02.022

[77] Fujishima A, Zhang A. Titanium dioxide photocatalysis: present situation and future approaches. Comptes Rendus Chimie. 2006;9:750-760. DOI: 10.1016/j.crci.2005.02.055

[78] Cheng Q, Li C, Pavlinek V, Saha P, Wang H. Surface-modified antibacterial TiO_2/Ag^+ nanoparticles: Preparation and properties. Applied Surface Science. 2006;252:4154-4160. DOI: 10.1016/j.apsusc.2005.06.022

[79] Trapalis C, Keivanidis P, Kordas G, Zaharescu M, Crisan M, Szatvanyi A, Gartner M. Nanostructured thin films with antibacterial properties. Thin Solid Films. 2003;433:186-190. DOI: 10.1016/S0040-6090(03)00331-6

[80] Zhang H, Wen D. Antibacterial properties of Sb-TiO_2 thin films by RF magnetron co-sputtering. Surface & Coating Technology. 2007;201:5720-5723. DOI: 10.1016/j.surfcoat.2006.07.109

[81] Yuranova T, Laub D, Kiwi J. Synthesis, activity and characterization of textiles showing self-cleaning activity under daylight irradiation. Catalysis Today. 2007;122:109-117. DOI: 10.1016/j.cattod.2007.01.040

[82] Mills A, Lepre A, Elliott N, Bhopal S, Parkin IP, O'Neill S. Characterisation of the photocatalyst Pilkington Activ (TM): a reference film photocatalyst? Journal of Photochemistry and Photobiology A: Chemistry. 2003;160:213-224. DOI: 10.1016/S1010-6030(03)00205-3

[83] Ashkarran A, Aghigh S, Kavianipour M, Farahani N. Visible light photo-and bioactivity of Ag/TiO_2 nanocomposite with various silver contents. Current Applied Physics. 2011;11:1048-1055. DOI: 10.1016/j.cap.2011.01.042

[84] Cotolan N, Rak M, Bele M, Cör A, Muresan LM, Milošev I. Sol-gel synthesis, characterization and properties of TiO_2 and Ag-TiO_2 coatings on titanium substrate. Surface & Coating Technology. 2016;307:790-799. DOI: 10.1016/j.surfcoat.2016.09.082

[85] Liu C, Geng L, Yu Y, Zhang Y, Zhao B, Zhao Q. Mechanisms of the enhanced antibacterial effect of Ag-TiO_2 coatings. Biofouling. 2018;34:190-199. DOI: 10.1080/08927014.2017.1423287

[86] Pilkington. self-clean glass [Internet]. 2021. Available from: https://www.pilkington.com/en-gb/uk/householders/types-of-glass/self-cleaning-glass. [Accessed: 2021-04-08]

[87] Kallio T, Alajoki S, Pore V, Ritala M, Laine J, Leskela M, Stenius P. Antifouling properties of TiO_2: Photocatalytic decomposition and adhesion of fatty and rosin acids, sterols and lipophilic wood extractives. Colloids and Surfaces A: Physicochemical and Engineering Aspects. 2006;29:162-176. DOI: 10.1016/j.colsurfa.2006.06.044

[88] Piranniemi M, Sillanpaa M. Heterogeneous water phase catalysis as an environmental application: a review. Chemosphere. 2002;48:1047-1060. DOI: 10.1016/S0045-6535(02)00168-6

[89] Carp O, Huisman CL, Reller A. Photoinduced reactivity of titanium

dioxide. Progress in Solid State Chemistry. 2004;32:33-177. DOI: 10.1016/j.progsolidstchem.2004.08.001

[90] Wang Y, Huang Y, Ho W, Zhang L, Zou Z, Lee S. Biomolecule-controlled hydrothermal synthesis of C–N–S-tridoped TiO_2 nanocrystalline photocatalysts for NO removal under simulated solar light irradiation. Journal of Hazardous Materials. 2009;169:77-87. DOI: 10.1016/j.jhazmat.2009.03.071

[91] Frank SN, Bard AJ. Heterogeneous photocatalytic oxidation of cyanide ion in aqueous solutions at titanium dioxide powder. Journal of the American Chemical Society. 1977;99:303-304. DOI: 10.1021/ja00443a081

[92] Frank SN, Bard AJ. Heterogeneous photocatalytic oxidation of cyanide and sulfite in aqueous solutions at semiconductor powders. Journal of Physical Chemistry. 1977;81:1484-1488. DOI: 10.1021/j100530a011

[93] Konstantinou IK, Albanis TA. TiO_2-assisted photocatalytic degradation of azo dyes in aqueous solution: kinetic and mechanistic investigations: A review. Applied Catalysis B: Environmental. 2004;49:1-14. DOI: 10.1016/j.apcatb.2003.11.010

[94] Molinari R, Pirillo F, Loddo V, Palmisano L. Heterogeneous photocatalytic degradation of pharmaceuticals in water by using polycrystalline TiO_2 and a nanofiltration membrane reactor. Catalysis Today. 2006;118:205-213. DOI: 10.1016/j.cattod.2005.11.091

[95] Li L, Zhu P, Zhang P, Chen Z, Han W. Photocatalytic oxidation and ozonation of catechol over carbon-black-modified nano-TiO_2 thin films supported on Al sheet. Water Research. 2003;37:3646-3651. DOI: 10.1016/S0043-1354(03)00269-0

[96] Fu W, Yang H, Chang L, Bala H, Li M, Zou G. Anatase TiO_2 nanolayer coating on strontium ferrite nanoparticles for magnetic photocatalyst. Colloids and Surfaces A: Physicochemical Engineering Aspects. 2006;289:47-52. DOI: 10.1016/j.colsurfa.2006.04.013

[97] Ismail AA, Bahnemann DW. Mesoporous titania photocatalysts: preparation, characterization and reaction mechanisms. Journal of Materials Chemistry. 2011;21:11686-11707. DOI: 10.1039/C1JM10407A

[98] Di Paola A, Carcia-Lopez E, Marci E, Palmisano L. A survey of photocatalytic materials for environmental remediation. Journal of Hazardous Materials. 2012;211-212:3-29. DOI: 10.1016/j.jhazmat.2011.11.050

[99] Fujishima A, Zhang X, Tryk DA. TiO_2 photocatalysis and related surface phenomena. Surface Science Reports. 2008;63:515-582. DOI: 10.1016/j.surfrep.2008.10.001

[100] Emeline AV, Kuznetsov VN, Rybchuk VK, Serpone N. Visible-light-active titania photocatalysts: the case of N-doped TiO_2-properties and some fundamental issues. International Journal of Photoenergy. 2008;Article ID 258394:19 pages. DOI: 10.1155/2008/258394

[101] Irie H, Watanabe Y, Hashimoto K. Carbon-doped anatase TiO_2 powders as a visible-light sensitive photocatalyst. Chemistry Letters. 2003;32:772-773. DOI: 10.1246/cl.2003.772

[102] Sakthivel S, Kisch H, Daylight Photocatalysis by Carbon-modified titanium dioxide. Angewandte Chemie International Edition. 2003;42:4908-4911. DOI: 10.1002/anie.200351577

[103] Morikawa T, Asahi R, Ohwaki T, Aoki K, Taga Y. Band-gap narrowing of titanium dioxide by nitrogen doping.

Japanese Journal of Applied Physics. 2001;40:L561-L563. DOI: 10.1143/JJAP.40.L561

[104] Primo A, Corma A, Garcia H. Titania supported gold nanoparticles as photocatalyst. Physical Chemistry Chemical Physics. 2011;13:886-910. DOI: 10.1039/C0CP00917B

[105] Smirniotis PG, Donovan AP, Uphade BS. Low-temperature selective catalytic reduction (SCR) of NO with NH_3 by using Mn, Cr, and Cu oxides supported on Hombikat TiO_2. Angewandte Chemie International Edition. 2001;40:2479-2482. DOI: 10.1002/1521-3773(20010702)40:13 <2479::AID-ANIE2479>3.0.CO;2-7

[106] Nakajima F, Hamada. The state-of-the-art technology of NOx control, Catalysis Today. 1996;29:109-115. DOI: 10.1016/0920-5861(95)00288-X

[107] Huang H, Long R, Yang R. A highly sulfur resistant Pt-Rh/TiO/Al_2O_3 storage catalyst for NOx reduction under lean-rich cycles, Applied Catalysis B: Environmental. 2001;33:127-136. DOI: 10.1016/S0926-3373(01)00176-X.

[108] Busca G, Lietti L, Ramis G, Berti F. Chemical and mechanistic aspects of the selective catalytic reduction of NOx by ammonia over oxide catalysts: A review. Applied Catalysis B: Environmental. 1998;18:1-36. DOI: 10.1016/S0926-3373(98)00040-X

[109] Dunn JP, Stenger HG, Wachs IE. Oxidation of sulfur dioxide over supported vanadia catalysts: molecular structure-reactivity relationships and reaction kinetics. Catalysis Today. 1999; 51:301-318. DOI: 10.1016/S0920-5861(99)00052-8

[110] Lai J, Wachs IE. A perspective on the selective catalytic reduction (SCR) of NO with NH_3 by supported V_2O_5-WO_3/TiO_2 catalysts. ACS Catalysis. 2018;8:6537-6551. DOI: 10.1021/acscatal.8b01357

[111] Boonstra AH, Mutsaers CA. Photohydrogeneration of ethyne and ethene on the surface of titanium dioxide. Journal of Physical Chemistry. 1975;79:2025-2027. DOI: 10.1021/j100586a009

[112] Yun C, Anpo M, Kodama S, Kubokawa Y. U.V. Irradiation-induced fission of a C=C or C≡C bond adsorbed on TiO_2. Journal of Chemical Society, Chemical Communications. 1980; 609-609. DOI: 10.1039/C39800000609

[113] Baba R, Nakabayashi S, Fujishima A, Honda K. Photocatalytic hydrogenation on the bimetal-deposited semiconductors powders. Journal of the American Chemical Society. 1987;109: 2273-2277. DOI: 10.1021/ja00242a007

[114] Kanno T, Oguchi T, Sakuragi H, Tokumaru K, Semiconductor-catalyzed photooxygenation of aromatic olefins, Tetrahedron Letters. 1980;21:467-470. DOI: 10.1016/S0040-4039(00)71435-3

[115] Joyce-Pruden C, Pross JK, Li Y. Photoinduced reduction of aldehydes on titanium dioxide. Journal of Organic Chemistry. 1992;57:5087-5091. DOI: 10.1021/jo00045a018

[116] Kohtani S, Yoshioka E, Saito K, Kudo A, Miyabe H. Photocatalytic hydrogenation of acetophenone derivatives and diary ketones on polycrystalline titanium dioxide. Catalysis Communications. 2010;11: 1049-1053. DOI: 10.1016/j.catcom.2010.04.022

[117] Mahdavi F, Brudon TC, Li Y. Photoinduced reduction of nitro compounds on semiconductor particulates. Journal of Organic Chemistry. 1993;58:744-746. DOI: 10.1021/jo00055a033

[118] Ohtani B, Goto Y, Nishimoto S, Inui T. Photocatalytic transfer hydrogenation of Schiff Bases with propan-2-ol by suspended semiconductor particles loaded with

platinum deposits. Journal of the Chemical Society, Faraday Transactions. 1996;92:4291-4295. DOI: 10.1039/FT9969204291

[119] Urbach F. The historical aspects of sunscreens. Journal of Photochemistry and Photobiology B: Biology. 2001;64: 99-104. DOI: 10.1016/S1011-1344(01)00202-0

[120] Gallagher RP, Lee TK. Adverse effects of ultraviolet radiation: A brief review. Progress in Biophysics and Molecular Biology. 2006;92:119-131. DOI: 10.1016/j.pbiomolbio.2006.02.011

[121] Mansoori GA, Mohazzabi P, McCormack P, Jabbari S. Nanotechnology in cancer prevention, detection and treatment: bright future lies ahead. World Review of Science, Technology and Sustainable Development. 2007;4:226-257. DOI: 10.1504/WRSTSD.2007.013584

[122] Shen B, Scaiano JC, English M. Zeolite encapsulation TiO_2-photosensitized ROS generation in cultured human skin fibroblasts, Photochemistry and photobiology. 2006; 82:5-12. DOI: 10.1562/2005-05-29-RA-551

[123] Dondi D, Albini A, Serpone N. Interactions between different solar UVB/UVA filters contained in commercial suncreams and consequent loss of UV protection, Photochemistry and Photobiology Science. 2006;5: 835-843. DOI: 10.1039/B606768A

[124] Morsella M, d'Alessandro N, Lanterna A, Scaiano J. Improving the sunscreen properties of TiO_2 through an understanding of its catalytic properties. ACS Omega. 2016;1:464–469. DOI: 10.1021/acsomega.6b00177

[125] Cai R, Hashimoto K, Itoh K, Kubota Y, Fujishima A. Photokilling of malignant cells with ultrafine TiO_2 powder. Bulletin of the Chemical Socie ty of Japan. 1991;64:1268-1273. DOI: 10.1246/bcsj.64.1268

[126] Rajh T, Dimitrijevic N, Bissonnette M, Koritarov T, Konda V. Titanium dioxide in the service of the biomedical revolution. Chemical Reviews. 2014;114:10177–10216. DOI: 10.1021/cr500029g

[127] Liu L, Miao P, Xu Y, Tian Z, Zou Z, Li G. Study of Pt/TiO_2 nanocomposite for cancer-cell treatment. Journal of Ph otochemistry and Photobiology B: Biology. 2010;98:207–210. DOI: 10.1016/j.jphotobiol.2010.01.005

[128] Çeşmeli S, Avci C. Application of titanium dioxide (TiO_2) nanoparticles in cancer therapies. Journal of Drug Targeting. 2018;27:762-766. DOI: 10.1080/1061186X.2018.1527338

[129] Ninomiya K, Ogino C, Oshima S, Sonoke S, Kuroda S Shimizu N. Targeted sonodynamic therapy using protein-modified TiO_2 nanoparticles. Ultrasonic s Sonochemistry. 2012;19:607-614. DOI: 10.1016/j.ultsonch.2011.09.009

[130] Yamaguchi S, Kobayashi H, Narita T, Kanehira K, Sonezaki S, Kudo N, Kubota Y, Terasaka S, Houkin K. Sonodynamic therapy using water-dispersed TiO_2-polyethylene glycol compound on glioma cells: Comparison of cytotoxic mechanism with photodynamic therapy. Ultrasonics Sonochemistry. 2011;18:1197-1204. DOI: 10.1016/j.ultsonch.2010.12.017

[131] Wang T, Jiang H, Wan L, Zhao Q, Jiang T, Wang B, Wang S. Potential application of functional porous TiO_2 nanoparticles in light-controlled drug release and targeted drug delivery, Acta Biomaterialia. 2015;13:354-363. DOI: 1016/j.actbio.2014.11.010

[132] Wang Q, Huang J, Li H, Chen Z, Zhao Z, Wang Y, Zhang K, Sun H, Al-Deyab S, Lai Y. TiO_2 nanotube platforms for smart drug delivery: a review. International Journal of Na nomedicine. 2016;11:4819-4834. DOI: 10.2147/IJN.S108847

Section 5

Properties of Titanium Dioxide Based Materials

Chapter 10

Effect of Titania Addition on Mechanical Properties and Wear Behavior of Alumina-10 wt.% Tricalcium Phosphate Ceramics as Coating for Orthopedic Implant

Rachida Barkallah, Rym Taktak, Noamen Guermazi and Jamel Bouaziz

Abstract

The aim of this study is to determine the effect of Titania on mechanical properties and wear behavior of Alumina-10 wt.% TCP ceramics and to evaluate the performance of Titania in improving their resistance to these effects. Al_2O_3–10 wt.% β-TCP mingled with TiO_2 to obtain a mixture which is considered as a bioactive coating that may be used in orthopedic implants. Representative bioceramic samples of such blends were prepared with different percentages of Titania and then tested using different methods and techniques. Mechanical properties, fracture toughness were evaluated using the modified Brazilian, semi-circular bending specimens. A pin-on-disk tribometer was retained to study the wear behavior. Based on the obtained results, it was found that the best mechanical properties and wear resistance was displayed for Alumina-10 wt.% TCP-5 wt.% Titania composite. This composite presents a good combination of flexural strength ($\sigma_f \approx 98$ MPa), compressive strength ($\sigma_c \approx 352$ MPa), fracture toughness ($K_{IC} \approx 13$ MPa $m^{1/2}$) and micro-hardness (Hv ≈ 8.4 GPa). In terms of tribological properties, the lowest wear volume and wear resistance was recorded for Al_2O_3–10 wt.% TCP – 5 wt.% TiO_2 composition.

Keywords: Titania, mechanical properties, wear behavior, fracture, biomaterial

1. Introduction

Biomaterial research is described through the introduction of biotechnology and advances in the comprehension of the biocompatibility of human tissues [1]. In this context, Tissue engineering applies several methods from materials engineering and life sciences in order to create artificial constructs and to achieve better and faster biological healing outcomes [2]. Bone tissue engineering researchers are interested to develop new synthetic biomaterials with similar properties to native bone [2]. Among the biomaterials, bioceramics which are widely used in medical applications and more precisely for implants in orthopedics [2].

Special attention has been given to β-tricalcium phosphate (β-$Ca_3(PO_4)^2$) (β-TCP) due to their bone-like chemical composition as well as excellent biological properties and its outstanding biological responses to physiological environments. The use of TCP has been limited in the human body due to its weak rupture resistance [3]. Therefore, much research has been interested in enhancing the mechanical resistance of β-TCP by the inclusion of several additives [4].

Alumina (Al_2O_3) have been widely studied due to their bioinert with human tissues, high wear resistance, fracture toughness and high strength [5]. The study by Sakka et al. [6] has recently been concerned with alumina/ β- tricalcium phosphate system with different percentages of β-TCP. These Al_2O_3/β-TCP composites have showed a good combination of tensile strength (26.69 MPa), compressive strength (173,468 MPa) and fracture toughness (8,762 MPa $m^{1/2}$).

The functions of additives for alumina have been often aimed to lower the sintering temperature, customize the microstructure as well as improve the product properties. In order to improve the mechanical and tribological resistance of these composites, it is essential to introduce a reinforcing agent: metallic dispersion or ceramic oxide. In this case, among the ceramic oxide agents, the addition of Titania (TiO_2) has been reported to promote the sintering. TiO_2 addition not only reduces the sintering temperature of alumina, but also influences the mechanical properties [7]. Due to its excellent wear resistance, its biocompatibility, its chemical inertness and its chemical stability in aqueous environments, we have chosen titania as the agent of reinforcement to be added to the Al_2O_3–10 wt.% TCP composites [8, 9]. The amounts of Titania were varied from 0 wt.% to 10 wt.%. Hence, this study aims to investigate the effects of TiO_2 addition to the Alumina-10 wt.% TCP composite mechanical and triboligical properties and evolution of microstructure.

Firstly, after implantation, the bone substitute suffers from several mechanical stresses notably bending and compressive stresses. Fracture, compression and bending tests are essential to ensure adequate resistance and compatibility with bone resistance and to evaluate the fracture behavior of the substitute used under tensile-shear loading [10–12].

Secondly, cracks and flaws which certainly exist in the sample reduce in a significant way the load-bearing capacity and then cause the substitute to break [13, 14]. The fracture toughness and stress intensity factor have been proposed to express the critical stress states in the vicinity of the crack tip, in the aim to analyze crack initiation and propagation [14].

Thirdly, the problem of wear due to friction in the prosthesis for the substitution of knee joints and hip has been addressed by many authors [15, 16]. This problem induces inflammatory responses in the tissues surrounding the joint which leading to a surgical Intervention. In the same vein, in artificial joints, the surfaces must be biocompatible and resisted to wear to reduce debris generation [17]. For that, the surface of the prosthesis must have sufficient mechanical and tribological stability when subjected to stresses associated with moving to avoid detachment of the surface of the implant.

In this context, our research was undertaken to discuss the influence of Titania on the densification, the mechanical and tribological behavior and microstructures of those composites as a coating for orthopedic implant. Within this context, we are interested in examining the effect of TiO_2 (1 wt.%, 2.5 wt.%; 3 wt.%; 4 wt.%; 5 wt.%; 7.5 wt.% and 10 wt.%) on the Al_2O_3–10 wt.% TCP composites sintered at 1600°C for 60 min. To reach this purpose, after sintering, flattened Brazilian discs (FBD) were used to determine the tensile strength (σt) and the elastic modulus (E). The semi-circular bending test was realized to study the σf and the mode I K_{IC} and the mode I stress intensity factor K_I was determined using the CSTBD specimen. Compression tests were conducted to determine the compressive strength (σc). Finally, the specimens were tested in sliding experiments to measure wear volume and friction coefficient. The characteristics were examined by X-ray diffraction and scanning electron microscopy. All those parameters are used to compare various formulations and then to retain the best formulation for testing samples.

2. Materials and methods

2.1 Preparation of ceramics

To produce biocomposites, the synthesis of β-tricalcium phosphate powder was conducted by solid-state reaction from calcium carbonate ($CaCO_3$) and calcium phosphate dibasic anhydrous ($CaHPO_4$). Stoichiometric amounts of high purity powders such as $CaCO_3$ (Fluka, purity ≥98.5%) and $CaHPO_4$ (Fluka, purity ≥99%), were sintered at 1000°C for 20 h according to the reaction reported in Ref [18]. While Alumina and Titania used are of commercial origin. They are high purity powders (Riedel-de Haien, purity ≥99%).

It is worthwhile to note that the size of particles of the powder was (2.53 ± 0.2 μm for Al_2O_3, 2.79 ± 0.2 μm for β-TCP and 0.11 μm for TiO_2).

The initial mixture was 90 wt% Al_2O_3 and 10 wt% TCP and was mixed with different amounts of TiO_2 (1 wt.%, 2.5 wt.%, 3 wt.%, 4 wt.%, 5 wt.%, 7.5 wt.% and 10 wt.%). The mixtures were homogeneously mixed in an agate mortar and were milled in ethanol utilizing an ultrasound machine then dried for 24 h at 80°C to eliminate the ethanol and generate a finely divided powder. The dried powder was pressed in metallic mold depending on the geometry of the specimens and uniaxially pressed using LLOYD model test (machine LR50K) at 67 MPa.

Subsequently, those compacted samples of all the compositions were sintered at 1600°C for 60 min within a programmable muffle furnace (Vecstar furnace model XF5) and in normal atmospheric conditions. All specimens were heated at a rate of 10°C min^{-1} and cooled at a rate 20°C min^{-1}.

The densification of the sintered samples was evaluated from measurements of the dimensions of the samples and the relative error of the densification value was about 1%.

2.2 Mechanical tests

At least six specimens were tested under each condition and then average values (E, σ_t, σ_f, σ_c and K_{IC}) were considered.

2.2.1 Modified Brazilian test

For the Flattened Brazilian Disc, the specimen would be separated into two halves. In fact, the crack initiation point is at the center of the specimen surface, the resulting crack would propagate in a plane normal to the loading direction. The tensile stress meets the maximum tensile strength criterion.

The FBD is more favorable to measure the tensile strength σ_t [19].

For FB test, the form of specimens is a cylindrical with a thickness of 5 mm, a diameter of 30 mm and the width of the flat 2b of 5 mm (2α = 20°) (**Figure 1**).

Using the Griffith strength criterion, for a loading angle 2 α = 20°, the tensile strength (σ_t) was determined by this equation [20]:

$$\sigma_t = 0.9644 \frac{2P_c}{\pi D t} \tag{1}$$

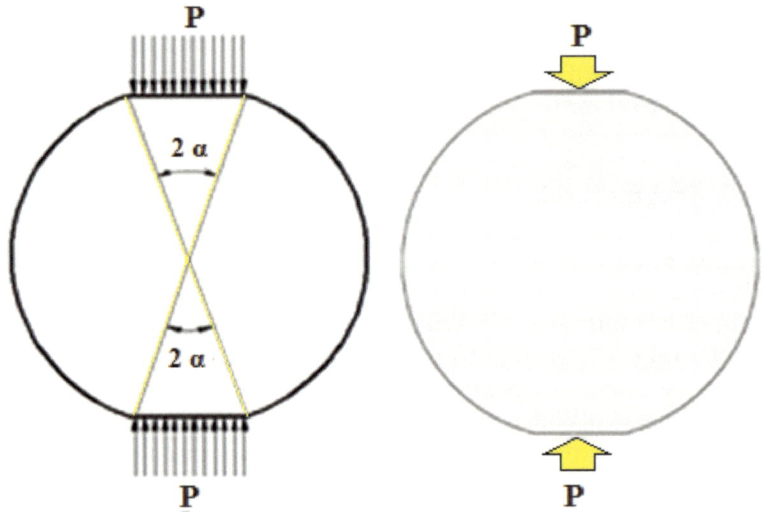

Figure 1.
Flattened Brazilian disc.

Where D is the specimen diameter, P_c is the tensile load, t is the specimen thickness.

2.2.2 Hardness measurements

In order to exclude the effect of Titania, the samples are evaluated mechanically by NanoIndenter (INNOVASTTEST). The nanoindentation experiments were performed using the instrumented indentation techniques. The indented area was measured by optical microscopy for hardness calculation.

2.2.3 Bending test

The mechanical properties of the samples were assessed using Semi Circular Bending tests to determine the flexural strength σ_f and the fracture toughness K_{IC}. The samples were positioned on the loading platform by 3-point compressive loading, at an uniform loading speed of 0.075 mm/min. The SCB specimen diameter is equal to 30 mm and 5 mm for thickness (**Figure 2a**).

The flexural strength σ_f is given according to Refs. [21–23]:

$$\sigma_f = \frac{P_{mac}}{\pi R t} Q \tag{2}$$

In which P_{max} is the maximum load, t is the specimen thickness and R is the specimen radius. Q is the corresponding components of the dimensionless stress tensor, for the isotropic case Q = 5.132 [24].

In terms of fracture toughness K_{IC}. The same specimen dimension was used by adding a crack of 4 mm in the semi disc, as shown in (**Figure 2b**). The crack-length-to diameter ratio S/D was 0.13.

Using the SCB specimen with straight crack, the fracture toughness K_{IC} was calculated with the following formula [21]:

$$K_{IC} = \frac{P_{\max}\sqrt{\pi a}}{2Rt} Y_I(a/R, S/R) \tag{3}$$

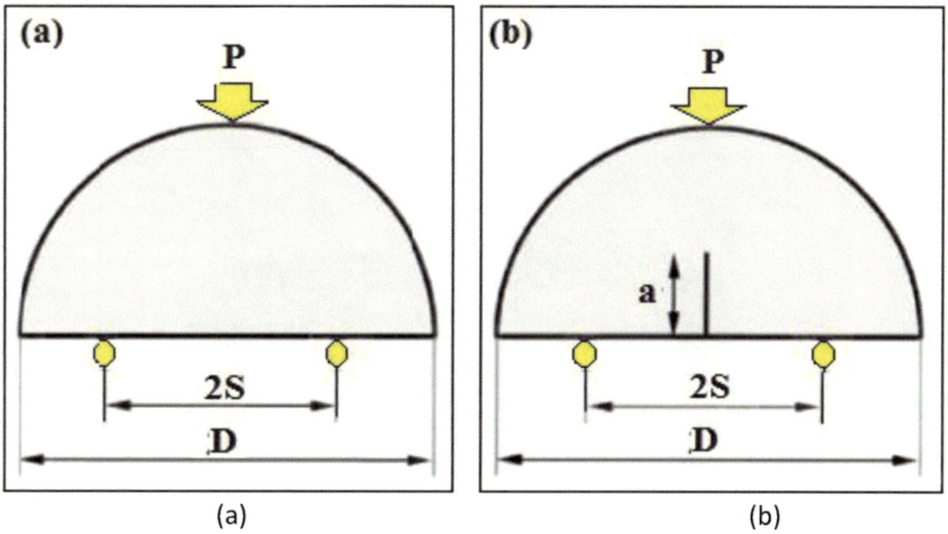

Figure 2.
Semi-circular bending Disc: (a) uncracked SCB specimen and (b) cracked SCB specimen.

Where a is the crack length, P_{max} is the maximum load, D is the cylindrical block diameter and Y_I is the geometry factor. The latter is a function of the ratio of the crack length (a) over the semi-disc radius (R) and the ratio of the half-distance between the two bottom supports (S) over the semi-disc radius (R) (**Figure 2b**). The geometry factor YI is expressed as follows [21]:

$$Y_I(a/R, S/R) = \frac{S}{R}\left(2.91 + 54.39\frac{a}{R} + 391.4\left(\frac{a}{R}\right)^2 + 1210.6\left(\frac{a}{R}\right)^3 - 1650\left(\frac{a}{R}\right)^4 + 875\left(\frac{a}{R}\right)^5\right) \tag{4}$$

2.2.4 Compression test

For the compressive test, D = 9 mm and l = 18 mm where D is the diameter and t is the length of the cylindrical specimen (**Figure 3**), as specified in ASTM standards [24, 25]. During the compressive test, the samples are positioned between compressive plates parallel and then compressed at a loading rate of 1 mm/min.

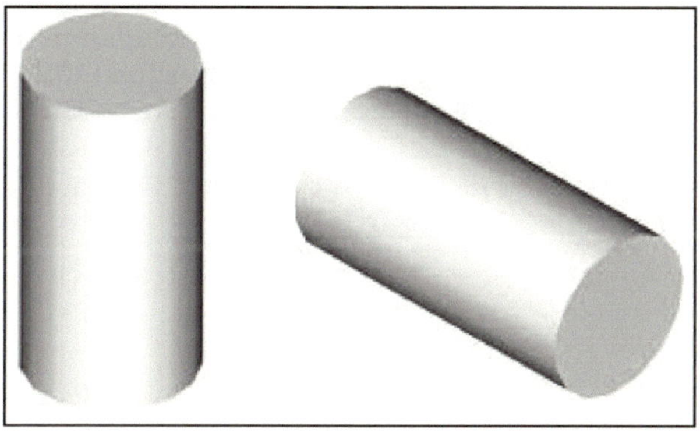

Figure 3.
Specimen for compressive test.

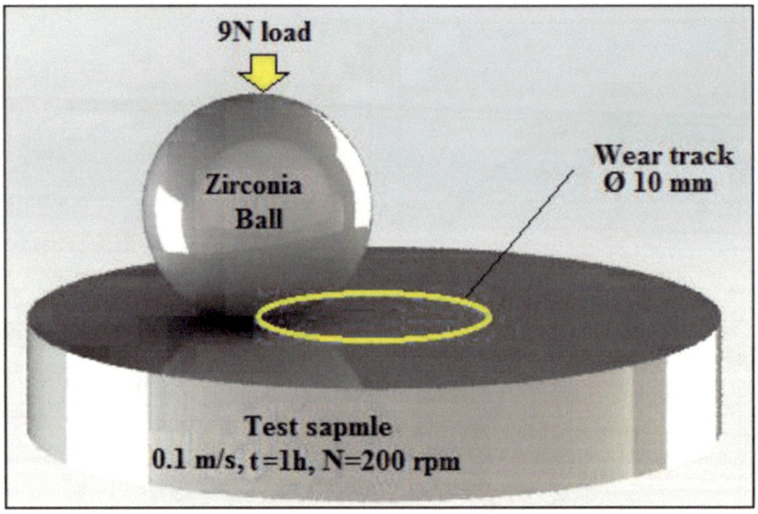

Figure 4.
Schematic representation of sliding (pion-on-disk configuration) test, showing experimental sliding parameters for Alumina- 10 wt.% TCP-TiO_2 samples (pressureless sintering at 1600°C for 1h) against ZrO_2 ball. The Sliding distance length is 377 m.

In terms of compressive properties, the compressive strength σ_c is given as follows:

$$\sigma_c = \frac{4P_{\max}}{\pi D^2} \tag{5}$$

Where D is the cylindrical block diameter and P_{max} is the maximum load.

2.3 Tribological tests

In order to evaluate the tribological properties of the Alumina-10 wt.% TCP-TiO_2 composite, sliding wear tests were carried out against sintered composites under 9 N normal load [26, 27]. Wear tests were performed using a rotating pion-on-disk tribometer, **Figure 4**, that was developed in the LGME lab [28]. A Zirconia ball of 10 mm diameter and 11 GPa hardness is fixed using a suitable device used to solicit our samples in sliding. All specimens are polished cylindrical discs of 30 mm diameter and 5 mm thickness. The parameters set for the sliding tests are a sliding velocity of 0.1 m/s. All the tests were achieved out without lubricated environment for the normal test duration of 3600 s, at room temperature (20°C) and with a relative humidity of 35 $\pm$ 5%. For each test, the friction coefficient and the wear rate are measured.

The coefficient of friction is determined from the tangential effort and the normal force by the following formula (6) [29]:

$$COF = \frac{Ft}{Fn} \tag{6}$$

Where COF is the coefficient of friction (dimensionless), *Fn* is the normal applied force (N) and *Ft* is the tangential effort (N).

The wear rate of all compositions is deduced from the measurement of mass loss. It is determined using this following equation (Eq. (7)) [30]:

$$W_r = \frac{(M_0 - M_t)}{M_0} * 100 \quad (7)$$

in which W_r is the wear rate (%), M_0 is the initial mass and M_t the mass of the specimen after sliding test.

In order to control the wear of Zirconia ball, the ball was weighed before and after each sliding test.

The wear volume of all compositions is determined from the measurement of worn surfaces of tested samples. The wear tracks were determined using 2D-profilometer.

To determine the wear volume, a profilogram (2D-profiles) is drawn in the perpendicular direction of the wear scars over a length of 8 mm [31, 32] with which we can determine the area (S_-) of the wear track (**Figure 5**). The wear volume is determined using by the following formula (8):

$$W_v = 2\,\pi r_t S_- \quad (8)$$

Where is the wear volume (mm^3), $S_-$is the area of the wear track and r_t is the radius of the wear track.

2.4 Physico-chemical characterization

The characterization of the samples after the tests is carried out using several techniques.

X-ray powder diffraction analysis (DIFFRAC SUITE, Brucker, Germany) was conducted in order to analyze the phase transformation in the different structures of each composites before and after the sintering process. The Xray radiance was created by using CuKα radiation (λ = 1.5406 Å) in the 2θ range 5–60° at a current of 40 mA, a voltage of 40 kv, and a scanning rate of 1.2°/min. The identification phase was identified out by comparing the experimental XRD-patterns with the standard files assembled by ICDD (the International Center for Diffraction Data).

Scanning electron microscopy (JEOL JSM-5400) was used to observe the surfaces of the fractured samples after the sintering process. It was equally used

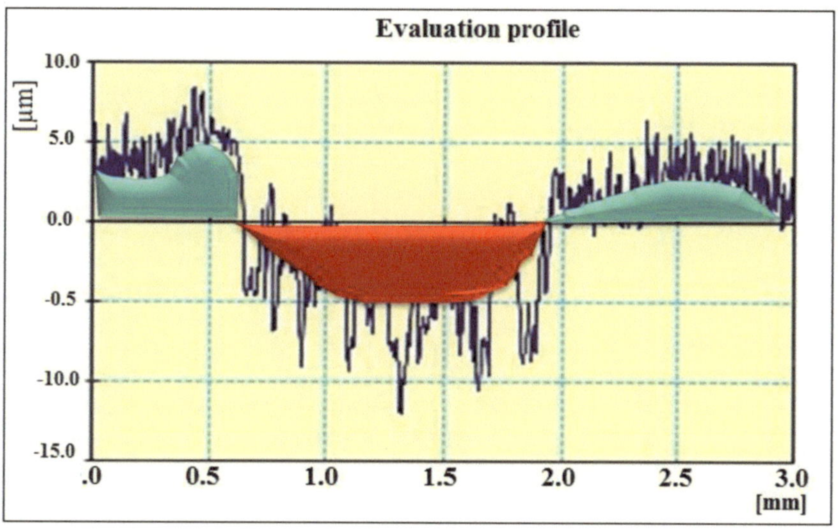

Figure 5.
Example of profilogram.

for the assessment of wear mechanisms, the microstructure of the sliding zone of the tested sample.

3. Results and discussion

In this section, we studied the effect of Titania on the mechanical behavior of Alumina-10 wt.% TCP composite with different additive amounts (1 wt.%; 2.5 wt.%; 3 wt.%; 4 wt.%; 5 wt.%; 7.5 wt.% and 10 wt.%).

The particle size analysis of the Al_2O_3–10 wt.% TCP-TiO_2 composites illustrated in **Figure 6** show that the particle distribution and the average grain size of the Al_2O_3–10 wt.% TCP-TiO_2 mixtures vary a function of the amount of Titania additive.

The particle size distribution of composites without titania is narrow, while mixtures containing TiO_2 exhibit a wide particle size dispersion. The median diameter (D_{50}) is around 0.34 μm.

3.1 Sintering behavior and densification effect

Figure 7 shows the densification behavior of various Al_2O_3–10 wt.% TCP- TiO_2 samples that contain different amounts of Titania.

It is noted that the addition of Titania has been effective in improving the densification and lowering the porosity. In fact, the addition of TiO_2 improves the density (by 30%) compared to that recorded without TiO_2. The maximum recorded densification rate is around 90% following an addition of 5% TiO_2 which is suitable for an optimum in the porosity of the order of 10%. Beyond this percentage, the density of these samples decreases slightly. So, a higher densification and a lower porosity for this composition indicates its good resistance.

3.2 Mechanical proprieties

In order to evaluate the material stiffness, ultrasound technique, is retained to estimate the elastic modulus E for different percentages of TiO_2. The evolution of elastic modulus of Al_2O_3–10 wt.% TCP- TiO_2 is illustrated in **Figure 8**. As can be seen, the elastic modulus increases with the amount of Titania additive until 194.96 GPa for 5 wt.% TiO_2. Beyond the 5 wt.% TiO_2, the overall stiffness falls

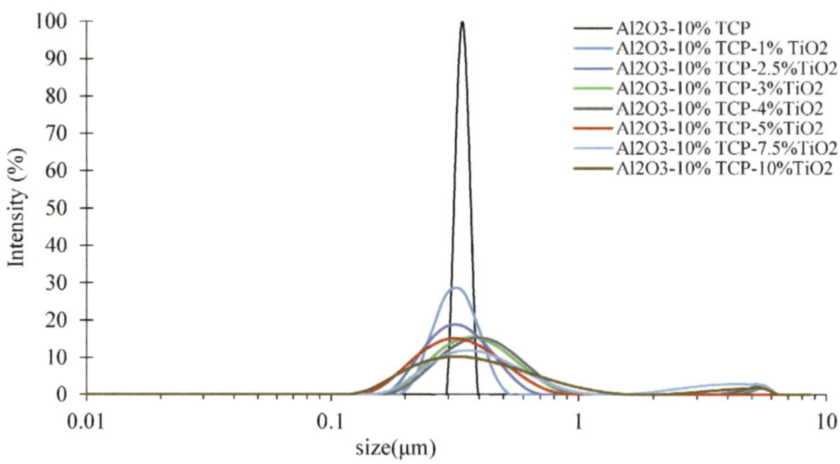

Figure 6.
Average particle size distribution plot composition for different amounts of TiO_2.

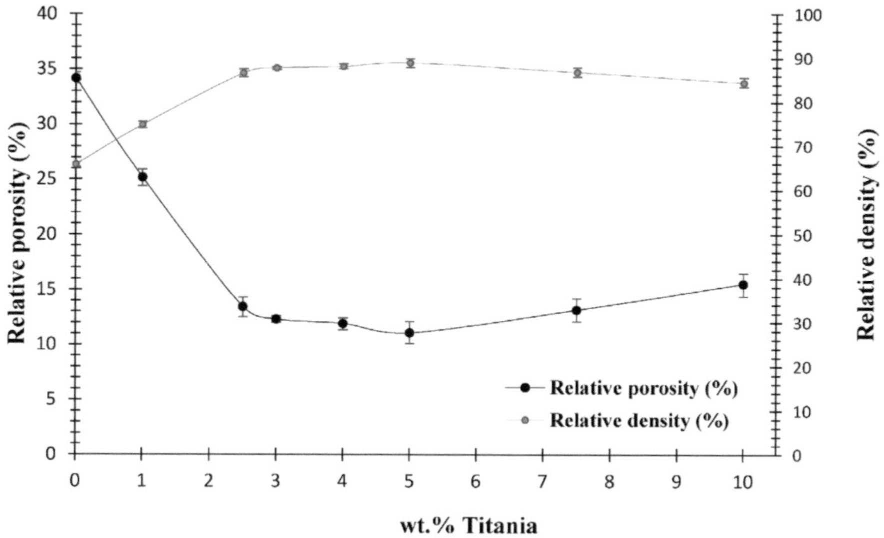

Figure 7.
Relative density and porosity of specimens added with different amounts of Titania.

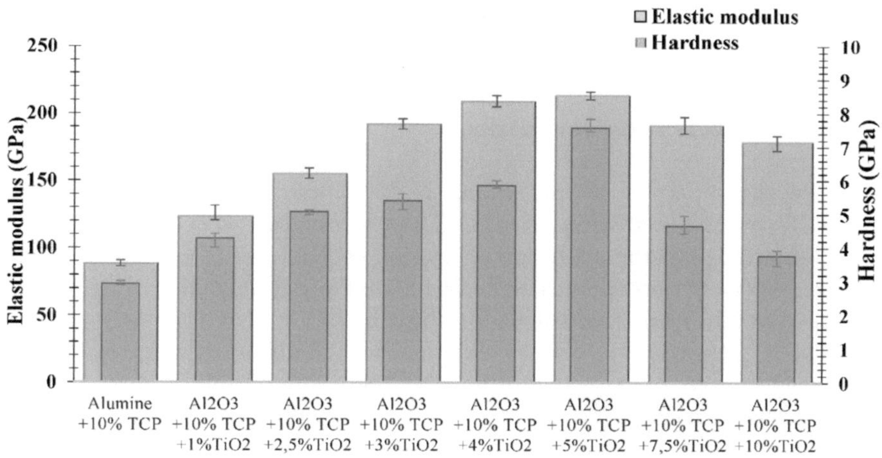

Figure 8.
Elastic modulus and hardness of Alumina-10 wt.% TCP-TiO_2 versus percentage of Titania sintered at 1600 °C for 1 h.

gradually. The Vickers indenter was applied to determine the hardness. On the other hand, at the same filler content of 5 wt % Titania, it was found that this composite provides the highest hardness values than those given by Alumina-10 wt% TCP (**Figure 8**).

Regarding the mechanical stresses, these were determined via bending, compression and Flattened Brazilian tests. The results are presented in **Figure 9**.

The tensile, flexural and compressive strength appears as being dependent on the amount of Titania additive.

It is observed that the tensile strength σ_t switches from 26.68 MPa for Alumina-10 wt.% TCP to 86.65 MPa with the addition of 5 wt.% TiO_2. For the results of the bending test, the flexural strength σ_f relative to the Alumina- 10 wt.%TCP-TiO_2 composite reaches 98 MPa.

On the other hand, and for the variation in the compressive strength σ_c of the elaborated specimens with different percentages of the Titania, one notices an

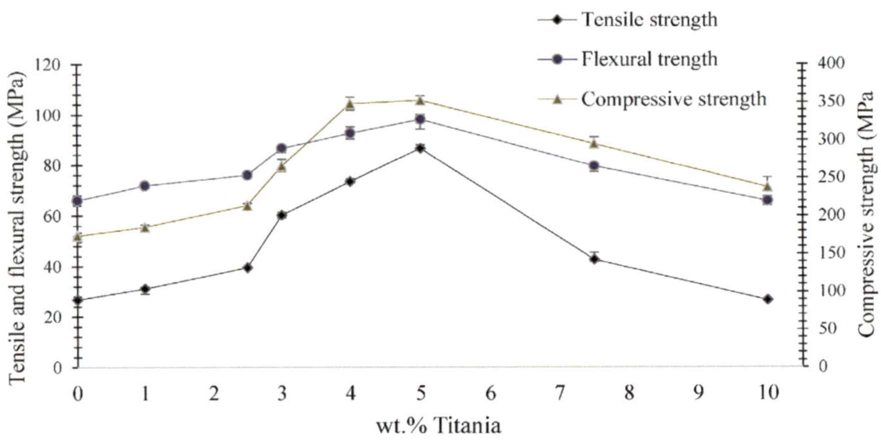

Figure 9.
Tensile strength, flexural strength and compressive strength Alumina-10 wt.% TCP-TiO$_2$ versus percentage of Titania sintered at 1600 °C for 1 h.

improvement in the compressive resistance σ_c whose maximum value reached is of the order of 352 MPa in the presence of 5 wt.% Titania (**Figure 9**).

Figure 10 shows the fracture toughness of different amounts of Titania added to Alumina-10 wt% TCP composite. As sintered at 1600°C/1 h, the toughness is improved to 13 MPa m$^{1/2}$ after 5 wt% TiO_2 adding.

3.3 Friction and wear behavior results

Likewise, the impact of adding Titania on the tribological properties of alumina-10 wt% TCP was explored with various TiO_2 percentages.

Representative plots the evolution of friction coefficients (COF) as a function of the time for different percentages of Titania tested under normal loading of 9 N at a sliding speed of 200 rpm are displayed in **Figure 11a**.

All friction curves display similar tendency for the different samples.

COF was found to increase abruptly at the beginning of the sliding test. This can be attributed to the topography of the initial surface roughness or fresh surface that has an influence on the contact intensity. Then, the variations in the curves become approximately constant and the COF values stabilized. This constant values

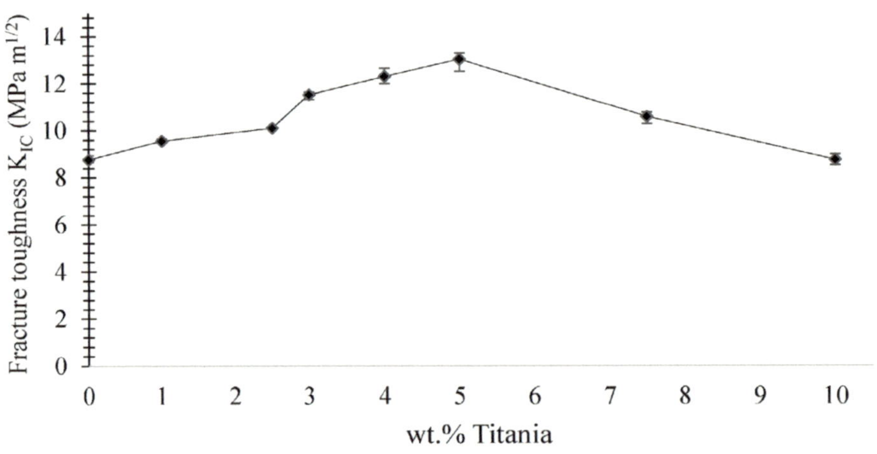

Figure 10.
Fracture toughness of different amounts of Titania added to alumina-10 wt.% TCP composite.

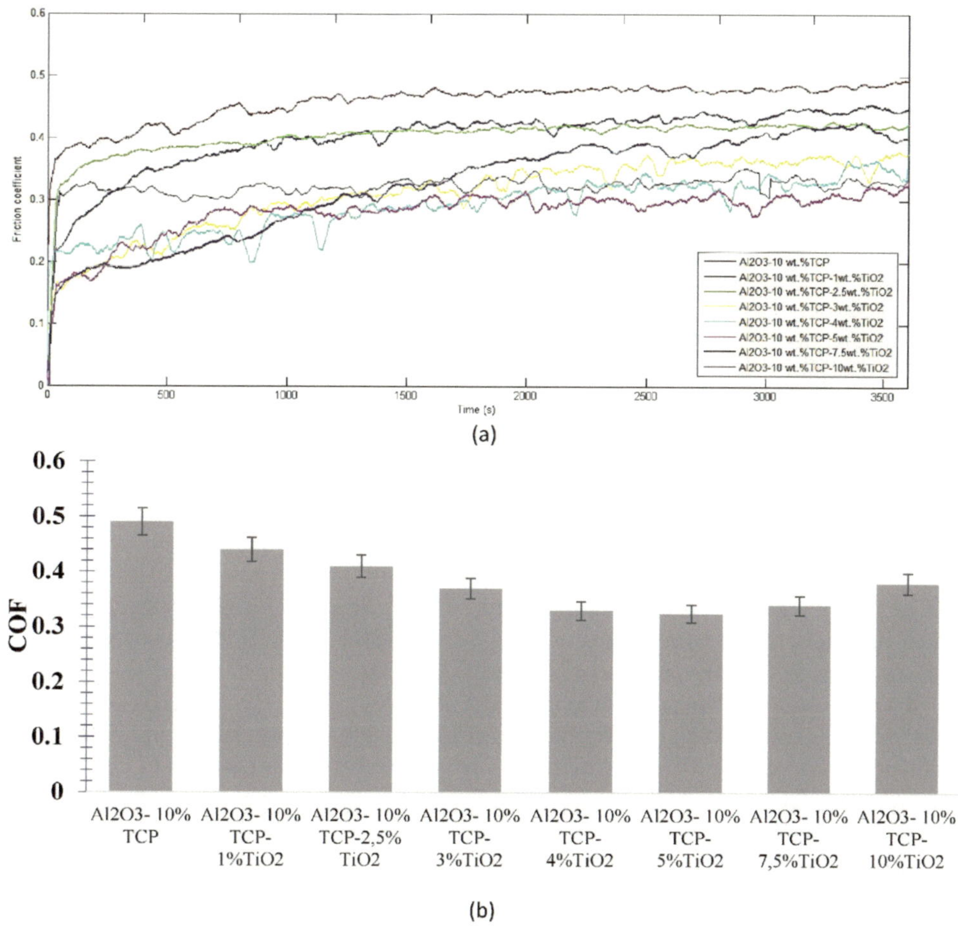

Figure 11.
(a) Plots of coefficient of friction (COF) versus time for Alumina-10 wt.% TCP-Titania composites in dry ambient condition. (b) The average steady COF as a function of % Titania.

corresponds to the average steady state value of the COF [33]. This step is named: stable wear stage.

Figure 11b presents the evolution of the friction coefficient versus the different compositions of samples during one hour (1 h) of sliding.

It is observed that a significant difference was noted between Alumina-10 wt.% TCP and Alumina-10 wt.% TCP- TiO_2 composites. Actually, the specimens were found to behave distinguishable depending on the percentage of titania under the explored conditions. In fact, the specimen of Alumina-10 wt.% TCP recorded maximum COF-values around 0.49, while Al_2O_3–10 wt.% TCP-TiO_2 specimens showed lowest COF-values about 0.325.

After each test, wear resistance and wear volume were estimated for different specimens.

The evolution of the wear rate and wear volume of Alumina-10 wt.% TCP-TiO_2 as a function of different Titania percentages is presented in **Figure 12**. According to this figure, it is noticed that the increasing of Titania content can lead to an increase in the wear resistance. The significant increase of this resistance is observed for 5 wt.% of Alumina- 10 wt.% TCP-TiO_2.

For the optimum composition, the wear volume decreases four times compared to Alumina- 10 wt.% TCP. Likewise, the lowest wear volume (0.26) was recorded for the Al_2O_3–10 wt.% TCP-5 wt.% TiO_2 composite.

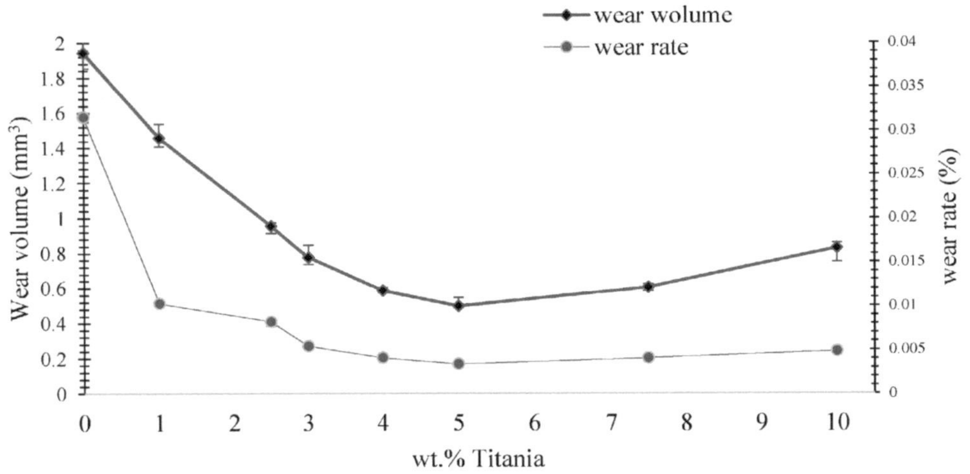

Figure 12.
Wear volume and wear rate of different samples (Alumina- 10 wt.% TCP- TiO_2) composites.

These low wear volumes are expected by the significantly higher values of the micro-hardness of the composites which go from 3800 MPa for Al_2O_3–10 wt.% TCP, to reach 9000 MPa, in the presence of 5% TiO_2 with a tenacity of the order of 13 MPa $m^{1/2}$. Indeed, the good densification of Al_2O_3–10% TCP-5% TiO_2 composite is due to mass transfers in liquid phase sintering. In this study, the liquid phase corresponds to an eutectic transformation between the initial powder (TCP) and the additive (TiO_2).

The tribological behavior of the biomaterials and specially the biocoating depends on the geometry of the contact, the tensile strength and also the surface roughness. The topography and surface roughness is an essential parameter during sliding contact which influences factors that govern sliding and wear behavior, the contact mode and the behavior of the interfacial medium.

The presence of pores in materials decreases the actual contact area between the two materials in contact, thus appears the cracks between the pores, causing wear debris to form. The wear resistance is thus reduced. The increase of the wear of these composites was probably due to the poor porosity of the Al_2O_3–10 wt% TCP–5 wt% TiO_2 (10%) (**Figure** 7).

To provide more information on wear mechanisms, microscopic observations of the worn surfaces are retained.

The specimens were analyzed by SEM after a duration of 1 hour and at a sliding speed of 200 rpm under a load of 9 N. **Figure 13** presents typical damages such as wear scars, micro-cracking, mechanical fracture and abrasion marks on the surfaces of all samples.

By co1mparing the **Figure 13a-c** and **e**, it is observed that the area of wear scar was smaller on the Al_2O_3–10 wt.%TCP-TiO_2 composite surface than Al_2O_3–10 wt.% TCP composite. This confirms that the addition of Titania to Al_2O_3–10 wt.%TCP enhances the wear resistance and tribological behavior [34]. For Al_2O_3–10 wt.% TCP-Titania composites, the debris became less deep and often compacted (**Figure 13b**, **d** and **f**).

3.4 Physicochemical characterization

The mechanical and tribological behavior confirms that a significant improvement in mechanical and tribological properties in the presence of 5 wt.% titania.

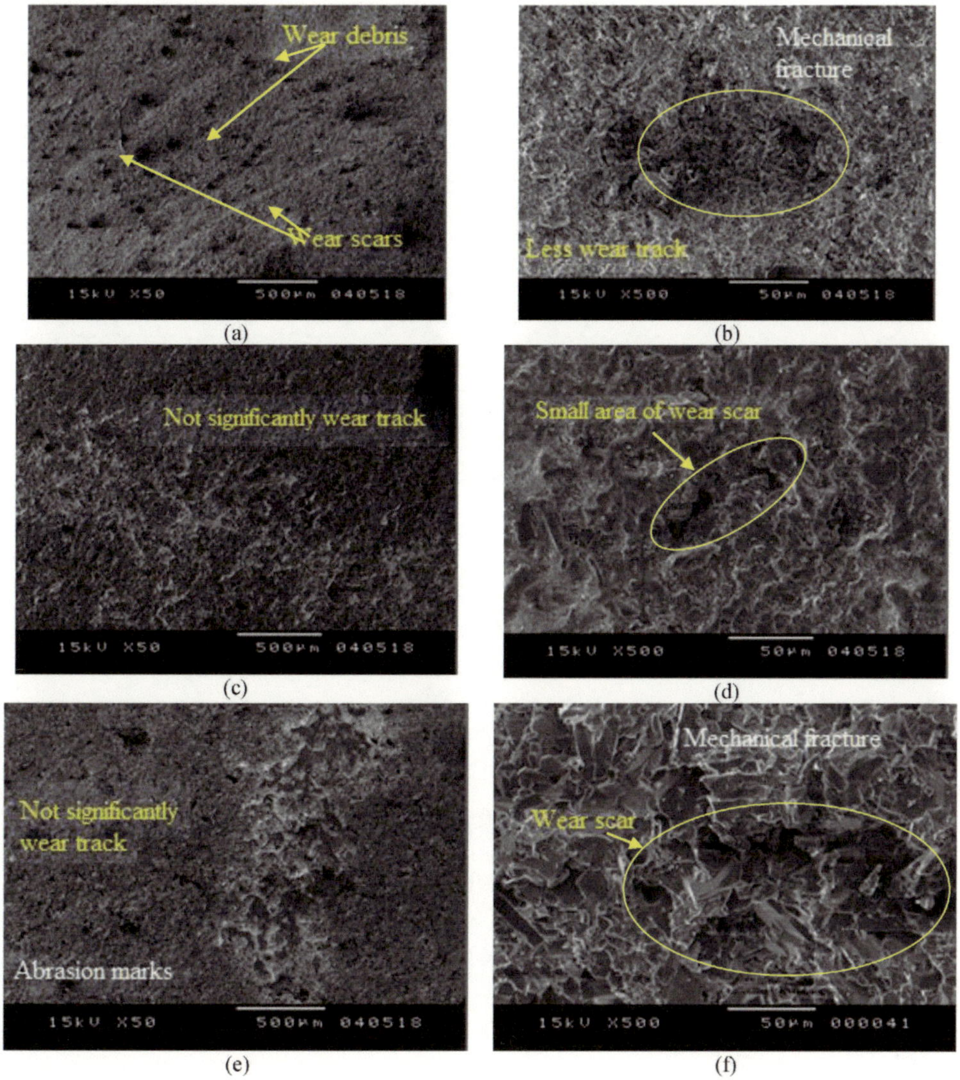

Figure 13.
SEM images of worn surfaces after testing against zirconia ball under 9N of (a) and (b) Al_2O_3- 10 wt.% TCP (c) and (d) Al_2O_3-10 wt.% TCP-5 wt.% TiO_2, (e) and (f) Al_2O_3-10 wt.% TCP-10 wt.% TiO_2.

In order to find an explanation for this improvement, an in-depth physico-chemical study was carried out to obtain more information, in particular, the microstructural changes of the samples.

Figure 14 exhibits XRD patterns of Al_2O_3–10 wt.% TCP, and Alumina-10 wt.% TCP-5 wt.% Titania composites sintered for 1 h at 1600°C. A new information was added about solid-state reactivity in the ternary system Al_2O_3-TCP-TiO_2. In fact, the spectra indicate the presence of traces of β-TCP, α-Al_2O_3, and new phases relative to β-Al_2TiO_5: aluminum titanate (**Figure 14**). This is an intermetallic compound relating to (βAl_2TiO_5) in the binary system Al_2O_3-TiO_2. It is a thermodynamically stable compound above 1280°C, formed as a result of a reaction between α- Al_2O_3 and TiO_2 in an oxidizing atmosphere. Note also that the intensity of the peak increases with the TiO_2 content in the Al_2O_3 composite – 10 wt.% TCP.

The micrograph (**Figure 15a**) displays a continuous phase and the spherical pores in the Al_2O_3–10 wt.% TCP composites. The microstructure of the Al_2O_3–10 wt.% TCP composites sintered with Titania show a high densification (**Figure 15b-d**).

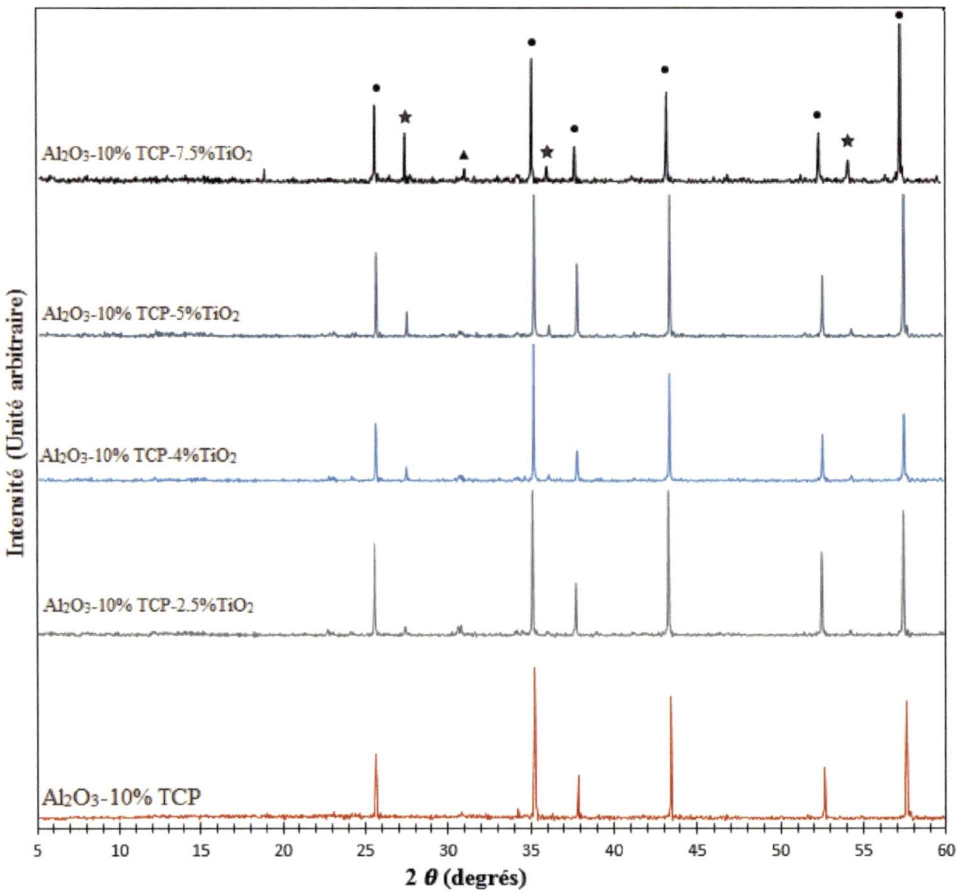

Figure 14.
XRD patterns of different composites Alumina-10 wt.% TCP-Titania versus percentage of Titania sintered at 1600 °C for 1 h.

For Al_2O_3–10 wt.% TCP-2.5 wt.% TiO_2 composite, we notice a high intergranular porosity (**Figure 15b**).

An important improvement of the characteristic performances of the Al_2O_3–10 wt.% TCP-5 wt.% TiO_2 composite was obtained by the addition of titania.

The microstructural analysis of the Al_2O_3–10 wt.% TCP-5 wt.% TiO_2 composite reveal high densification (90%) of the specimens. The creation of a liquid phase (between TCP and TiO_2) (**Figure 15c**) that can induce a higher wear and mechanical strength through the transition from solid sintering to liquid sintering. In summary, a small amount of liquid phase due to the presence of a TCP-TiO_2 eutectic caused the densification of the composites and the improvement of the mechanical properties of the Al_2O_3–10 wt.% TCP-5 wt.% TiO_2 composite.

On the other hand, the addition of high percentages of TiO_2 hinders densification was found that the intragranular pores and the growth of the grains are more important, consequently decreases in mechanical properties. For the composite Al_2O_3–10 wt.% TCP-10 wt.% TiO_2 composite (**Figure 15d**), it was found that there are intragranular pores and grain growth.

When increasing TiO_2 content in Al_2O_3–10 wt.%TCP composites, the mechanical and wear resistance of the Alumina-10 wt.% TCP composites were enhanced. In fact, the wear volume and the mechanical stresses of the Al_2O_3–10 wt.% TCP composites decreased to a minimum value with 5 wt.% Titania. In our research, the mechanical and tribological behavior of Alumina-TCP- TiO_2 bioceramics is

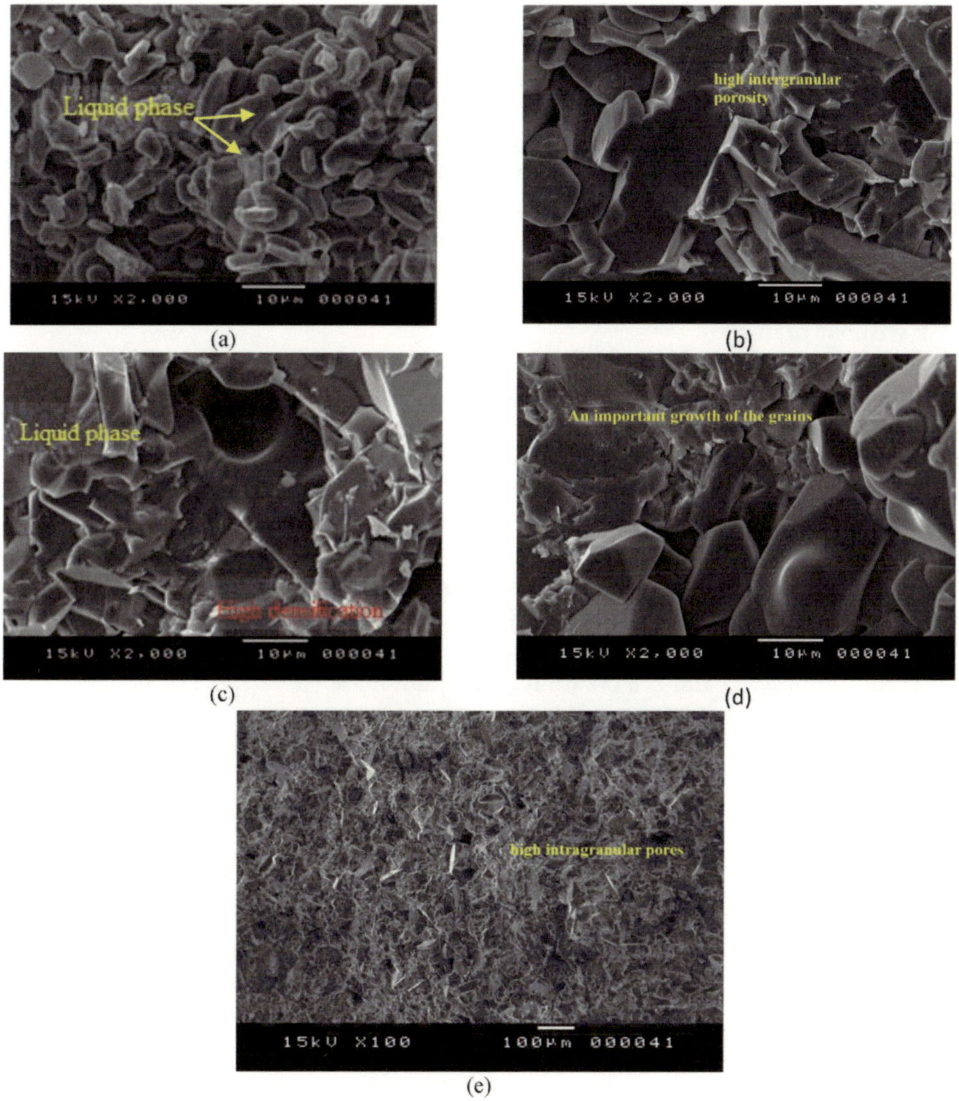

Figure 15.
SEM micrographs of the Alumina-10 wt.% TCP-Titania composites sintered at 1600 °C for 1 h: (a) Al_2O_3- 10 wt.% TCP, (b) Al_2O_3- 10 wt.% TCP-2.5 wt.% TiO_2 (c) Al_2O_3-10 wt.% TCP-5 wt.% TiO_2, (d) and (e) Al_2O_3-10 wt.% TCP-10 wt.% TiO_2.

enhanced by incorporating TiO_2, that's agrees well with other similar works [34, 35].

An important improvement of the characteristic performances of the Al_2O_3–10 wt.% TCP-5 wt.% TiO_2 composites were successfully obtained by the addition of titania oxide.

This important improvement can be explained by the high densification of the samples (90%), the creation of a liquid phase (between TCP and TiO_2) (**Figure 15c**).

4. Conclusion

In this chapter, we have focused on the effect of Titania on mechanical and tribological characterizations of Alumina-10 wt.% TCP biomaterials manufactured as coating for orthopedic implant.

After the sintering process, Alumina-10 wt.% TCP-Titania composites have been characterized by using X-ray diffraction and SEM analysis. The mechanical properties have been investigated by the Flattened Brazilian test, compression test, semi-circular bending test and NanoIndenter. A pin-on-disk tribometer was used to ensure sliding and wear experiments under dry condition and against zirconia ball. A 2D profilometer was retained to measure the wear volume.

The main results are summarized as follows:

The produced Al_2O_3–10 wt.% TCP-TiO_2 composites with different percentages of Titania (1 wt.%; 2.5 wt.%; 3 wt.%,4 wt.%,5 wt.%,7.5 wt.% and 10 wt.%) exhibited much better mechanical properties than the reported values of Alumina-TCP without titania.

After the sintering process at 1600°C for 1 hour, the Al_2O_3–10 wt.% TCP composites showed a higher elastic modulus, compressive strength, flexural strength, and fracture toughness which certainly increased with the Titania content and reached the optimum value with 5 wt.%. However, no cracks were observed in the microstructure of this composition.

In terms of tribological properties, the composite Alumina-10 wt.% TCP-5 wt.% Titania presents excellent wear rate and wear volume. Accordingly, the lowest wear volume (0.003%) was recorded for this composite. The mechanical properties and wear behavior of Alumina-TCP-TiO_2 biomaterial is enhanced by incorporating Titania. This important improvement can be explained by the high densification of the composite (90%) and the creation of a liquid phase (between Tricalcium Phosphate and Titania).

Author details

Rachida Barkallah[1*], Rym Taktak[1], Noamen Guermazi[2] and Jamel Bouaziz[1]

1 Laboratoire des Matériaux Avancés (LMA), Ecole Nationale d'Ingénieurs de Sfax (ENIS), Université de Sfax, Tunisia

2 Laboratoire de Génie des Matériaux et Environnement (LGME), Ecole Nationale d'Ingénieurs de Sfax (ENIS), Université de Sfax, Tunisia

*Address all correspondence to: barkallah_rachida1@live.fr

References

[1] Anderson J.M, Biological responses to materials. Annual review of materials research.2001;31(1),81-110. doi.org/10.1146/annurev.matsci.31.1.81

[2] EBRAHIMI Mehdi, BOTELHO, Michael G., et DOROZHKIN, Sergey V. Biphasic calcium phosphates bioceramics (HA/TCP): Concept, physicochemical properties and the impact of standardization of study protocols in biomaterials research. Materials Science and Engineering: C.2017;71, 1293-1312. doi:10.1016/j.msec.2016.11.039. Epub 2016 Nov 12

[3] D Williams, An introduction to medical and dental materials, in: D. Williams (Ed.), Concise Encyclopedia of Medical & Dental Materials, Pergamon Press and The MIT Press. 1990; pp. xvii–xx

[4] Sprion S, Guicciardi S, Dapporto M, Melandri C, Tampieri A. Synthesis and mechanical behavior of β-tricalcium phosphate/titania composites addressed to regeneration of long bone segments. J MechBehav Biomed Mater. 2013;17:1–10. doi.org/10.1016/j.jmbbm.2012.07.013

[5] Barker LM. A simplified method for measuring plane strain fracture toughness. EngFract Mech. 1977;9(2):361–364. doi.org/ 10.1016/0013-7944(77)90028-5

[6] Sakka S, Bouaziz J, Ben Ayed F. Sintering and mechanical properties of the alumina-tricalcium phosphate-titania composites. Mater. Sci. Eng. C. 2014;40:92–101. doi.org/10.1016/j.msec.2014. 03.036

[7] Chih-Jen Wang, Chi-Yuen Huang. Effect of TiO_2 addition on the sintering behavior, hardness and fracture toughness of an ultrafine alumina. Materials Science and Engineering: A.2008: 492(1-2), 06-310. doi.org/10.1016/j.msea.2008.04.048

[8] KIM, Hyun-Min, MIYAJI, Fumiaki, KOKUBO, Tadashi, et al. Preparation of bioactive Ti and its alloys via simple chemical surface treatment. Journal of Biomedical Materials Research: An Official Journal of the Society for Biomaterials and The Japanese Society for Biomaterials. 1996;32(3),409-417. doi: 10.1002/(SICI)1097-4636(199611)32:3<409::AID-JBM14>3.0.CO;2-B.

[9] KANEKO Hideki, UCHIDA, Masaki, KIM, Hyun Min, et al. Process of apatite formation induced by anatase on titanium metal in simulated body fluid. In: Key Engineering Materials. Trans Tech Publications Ltd. 2002; p. 649-652. doi:10.1016/j.biomaterials.2005.01.048

[10] MEHMANPARAST, H. N., MAC-THIONG, J. M., et PETIT, Yvan. Compressive properties of a synthetic bone substitute for vertebral cancellous bone. International Journal of Biomedical and Biological Engineering. 2012:6(5),144-147. doi.org/10.5281/zenodo.1077803

[11] LIU, Chen, WAN, Peng, TAN, Li Li, et al. Preclinical investigation of an innovative magnesium-based bone graft substitute for potential orthopaedic applications. Journal of Orthopaedic Translation, 2014; 2(3)139-148. https://doi.org/10.1016/j.jot.2014.06.002

[12] Liu H, Li H, Cheng W, Yang Y, Zhu M, Zhou C. Novel injectable calcium phosphate/chitosan composites for bone substitute materials. Acta Biomater. 2006;2(5):557–565. DOI: 10.1016/j.actbio.2006.03.007

[13] Ayatollahi MR, Aliha MRM, Saghafi H. An improved semicircular bend specimen for investigating mixed mode brittle fracture. EngFract Mech. 2011;78(1):110–123. doi.org/10.1016/j.engfracmech.2010.10.001 11.

[14] XIE, Yousheng, CAO, Ping, JIN, Jin, et al. Mixed mode fracture analysis of semi-circular bend (SCB) specimen: a numerical study based on extended finite element method. Computers and Geotechnics. 2017;82:157–172. doi.org/10.1016/j.compgeo.2016.10.012

[15] KUMAR, Alok, BISWAS, Krishanu, et BASU, Bikramjit. Fretting wear behaviour of hydroxyapatite–titanium composites in simulated body fluid, supplemented with 5 g l− 1 bovine serum albumin. Journal of Physics D: Applied Physics. 2013;46 (40),404004. doi:10.1088/0022-3727/46/40/404004

[16] FELLAH, Mamoun, SAMAD, Mohammed Abdul, LABAIZ, Mohamed, et al. Sliding friction and wear performance of the nano-bioceramicα-Al_2O_3 prepared by high energy milling. Tribology international. 2015; 91,151-159. doi.org/10.1016/j.triboint.2015.07.006

[17] Nine MJ, Choudhury D, Hee AC, Mootanah R, Osman NAA. Wear debris characterization and corresponding biological response: artificial hip and knee joints. Materials. 2014;7(2),980–1016. doi: 10.3390/ma7020980

[18] SAKKA, S., AYED, F. Ben, et BOUAZIZ, J. Mechanical properties of tricalcium phosphate–alumina composites. In: IOP Conference Series: Materials Science and Engineering. IOP Publishing, 2012;012028. DOI: 10.1088/1757-899X/28/1/012028

[19] MAZEL, Vincent, GUERARD, Sandra, CROQUELOIS, Benjamin, et al. Reevaluation of the diametral compression test for tablets using the flattened disc geometry. International journal of pharmaceutics, 2016; 513,669-677. DOI : 10.1016/j.ijpharm.2016.09.088

[20] WANG, Q. Z., JIA, X. M., KOU, S. Q., et al. The flattened Brazilian disc specimen used for testing elastic modulus, tensile strength and fracture toughness of brittle rocks: analytical and numerical results. International Journal of Rock Mechanics and Mining Sciences. 2004;41(2) 245-253. doi.org/10.1016/S1365-1609(03)00093-5

[21] KURUPPU, Mahinda D. et CHONG, Ken P. Fracture toughness testing of brittle materials using semi-circular bend (SCB) specimen. Engineering Fracture Mechanics. 2012; 91,133-150. doi.org/10.1016/j.engfracmech.2012.01.013

[22] DAI, F., XIA, K., ZUO, J. P., et al. Static and dynamic flexural strength anisotropy of Barre granite. Rock Mechanics and Rock Engineering. 2013; 46(6),1589-1602. DOI: 10.1007/s00603-013-0390-y

[23] LIM, I. L., JOHNSTON, I. W., CHOI, S. K., et al. Fracture testing of a soft rock with semi-circular specimens under three-point bending. Part 1—mode I. In: International journal of rock mechanics and mining sciences & geomechanics abstracts. Pergamon. 1994;185-197. DOI: 10.1016/0148-9062(94)90463-4

[24] LIU, Hua, LI, Hong, CHENG, Wenjun, et al. Novel injectable calcium phosphate/chitosan composites for bone substitute materials. Acta Biomaterialia. 2006;2(5)557-565. doi.org/10.1016/j.actbio.2006.03.007

[25] ASTM C1424-15, Standard Test Method for Monotonic Compressive Strength of Advanced Ceramics at Ambient Temperature, ASTM International, West Conshohocken, PA, 2015C.

[26] RAHMATI, Maryam et MOZAFARI, Masoud. Biocompatibility of alumina-based biomaterials–A review. Journal of cellular physiology. 2019;234(4)3321-3335. doi.org/10.1002/jcp.27292

[27] NATH, Shekhar, UMMETHALA, Raghunandan, et BASU, Bikramjit. Fretting wear behavior of calcium phosphate-mullite composites in dry and albumin-containing simulated body fluid conditions. Journal of Materials Science: Materials in Medicine. 2010;21 (4)1151-1161. doi.org/10.1007/ s10856-009-3983-y

[28] KHALLADI, A. et ELLEUCH, K. Tribological behavior of wheel–rail contact under different contaminants using pin-on-disk methodology. Journal of Tribology, 2017;139(1). doi.org/ 10.1115/1.4033051

[29] JENG, Yeau-Ren, LIN, Tsung-Ting, HSU, Hsiu-Ming, et al. Human enamel rod presents anisotropic nanotribological properties. Journal of the mechanical behavior of biomedical materials. 2011;4(4)515-522.DOI: 10.1016/j.jmbbm.2010.12.002

[30] KARAVAEV, M. G. et KUKAREKO, V. A. Automated tribometer with reciprocating. Nadezhnostmashin-itekhnicheskikhsistem-Reliability of machines and technical systems. 2001 (1),37-39.

[31] Sharma S K, Kumar B V M, Kim Y W. Effect of WC addition on sliding wear behavior of SiC ceramics. Ceram Int. 2015;41(3),3427-3437. doi.org/ 10.1016/j.ceramint.2014.10.144

[32] Guermazi N, Elleuch K, Ayedi HF, Fridrici V. Kapsa P, Tribological behaviour of pipe coating in dry sliding contact with steel. Mater Des. 2009;30(8):3094–104. doi:10.1016/j.matdes.2008.12.003

[33] KALIN, Mitjan, JAHANMIR, Said, et IVES, Lewis K. Effect of counterface roughness on abrasive wear of hydroxyapatite. Wear. 2002;252(9-10) 679-685. DOI: 10.1016/S0043-1648(02) 00028-5

[34] SUN, X. J. LI, J. G. Friction and wear properties of electrodeposited nickel–titania nanocomposite coatings. Tribology Letters. 2007;28(3)223-228. DOI: 10.1007/s11249-007-9254-5

[35] Miao. X, Sun. D, Hoo. P. W, Liu. J, Hu. Y, Chen. Y, Effect of titania addition on yttria-stabilised tetragonal zirconia ceramics sintered at high temperatures. Ceramics international. 2004;30(6)1041-1047. DOI: 10.1016/j. ceramint.2003.10.025

Chapter 11

Application of Titanium Dioxide in the Synthesis of Mesoporous Activated Carbon Derived from Agricultural Waste

Ashok Kumar, Kaman Singh and Rayees Ahamad Bhat

Abstract

Adsorption is an important technique that significances the characteristics of porous solid materials and fine powders. The importance of porous solid materials and fine powders has been recognized when porous coal used for various applications such as catalysis, separation, isolation, sensors, chromatography, etc. Herein, the synthesis of mesoporous activated carbon derived from agricultural waste using TiO_2. The TiO_2-modified carbon was characterized employing scanning electron microscope (SEM), attenuated total reflection-Fourier transform infrared (ATR-FTIR) spectroscopy, powder X-ray diffraction (pXRD), Brunauer–Emmett–Teller (BET) surface area analyzer and X-ray photoelectron spectroscopy (XPS). The obtained results suggested that the TiO_2-modified carbon could be a potential material for various application like dye removal, metal removal and allied areas. This book chapter describes the commonly used classifications of porous bulk materials and also reported here the characterization of porous solid materials and fine powders with special reference to the evaluation of the surface area, pore size distribution and thermodynamic parameters of the different mesoporous material, at various scales of resolution using relevant techniques. These materials comprise several levels of structures that of the mesopores, micropores as well as macropores. The apparent topography analysis of these materials, of various pore diameters, synthesized in our laboratory has been determined at various scales with the help of various characterization techniques.

Keywords: Titanium Dioxide, Mesoporous Materials, N_2 adsorption–desorption isotherm, Capillary (or pore) condensation

1. Introduction

Since ancient times, the importance of porous solid materials and fine powders has been recognized when porous coal used for its medicinal properties [1]. Research on porous solid materials and fine powders, which have various applications such as catalysis, separation, isolation, sensors, chromatography, etc., have reinforced the global interest [2] in environmental sustainability and energy storage. In the above - referenced applications, the outperformance of different porous solid materials and fine powders is extremely dependent on each material's internal porous structure. Therefore, the internal geometry, size, connectivity, etc. such as porous materials of different properties were fully characterized to better understand a particular physical process taking place in a porous medium [3]. Gas adsorption is one of many experimental methods available to characterize porous materials by the surface and pore size. These include ray and neutron scattering of small angles (SAXS and SANS), porosimetry of mercury, electron microscopy (scanning and transmission), thermoporometry, nuclear magnetic resonance methods, and others. Each method has a limited application scale for analyzing pore size. IUPAC recently provided an overview of different methods for characterizing pore size and their range of application [4]. Among these methods, gas adsorption is the most popular because it enables the evaluation of a wide range of pore sizes (from 0.35 nm to >100 nm), including the full range of micro-and mesopores and even macropores. Moreover, the techniques of gas adsorption are convenient to use and are not that cost-intensive compared to some of the other methods [5].

Adsorption finds very extensive potential application in the research laboratory and industry. Adsorption plays a very important role in various aspects of the catalysis of gases reactions by solid surfaces. As applications of adsorption from solutions may be mentioned the clarification of sugar liquors by charcoal, the removal of coloring materials from various types of solutions, and recovery of dyes from dilute solutions in several solvents. Adsorption has also been employed for the recovery and concentration of vitamins and other biological substances and finds immense utility in the chromatographic analysis [6]. The importance of the adsorption can be assessed in the sense that Langmuir was awarded Nobel Prize in Chemistry for his seminal contribution to surface science. Since then, two more Nobel Prizes Has been given in this area (Ertl in 1992 for his studies of chemical processes on solid surfaces and Marcus in 1992 for his theory and mechanism of electron transfer reactions on surfaces).

The materials comprising of pore size lying between Micro- and Macro-porous materials are regarded as mesoporous materials [7–9]. Mesoporous materials are the materials that have their pore size in between Micro- and Macro-porous materials. Meso a Greek prefix – "in-between" - micro and macroporous system. Following the IUPAC standard, microporous material has pore smaller than 2 nm while macroporous material possesses pores larger than 50 nm. Mesoporous material has high specific surface area due to high porosity within the mesopore range which forms the basis of their applications in varying fields. They possess high surface area -400-1000 m^2/g, large pore volume and high stability –500-600°C. A very common mesoporous material is customary activated carbon which is typically composed of a carbon framework having both mesoporosity and microporosity depending on the conditions under which it was manufactured. According to IUPAC, a mesoporous material can be deliberately ordered or disordered in a mesostructure [10–13]. In crystalline inorganic materials, mesoporous structure noticeably limits the number of lattice units, and this significantly changes the physics and chemistry. For example, the battery performance of mesoporous electroactive materials is significantly different from that of their bulk structure. "The overwhelming tendency for solids to minimize void space within their structure" is inherent, porous materials are difficult to make naturally. But Einstein says "In the middle of difficulty lies opportunity". The above statement was made true by the Mobil scientist in the year 1992 by successfully synthesizing the Mesoporous materials i.e. Mobil Crystalline Materials (MCM-41 and MCM-48) employing soft template strategy. This has opened new potentialities for Mesoporous materials, and extensive research has been contributed in this field.

Material can be classified as porous if its internal voids can be filled with gases. The history of porous materials began with the zeolites having an aluminosilicates framework which was synthesized employing a single template molecule with a small pore. Usually, they are synthesized by the use of the soft template method. Since then, research and development in this field have grown steadily. Notable examples of prospective industrial applications are catalysis, sorption, gas sensing, ion exchange, optics, and photovoltaics. Gas adsorption is the most effective and corroborative method for the characterization of the texture of porous solids and fine powders. The analysis was done on "Reporting Physisorption Data for Gas/ Solid Systems" on the surface area and porosity [14, 15].

2. Materials and methods

The agriculture waste namely mustard cake (sample) was purchased from the local mill (Sai Enterprise, Lucknow, India). Sodium hydroxide (98%), hydrochloric acid (35.5%), potassium chloride (99.5%), sulfuric acid (98%) and TiO_2-Degussa P25 (ca.80% anatase, 20% rutile) chemicals were used and supplied by Bionic Enterprises, Lucknow, India. High-purity grade nitrogen gas (99.999%) and helium gas (99.9999%) were used for physicochemical characterization of the sample supplied by Krishna Gas Agencies, Lucknow, India. The double-distilled water (DDW) was prepared using a double distillation unit (Glassco Laboratory Equipment Pvt. Ltd., Ambala Cantt., India) and used for the preparation of modified material. All the reagents used for synthesis and experimental studies were of analytical and laboratory grade.

The agricultural waste sample was crushed and washed with double distilled water (DDW) and then sun-dried. This material was treated with 20 wt% H_2O_2 at 60°C for 24 h to oxidize the adhering organic matter and it was washed several times using DDW. This material was calcinated in the presence of O2 gas at 715°C for 25 min. The material, powdered activated carbon (PAC) was grounded and sieved to desired particle sizes. To get TiO_2-MMC, TiO_2-Degussa P25 (ca.80% anatase, 20% rutile) was added to PAC in a 5:1 weight ratio, in NaOH 2 N alkaline solution; in the 1000 ml, volumetric flask with a reflux condenser, the new potential material (TiO_2-MMC) was obtained from the slurry under stirring (300 ppm) for 24 h, at atmospheric pressure and 100°C, on a thermostat heating plate. After filtration, washing and drying at 105–120°C till a constant mass, the dried TiO_2-MMC was ground and passed through a 100 mesh sieve and stored in a desiccator. The material was characterized employing ATR- FTIR, SEM, XRD, BET and XPS [6, 7].

For the functional group analysis, the sample was directly (without any preparation) scanned in transmittance mode with a wave region ranging from 4000 to 500 cm^{-1} by ATR-FTIR (Bruker-Tensor 27) [16, 17]. The particle morphologies of samples were studied using SEM (JEOL JSM 6490LV). The sample was mounted on an aluminum stub with the help of double-sided tape. Mounted stubs were coated with gold–palladium for analysis using a Polaron sputter coater [16, 17]. The specific surface area, size and pore diameter of the materials were measured using a surface area analyzer (Belsorp-Mini-II, Japan). A definite amount of the adsorbent materials (0.20 g) were heated under pre-treatment at 120°C under vacuum for approximate 5–6 hrs. Subsequently, the process of N_2 adsorption and desorption was performed. The surface area and pore diameter were calculated using the standard Brunauer–Emmett–Teller (BET) equation and pore size was obtained employing a method of Barrett–Joyner–Halenda (BJH) and T-plot [16, 17]. The crystalline nature of the sample was examined by powder X-ray diffraction ($_P$XRD) (PAN analytical X'pert, Powder, Malvern Panalytical, UK) in 2-theta(degree) angle range 5°-80° at 45 kV using CuK radiation with 148.92 times per step [6, 7]. The XPS experiments were performed in an ultrahigh vacuum (UHV) using a high-resolution X-ray photoelectron spectrophotometer (HR-XPS), (PHI 5000 Versa Prob II). All samples were dehydrated under vacuum before XPS analysis. All Data of XPS analysis were plotted using Casa XPS Version 2.3.17PRI. I and Origin Pro 8 SRO v8.0724 (B724) software [16, 17].

3. Result and discussion

3.1 Functional group analysis

Vibrational spectroscopy in the infrared region was used to perform the characterization of the molecular structure of the adsorbent from the vibrational modes related to functional groups present in the sample. ATR FTIR spectroscopy provides evidence for specific functional groups on the surface of the TiO_2-modified mesoporous carbon (TiO_2-MMC). Several characteristics bands were observed in the ATR FTIR spectrum (**Figure 1**). Many existing groups in carbon derived from agriculture waste (CDAW) (blank) [18] are also present in the TiO_2-MMC, showing that modified material exhibits many of the same functional groups of CDAW (blank).

It was observed that the infrared spectra of TiO_2-MMC which overlapping with CDAW (source material) so that there are new peaks as observed in 3832.20–3425.4, 1632.21, 1570, 1046.50, 600.23, 500.30 cm^{-1}. The broad and strong bands at 3832.20–3425.4 cm^{-1} have been assigned because of bonded hydroxyl (-OH) or amine (-NH_2) groups. The band 1632.21 can be ascribed to –C=O stretching. The band around 1570 cm^{-1} can be ascribed to –C=C. The band at 1046.50 cm^{-1} is due to the C-O of carbonyl groups. The stretching band corresponding to Ti-O-Ti is clearly in the region of 509 cm^{-1} [18]. The band around 620.23 cm^{-1} is attributed to the Ti-O-C vibration which indicates that TiO_2 (Dugasa P25) were chemically bonded with carbonized carbon in TiO_2-MMC [16, 17].

3.2 Microstructural and elemental composition analysis

The SEM images of CDAW (**Figure 2a**) [16] and TiO_2-MMC are shown in **Figure 2b**, which demonstrate the porosity and texture of TiO_2-MMC. It has been observed that TiO_2-MMC retain the surface characteristics of its precursor (CDAW

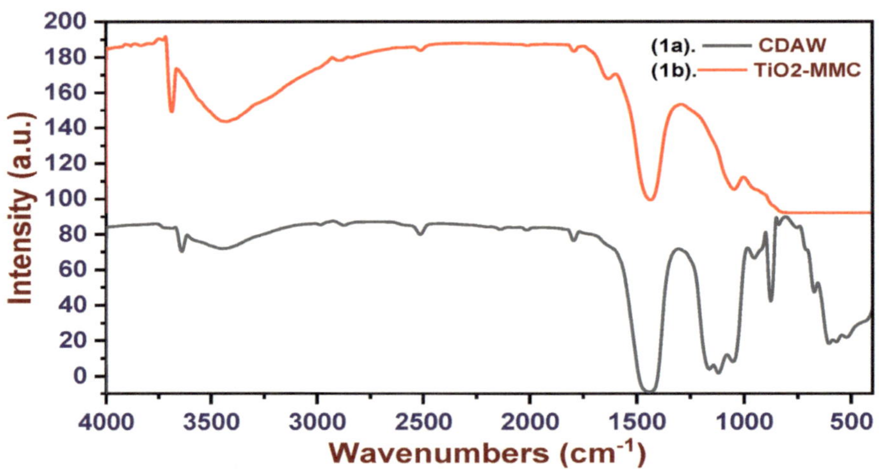

Figure 1.
FTIR spectra of (1a). CDAW and (1b) TiO2-MMC.

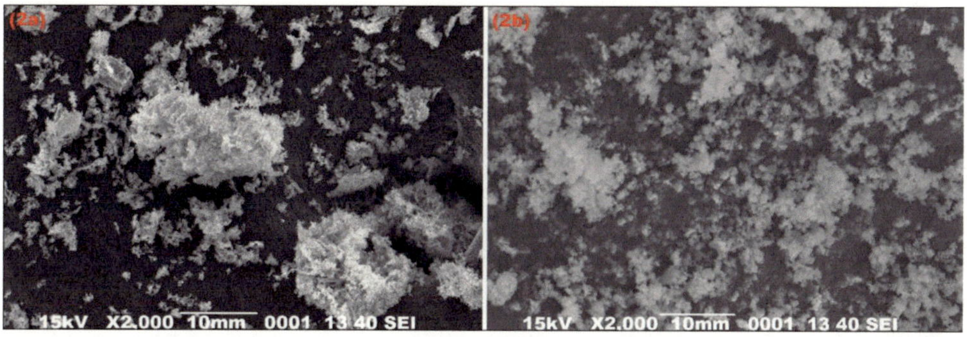

Figure 2.
SEM micrographs of (2a). CDAW (blank) and (2b) TiO2-MMC.

and TiO_2) as shown in **Figure 2**. However, with some remarkable modifications, the results (**Figure 2b**) obtained showed that there was displacement as the temperature of thermal decomposition of the modified adsorbent suggesting that the modification caused a difference in structure. The SEM micrographs (**Figure 2b**) show a spongy aspect, fibrous surface with irregular and heterogeneous structure with many pits and fissures through the surface prominent possibly due to powerful oxidizing agent H_2O_2, which caused some changes in the surface of TiO_2-MMC.

3.3 Specific surface area, particle size measurement

The most applicable and accessed procedure for evaluating the surface area of porous and finely-divided material is the Brunauer–Emmett–Teller method [6]. Given the specific controlled conditions, the BET-area of a nonporous, macroporous or mesoporous solid can be referred to as the 'probe accessible area' (which is the

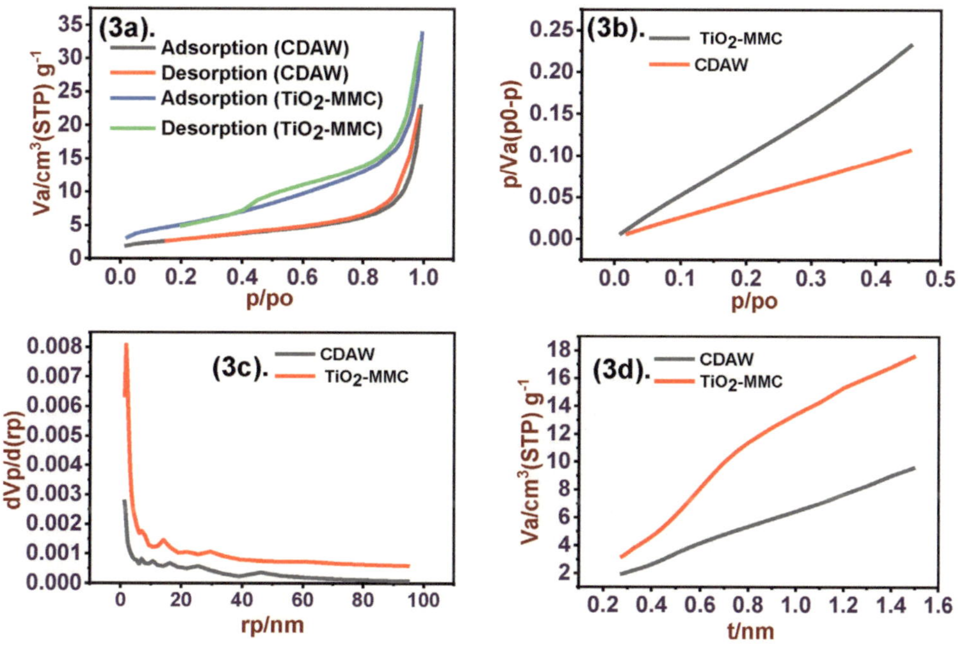

Figure 3.
BET analysis for CDAW and TiO2-MMC, (3a). N2 adsorption and desorption, (3b). BET, (3c). BJH and (3d). T-plot respectively.

effective area available for the adsorption of the specified adsorptive). There are two particular stages involved in the application of the BET method. It is important to transform a physisorption isotherm into the 'BET plot'. The calculated value of the BET area is dependent on (i). The adsorptive and operational temperature and (ii). The procedure used to locate the pressure range in applying the BET equation.

The N_2 adsorption–desorption of both CDAW and TiO_2-MMC was conducted (**Figure 3a**). BET analysis (**Figure 3**) confirms that TiO_2-MMC falls on the mesoporous domain. The typical specific surface area of TiO_2-MMC was found as S_{BET} = 18.845 m^2g^{-1} whereas the surface area of the CDAW was typically S_{BET} = 10.251 m^2g^{-1} (**Figure 3b**). In the case of fine and especially, free-floating powdered samples can present an additional difficulty to the during BET analysis. Particles may also be deposited on valve seats causing problematical leaks. This loss of powder from the confinement of the sample bulb, known as elutriation, should be controlled and eliminated as far as is reasonably practicable.

The adsorption/desorption isotherm (**Figure 3a**) shows the relationship between the amount of adsorbed/desorbed gas (y-axis) and the pressure of adsorptive (x-axis) at the constant temperature. Nitrogen (at the boiling temperature of 77 K) was the conventional and usual choice for the adsorptive to obtain (BET), with $\sigma m(N_2)$ assumed to be 0.162 nm^2 and being a close-packed monolayer. Moreover, liquid nitrogen was readily available and also nitrogen isotherms on many adsorbents were found to exhibit a well-defined Point B. Though, present studies have shown that due to its quadrupole moment, the orientation of a nitrogen molecule is dependent on the surface chemistry of the adsorbent. And, this could lead to the tentative reading of the value of σm (N_2) – possibly ~ 20% for some surfaces [7].

A more suitable alternative adsorptive for surface area analysis could be Argon. Argon does not have a quadrupole moment and is less reactive than the diatomic nitrogen molecule. Argon, though, at 77 K is, possibly, less reliable than Nitrogen. Argon adsorption at 87 K (liquid argon temperature) is the possible alternative. At 87 K, a cross-sectional area, $\sigma m(Ar)$, of 0.142 nm^2 is assumed. Due to the absence of a quadrupole moment and the higher temperature, σm (Ar) is less prone to the differences in the structure of the adsorbent surface. Argon adsorption at 87 K facilitates the micropore analysis. Analysis at 87 K can be done by using either liquid argon (replacing liquid nitrogen) or a cryostat (or cryocooler) [1, 7]. Manometric adsorption equipment of maximum accuracy can effectively assess the surface areas as low as (~ 0.5–1) m^2 with nitrogen or argon as the adsorptive. For lower surface areas, krypton adsorption at 77 K is the preferred adsorptive. Though krypton at 77 K is similar to argon at 77 K. Hence, the standard thermodynamic state of the adsorbed layer is not accurately gauged.

These methods (**Figure 3**) used to calculate the micropore radius (cylindrical shape), micropore diameter (cylindrical shape), area distribution, volume distribution, Integral curve, pore specific surface area and pore volume. Methods for mesopore size analysis have been proposed by Barrett, Joyner, and Hlenda (BJH) (**Figure 3c**) [19, 20]. The preadsorbed multilayer film can be accounted for with the integration of the Kelvin equation with a standard isotherm (the t-curve) to analyze the specific nonporous solids. Still, for analyzing the accurate size of narrow mesopores, the standard t-curve is not a convincing procedure (**Figure 3d**), the validity of the Kelvin equation also provides an approximation only because the mesopore width is reduced in the procedure. Studies have proven that the Kelvin equation-based procedures, such as the BJH method, underestimate the pore size for narrow mesopores.

The IUPAC classification, where pores are classified in macro-, meso-, and micropores, is mostly based on the different mechanisms that occur in these pores

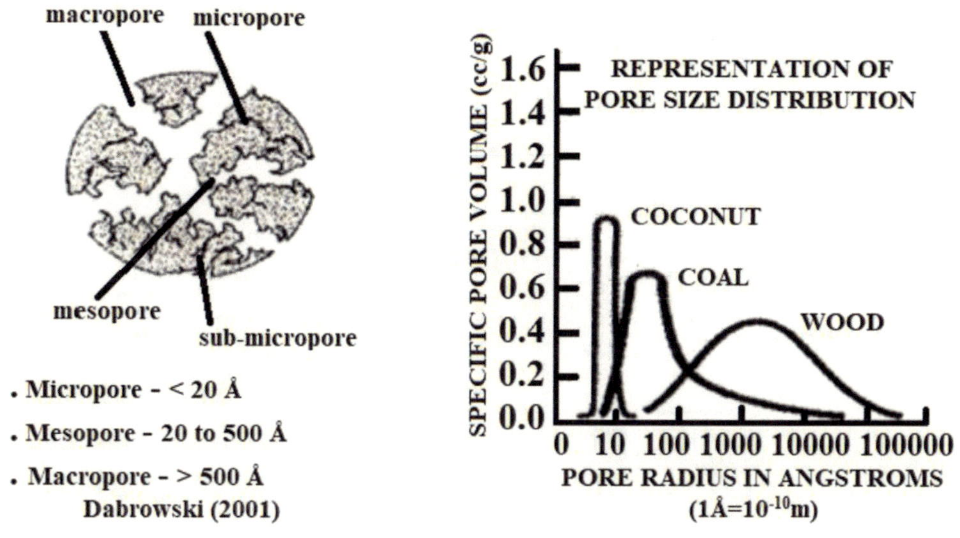

Figure 4.
IUPAC classification of pores. Figure adapted from reference [1]. © 2015 IUPAC & De Gruyter.

during 77 K and 1 atm isothermal adsorption of N_2 Pressure (**Figure 4**). Multilayer adsorption, capillary condensation, and microporous filling, respectively, are processes related to macropores, mesopores, and micropores. Depending on the different relative pressures p/p° ratio, the pore width classes correspond to the application of the capillary condensation theory. The 50 nm pore width is associated with 0.96 relative pressures. Above this value, isothermic adsorption experiments are quite difficult to interpret and the capillary condensation theory of applicability has not been adequately tested [21–25].

3.4 Phase determination

The pXRD spectra shown in **Figure 5** illustrates that the treated CDAW and modified with TiO_2 under optimum preparation condition is crystalline in structure. There are seven broad peaks centred on a 2θ value of 250, 300, 370, 470, 520, 540 and 630. 2θ = 68° is not clearly described [21–25].

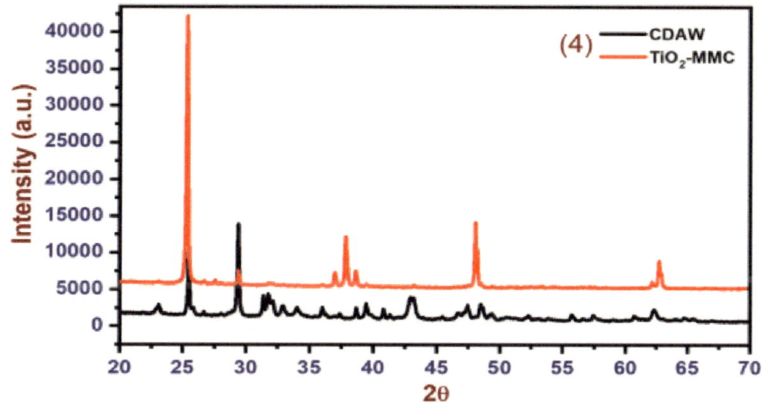

Figure 5.
XRD spectra of CDAW and TiO2-MMC.

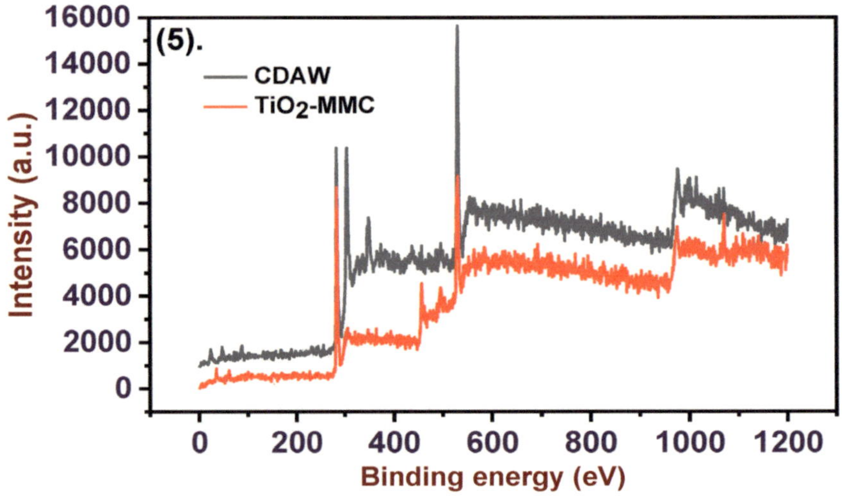

Figure 6.
XPS survey AlKα PES of CDAW and spectra CDAW.

3.5 XPS analysis

To determine the chemical composition of the CDAW and TiO_2-MMC XPS analysis was performed. The surveys scan of CDAW in (**Figure 6**) illustrated that its oxygen content is around 17.4%, suggesting the containing of abundant surface oxygenic functional groups. In the C 1 s XPS spectra of CDAW (**Figure 6**), the peaks at 284.8, 286.7, and 288.9 eV are ascribed to the sp2 hybridized (C–C), alcoholic (C–OH) and carbonyl (C=O) carbon atoms, respectively.

The XPS survey scans of TiO_2-MMC (**Figure 6**) show the typical spectra of TiO_2-MMC (P25), and the more intense feature peak of C1s in TiO_2-MMC was attributed to the layer of the hybrid. Concerning the C1s XPS spectra of TiO_2-MMC, aside from the above listed characteristic peaks, it was noticed that an additional peak emerges at the binding energy of 283.5 eV, which is higher than that of titanium carbides ($Ti4^{+}$–C) and should be assigned to the carbon bonded with oxygen-deficient titanium ion ($Ti3^{+}$–C) revealing the distinctive chemically bonded carbon/TiO_2-MMC. Besides, the atomic ratios of O: C for PAMC and TiO_2-MMC are 0.21 and 0.17, respectively, suggesting the decreased oxygen content of coated PAMC TiO_2-MMC [21–25].

4. Conclusions

- This chapter briefly introduced the synthesis of impregnated titanium dioxide mesoporous carbon derived from agricultural waste and high-resolution characterization techniques and reliable commercial instrumentation.

- The AT-FTIR analysis revealed the specific functional groups present in the TiO_2-MMC. The microstructural analysis confirmed the chemical composition and surface nature.

- The initial IUPAC classifications of physisorption isotherms and hysteresis loops have encapsulated the recent characteristic types the Brunauer–Emmett–Teller (BET) method for the analysis of surface area needs careful attention.

- The nitrogen adsorption isotherms, pores size distribution and scanning electron micrographs of TiO_2-MMC suggest that it could be very good mesoporous materials for various applications.

- Relying on the standard methods for the analysis of the surface area and pore size can lead to inconclusive BET surface areas and pore size distributions. Such gaps have necessitated the introduction of more convincing and validated pore models based on the non-rigid nature of the adsorbent.

- The study highlights that the pore size analysis of narrow mesopores remains inconclusive by enlisting the procedures derived from the Kelvin equation, as the Barrett–Joyner–Halenda (BJH) method.

- The classification of non-rigid adsorbents poses another area for further investigations. Enhanced procedures for standard analysis and development of the latest certified reference materials are the imminent need of research in this domain.

Nomenclature

SEM	Scanning Electron Microscope
ATR-FTIR	Attenuated Total Reflection-Fourier Transform Infrared Spectroscopy
$_{P}$XRD	Powder X-ray Diffraction
BET	Brunauer–Emmett–Teller
XPS	X-ray Photoelectron Spectroscopy
SANS	Small-Angle Neutron Scattering
SAXS	Small-Angle X-ray Scattering
IUPAC	International Union of Pure and Applied Chemistry
MCM	Mobil Crystalline Materials
PAC	Powdered Activated Carbon
TiO_2-MMC	Titanium Dioxide-Modified Mesoporous Carbon
BJH	Barrett–Joyner–Halenda Method
S_{BET}	Specific Surface Area
Ultrahigh Vacuum	UHV
HR-XPS	High-resolution X-ray Photoelectron Spectrophotometer
CDAW	Carbon Derived from Agriculture Waste

Author details

Ashok Kumar, Kaman Singh* and Rayees Ahamad Bhat
Department of Chemistry, Babasaheb Bhimrao Ambedkar University (A Central University), Lucknow, U.P., India

*Address all correspondence to: singh.kaman@bbau.ac.in; drkamansingh@yahoo.com

References

[1] Nicholson D, Using Computer Simulation To Study The Properties Of Molecules In Micropores, J. Chem. Soc., Faraday Trans. 1996;92: 1-9. DOI: 10.1039/FT9969200001

[2] Menon S, Pidugu VC, Goworek R, Stefaniak J, Temperature Programmed Desorption Vs. N2 Desorption In Determining Pore-Size Distribution Of Mesoporous Silica Molecular Sieves, J. Porous Mat., 1996;3: 115-119. DOI: 10.1007/BF01186041

[3] Rigby SP, Fletcher RS, Riley SN, Characterisation Of Porous Solids Using Integrated Nitrogen Sorption And Mercury Porosimetry, Chem. Eng. Sci., 2004;59:41-51. DOI:10.1016/j.ces.2003.09.017

[4] Sing KSW, Reporting Physisorption Data For Gas/Solid Systems With Special Reference To The Determination Of Surface Area And Porosity (Recommendations 1984). Pure & App. Chem.1985; 57:603-619. DOI: 10.1351/pac198557040603

[5] Zdravkov BD, Cermak JJ, Sefara M, Janku J, Pore Classification in the Characterization of Porous Materials: A Perspective. Central Europ. J. Chem. 2007; 5:385-395. DOI: 10.2478/s11532-007-0017-9

[6] Sing KSW, Reporting Physisorption Data For Gas/Solid Systems With Special Reference To The Determination Of Surface Area And Porosity (Recommendations 1984). Pure & App. Chem., 1985;57:603-619. DOI: 10.1351/pac198557040603

[7] Rouquerol J, Rouquerol F, Sing KSW, Llewellyn P, Maurin G, Adsorption by Powders and Porous Solids: Principles, Methodology and Applications. 2nd ed. Academic Press; 2014.646p .eBook ISBN: 9780080970363

[8] CohenE R, Cvitas T, Frey J G, Holmström B, Kuchitsu K, Marquardt R, Mills I, Pavese F, Quack M, Stohner J, Strauss HL, Takami M, ThorAJ, Quantities, Units and Symbols in Physical Chemistry. IUPAC Green Book, 3rd Edition, 2nd Printing, IUPAC & RSC Publishing, Cambridge, 2008.250p. ISBN: 978-0-85404-433-7

[9] Lowell S, Shields JE, Thomas MA, Thommes M, Characterization of Porous Solids and Powders: Surface Area. Porosity and Density. Springer, 2004.XIV, 350p.DOI: 10.1007/978-1-4020-2303-3

[10] Silvestre-Albero J, Silvestre-Albero AM, Llewellyn PL, Rodriguez-Reinoso F, High-resolution N2 Adsorption Isotherms At 77.4K: Critical Effect Of The He Used During Calibration. J. Phys. Chem. C, 2013; 117:16885-16889.DOI: 10.1021/jp405719a

[11] Sing KSW, Everett DH, Haul RAW, Moscou L, Pierotti RA, Rouquerol J, Siemieniewska T, Reporting Physisorption Data For Gas/Solid Systems With Special Reference To The Determination Of Surface Area And Porosity. Pure Appl. Chem. 1985:57:603-619. DOI:10.1515/iupac.57.0007

[12] Broekhoff JCP, Mesopore determination from nitrogen sorption isotherms: Fundamentals, scope, limitations. Stud. Surf. Sci. Catal. 1979; 3:663-684. DOI: 10.1016/S0167-2991(09)60243-3

[13] Shields JE, Lowell S, Thomas MA, Thommes M, Characterization of Porous Solids and Powders: Surface Area, Pore Size And Density; Kluwer Academic Publisher: Boston, MA, USA, 2004; pp. 43-45.ISBN:978-4020-2303-3

[14] Thommes M, Cychosz KA, Physical Adsorption Characterization of

Nanoporous Materials: Progress and Challenges. Adsorption. 2014; 20:233-250.DOI 10.1007/s10450-014-9606-z

[15] Monson PA, Understanding Adsorption/Desorption Hysteresis For Fluids In Mesoporous Materials Using Simple Molecular Models And Classical Density Functional Theory. Microporous and Mesoporous Materials. 2012; 160:47-66. DOI:10.1016/j.micromeso.2012.04.043

[16] Landers J, Gor GY, Neimark AV, Density Functional Theory Methods For Characterization Of Porous Materials. Colloids and Surfaces A: Physicochem. Eng. Aspects. 2013; 437:3-32. DOI:10.1016/j.colsurfa.2013.01.007

[17] Singh K, Mohan S, Kinetic studies of the sucrose adsorption onto an alumina interface. App. Surf. Sci. 2004; 221:308-318. DOI: 10.1016/S0169-4332(03)00950-4

[18] Brunauer, S.; Emmett, P.H.; Teller, E. Adsorption of Gases in Multimolecular Layers. J. Am. Chem. Soc. 1938, 60, 309-319. DOI: 10.1021/ja01269a023

[19] Barrett E.P.; Joyner L.G.; Halenda PP. The Determination of Pore Volume and Area Distributions in Porous Substances. I. Computations from Nitrogen Isotherms. J. Am. Chem. Soc. 1954, 73, 373-380. DOI: 10.1021/ja01145a126

[20] Joyner L.G.; Barrett E.P.; Skold, R. The Determination of Pore Volume and Area Distributions in Porous Substances. II. Comparison between Nitrogen Isotherm and Mercury Porosimeter Methods. J. Am. Chem. Soc. 1951, 73, 3155-3158. DOI: 10.1021/ja01151a046

[21] Guo, M.; Wang, H.; Huang, D.; Han, Z.; Li, Q.; Wang, X.; Chen, J. Amperometric Catechol Biosensor Based On Laccase Immobilized On Nitrogen-Doped Ordered Mesoporous Carbon (N-OMC)/PVA Matrix. Sci. Tech. Adv. Mat. 2014, 15, 035005. DOI: 10.1088/1468-6996/15/3/035005.

[22] Somasundaran P. Encyclopedia of Surface and Colloid Science, CRC press; Boca Raton, Florida. 2006, 6, pp6675. ISBN: 0849396158, 9780849396151

[23] Dabrowski. A. Adsorption from Theory to Practice. Adv. Coll. Interf. Sci. 2001, 93, 135-244. DOI: 10.1016/S0001-8686(00)00082-8

[24] McCusker, L.; Liebau, F.; Engelhardt, G. Nomenclature of Structural and Compositional Characteristics of Ordered Microporous and Mesoporous Materials with Inorganic Hosts (IUPAC Recommendations 2001). Pure Appl Chem. 2001, 73, 381-394. DOI: 10.1351/pac200173020381

[25] McCusker L, Liebau F, Engelhardt G. Nomenclature of structural and compositional characteristics of ordered microporous and mesoporous materials with inorganic hosts: (IUPAC recommendations 2001). Microporous Mesoporous Mater. 2003; 58:3-13. DOI: 10.1351/pac200173020381